Cereals and Cereal-Based Foods: Nutritional, Phytochemical Characterization and Processing Technologies

Cereals and Cereal-Based Foods: Nutritional, Phytochemical Characterization and Processing Technologies

Guest Editors

Donatella Bianca Maria Ficco

Grazia Maria Borrelli

MDPI

Basel • Beijing • Wuhan • Barcelona • Belgrade • Novi Sad • Cluj • Manchester

Guest Editors
Donatella Bianca Maria Ficco
Centro di Ricerca Cerealicoltura e Colture Industriali
Consiglio per la Ricerca in Agricoltura e l'Analisi dell'Economia Agraria
Foggia
Italy

Grazia Maria Borrelli
Research Centre for Cereal and Industrial Crops
CREA
Foggia
Italy

Editorial Office
MDPI AG
Grosspeteranlage 5
4052 Basel, Switzerland

This is a reprint of the Special Issue, published open access by the journal *Foods* (ISSN 2304-8158), freely accessible at: https://www.mdpi.com/journal/foods/special_issues/4IK67SHLVY.

For citation purposes, cite each article independently as indicated on the article page online and as indicated below:

Lastname, A.A.; Lastname, B.B. Article Title. *Journal Name* **Year**, *Volume Number*, Page Range.

ISBN 978-3-7258-5695-4 (Hbk)
ISBN 978-3-7258-5696-1 (PDF)
https://doi.org/10.3390/books978-3-7258-5696-1

Contents

Contents

About the Editors

Donatella Bianca Maria Ficco

Donatella Bianca Maria Ficco, senior researcher in the Council for Agricultural Research and Economics, Research Centre for Cereal and Industrial Crops in Foggia. Degree in Food Science and Technologies and PhD in "Food Science and Pediatric Nutrition". Scholarship for a 4800-hour training for "experienced specialists in Plant Breeding / Biotechnology in support of the National Durum Wheat Industry (project AGROGEN-Laboratory of Genomics for characters of agronomic importance in durum wheat: identification of useful genes, functional analysis, marker-assisted selection and the development of the national seed industry)" in 2010. Scholarship for a High Formation of 1600 hours in "Research, Technological Development and High Formation". Project title: "Innovative methodologies to the competitiveness of cereal farm" with stages in Germany and Spain. Italian patent RM2011A000487 on a "protective decapeptide" isolated from wheat carrying wheat-rye chromosomal translocations 1BL.1RS capable of interfering with K562(S) cell agglutination and intestinal mucosa immune activation induced by gliadin".

Member of the Expert Working Group on Durum Wheat Genomics and Breeding in the frame of the Wheat Initiative. Author of 41 publications in international journals with an IF. Referee for international journals with an IF. Her research activity is focused on the characterization of cereal grain quality, evaluating traits linked to health benefits (nutraceutical characters) and optimizing cereal processing techniques to enhance the quality of cereal-based foods.

Grazia Maria Borrelli

Grazia Maria Borrelli, senior researcher on the Council for Agricultural Research and Economics, at the Research Centre for Cereal and Industrial Crops in Foggia. Degree in Biological Sciences and PhD in "Sustainable Agricultural Ecosystems". Study grant "Agronomic aspects of *Azospirillum* use" and research grant on "Genetic improvement of durum wheat with innovative methods with the aim to develop resistance to biotic and abiotic stress" at the Section of Foggia of the Experimental Institute for Cereal Research, under the control of the Minister of Agricultural and Forest Policies. Supervisor of two degree theses and teacher in several courses for agricultural technicians and researchers. Co-author of 36 papers in scientific indexed journals. Reviewer for international journals with an IF. Member of the "Expert working group on durum wheat genomics and breeding in the frame of the Wheat Initiative" and member of the "Italian Society of Agricultural Genetics (SIGA)". Her research activity mainly concerns durum wheat and falls within three main topics: the study of lipoxygenase's role in pasta color, cereal processing, pigmented wheats for functional foods; in vitro culture and genetic transformation, including transgenesis and new breeding technologies, for traits of agronomical and quality interest; and multiple plant responses to abiotic stresses.

Preface

Consumers increasingly desire cereal-based products that offer added health benefits beyond basic nutrition, driving a diversification of products with features like whole grains, fortified ingredients, fermented components, and functional extracts. This trend stems from growing awareness of the link between whole grains and reduced risk of chronic diseases like cardiovascular issues and diabetes, as well as the desire for natural, plant-based, and probiotic-rich foods. The challenge is to improve cereal-based foods not only for health, but also for taste to guarantee consumer's acceptance, and the economic potential of the new products. This aim can be achieved using different strategies, with some based on breeding programs, and others focusing on the upgrade of milling processes or on using new foods formulations.

Donatella Bianca Maria Ficco and Grazia Maria Borrelli
Guest Editors

foods
MDPI

Editorial

Cereals and Cereal-Based Foods: Nutritional, Phytochemical Characterization and Processing Technologies

Grazia Maria Borrelli * and Donatella Bianca Maria Ficco *

Consiglio per la Ricerca in Agricoltura e l'Analisi dell'Economia Agraria—Centro di Ricerca Cerealicolturae Colture Industriali, 71122 Foggia, Italy
* Correspondence: graziamaria.borrelli@crea.gov.it (G.M.B.); donatellabm.ficco@crea.gov.it (D.B.M.F.)

Cereals have historically been recognized as an important part of the human diet. They include a variety of grains, such as wheat, durum wheat, maize, oats, rice, rye, barley, and sorghum, representing two-thirds of human food consumption. Cereal-based foods can be cooked and eaten as whole grains, ground into flour to make a variety of products like bread, pasta, and noodles, or used as ready-to-eat breakfast cereals and snacks [1].

With cereals being an important source of energy and compounds with nutritional and healthy properties, the development of cereal food products with enhanced features presupposes the exploitation of genetic materials with good levels of endogenous nutritional and phytochemical compounds.

This Special Issue comprises a collection of eleven peer-reviewed articles focused on the study of the nutritional/functional and physico-chemical features of raw materials for the improvement of derived cereal-based foods. In total, 75 authors contributed to this collection, which consists of ten research articles and one review article that addresses the most recent advances on these topics. The articles published in this Special Issue covered the characterization of the nutritional and technological properties of cereals to be used for the production of targeted foods (five articles), the study of primary and secondary processing (four articles), and the potential reuse of by-products as new ingredients in formulations for cereal food development (two articles).

Some cereals play an important role in the nutrition of subjects affected by particular pathologies, such as gluten-related disorders or diabetes. Gluten-related disorders, particularly celiac disease, are triggered by the ingestion of gluten proteins naturally present in grains such as rye, wheat, and barley. Patients suffering from celiac disease have a reduced quality of life due to the need to adopt a strictly gluten-free diet, with all the negative nutritional implications due to the characteristics of traditional gluten-free products that are often rich in lipids and poor in vitamins, antioxidants, and fibers [2]. To meet their dietary needs, much research is aimed at investigating other gluten-free cereal species to identify genotypes with good baking performance and good acceptability and technological quality for developing novel and healthier gluten-free products. Among these, sorghum, oat, and maize represent valuable gluten-free raw materials for the development of products with improved nutritional value.

Sorghum is an excellent source of proteins, carbohydrates (starch), vitamins, minerals (such as iron and potassium), other bioactive compounds (especially phenolic and flavonoid compounds), oligo-elements (mainly magnesium and manganese), and fatty acids (ω-6 and 3) [3,4]. Moreover, it has a low glycemic index, making sorghum-derived products beneficial to diabetics, in addition to for people affected by celiac disease or gluten intolerances [4]. Oat is rich in substances beneficial to the human body, including vitamin E

https://doi.org/10.3390/foods14071234

(tocols), phenolic compounds, β-glucans, and insoluble fiber [5]. Maize is a good source of starch, proteins, and lipids, and of several bioactive compounds. Different types of maize are characterized by grain color, varying from white to yellow, pink, red, blue or black, and these pigmented maizes are richer in many secondary metabolites, such as phenolic compounds, anthocyanins, carotenoids, and tocols [6]. The nutritional and technological traits of sorghum hybrids, hulled/naked oat varieties, and white/pigmented maize genotypes and landraces were analyzed by Gazza et al. (Contribution 1) to identify species and cultivars expressing better characteristics intended for various processed foods. Differences in the content of protein, total and resistant starch, amylose, and the total dietary fiber, particularly β-glucan, as well as in the content of phenolic compounds and antioxidant capacities were found in the three species. Although all the three species possess technological properties suitable for food processing, their pasting properties, related to the structural changes affecting the starch granules during gelatinization and retrogradation, provided more precise information regarding their aptitude for making particular products. In fact, some oat genotypes showed characteristics suitable for formulations of gluten-free pasta or bakery products, while some maize and sorghum genotypes could be adapted for baking goods. Moreover, other sorghum genotypes were found to be more appropriate for food products that require stable thickening after heating and cooling, like in sauce production.

Among the minor cereals, sorghum is the most investigated, as it can grow in semi-arid conditions, as well as having nutritional peculiarities. This led to much research investigating sorghum genotype variability. Pontieri et al. (Contribution 2) assessed the nutritional and bioactive components, including proteins, carbohydrates, dietary fiber, unsaturated fatty acids, and minerals, of nine inbred varieties grown under typical Mediterranean conditions with the aim of identifying varieties with superior nutritional attributes to utilize in breeding programs.

However, sorghum's nutritional components, particularly its proteins and minerals, are less available for humans and animals due to presence of anti-nutritional factors, like tannins and phytic acid. In fact, tannins interact with proteins, creating complexes that inhibit protein/enzyme and starch digestibility. Moreover, tannins, similarly to phytate, can also bind to minerals, thus reducing their bioavailability. Therefore, various approaches such as fermentation, germination, and hydrolysis are employed to cope with these limits. Among these, germination represents a cost-effective and efficient technology that facilitates structural modifications of bound phenolic compounds into free forms, improving the nutritional value, generating various bioactive compounds, and also reducing the anti-nutritional factors (tannins, phytates, and protease inhibitors) [7,8]. Kaisoglu et al. (Contribution 3) demonstrated the effectiveness of germination over a period of 7 days in increasing the nutritional and functional qualities of red and white sorghum grains, currently limited to animal feed use, in comparison to native sorghum. They found increased protein, lipid, fiber, minerals and lignin content. The authors also observed a decrease in starch content and modifications to the molecular structure of starch granules in both sorghum varieties that could pave the way for their possible use in human nutrition.

Besides being a source of food for celiacs, maize can have many other fields of food industry application related to the production of Resistant Starch type 5 (RS5). RS5 is highly interesting in the food industry for its versatility and functional properties, having characteristics of non-toxicity, biodegradability, and an ability to selectively release bioactive compounds in the gastrointestinal tract [9]. Consequently, it is important to explore cost-effective and efficient methods to produce it. The formation of RS5, primarily associated with amylose–lipid complexes, is typical of starches with a high amylose content due to their affinity for lipid compounds [10]. However, the parameters governing gelatinization are critical for the successful formation of RS5 [11]. Recently, some studies have also investigated the potential of amylopectin-rich starches to form amylopectin–lipid complexes (ALCs), broadening RS5 sources. Starches rich in amylopectin may be a more accessible source for RS5 formation, as the size of the chains obtained by amylopectin debranching significantly affects their ability to interact with lipids. The type of fatty acid, influencing the formation of complexes with starch molecules, also affects the content of RS5, the enzymatic hydrolysis rate, and the thermal properties of the starch. Specifically, complexes formed between oleic acid and amylopectin-rich (waxy) starch under certain conditions lead to a higher content of RS5 [12]. In this context, Castro-Campos et al. (Contribution 4) analyze the potential of waxy maize starch, which is rich in amylopectin, to develop amylopectin–lipid complexes, with oleic acid (10% *w*/*w*) under different thermal and mechanical conditions. They identify boiling conditions with 45 min of stirring as the optimal conditions for the formation of stable ALCs that, restricting the access of enzymes, showed the highest resistance to the digestion process.

Overall, to obtain superior foods with high nutritional and health value, a decisive contribution derives from the degree of refinement of flours. Cereal wholemeals provide many important nutrients and bioactive compounds with an important role in reducing the risk of developing various chronic diseases and age-related conditions [13]. Dragičević et al. (Contribution 5) evaluated the content of important macro- (protein) and micro-nutrients (mineral elements), as well as bioactive compounds, such as soluble and insoluble functional dietary fiber prebiotic dietary fiber (e.g., cellulose, arabinoxylan, β-glucan, xyloglucan, and fructan) and some antioxidants (glutathione, yellow pigment, and the specific profile of phenolic compounds), in wholemeals of cereals, such as barley, durum wheat, bread wheat, ancient wheats, such as spelt and emmer wheat, and minor crops, such as triticale, sorghum, oat, and rye. The results confirmed the importance of these meals as nutraceuticals. The unique chemical profile of oat and barley reiterates their utility in the production of functional foods. Regarding wheats, some of these compounds are mostly concentrated in ancient genotypes. Despite the lower yields, compared to the modern wheat genotypes, their value is associated with their sustainability, as they are better adapted to low-input and organic agriculture, growing on marginal soils and in dry conditions [14,15].

The second part of this Editorial is mainly devoted to evaluating the effect of processing on the nutritional and technological quality of cereal-based foods. Final quality results from the interaction among many different factors including species, genotype, environment, and primary (milling) and secondary processing steps are presented. For instance, in wheat, milling has an important effect beyond reducing nutritional compounds [16], also altering the structure and functionality of starch granules, thus producing damaged starch (DS) [17]. The content of DS influences the physicochemical and rheological properties of the flours, directly affecting their industrial performance [18]. The degree of starch damage during milling is influenced by the composition of the starch (amylose/amylopectin ratio), the hardness of the grain, and the milling conditions, increasing with the severity of the milling process. Damaged starch granules increased their water absorption capacity during dough preparation, affecting dough mixing and the rheological properties of flour. Furthermore, DS granules are more accessible to endogenous β-amylase hydrolysis, which results in the production of maltose and dextrin [18]. In conventional baking procedures, maltose is used by yeast to ferment to carbon dioxide, causing loaves to rise. An understanding of how the DS content of flour affects the performance of the product has been a very important requirement in the baking industry. Teobaldi et al. (Contribution 6) assessed the morphology of wheat starch granules with different DS, correlating it with the rheological and thermal behavior of starch-based systems. They demonstrated that higher levels of DS determine a decrease in the driving force for water attachment to the starch granules allowing to a weaker water–starch polymer chain interactions and to a reduction in the evaporation temperature that could affect bread quality.

It is also interesting to account for the effect of milling on wheat protein in an industrial setting. In fact, milling also affects protein content in the kernel. Wheat is the main source of protein worldwide. Within the starchy endosperm, protein content and composition have a positive gradient from the center of the endosperm towards the aleurone layer [19,20]. The extent of this gradient highly depends on the cultivar and the level of nitrogen fertilization (N-fertilization) [20]. Milling separates the starchy endosperm from the aleurone layer and bran and grinds this into white flour. It was reported that the last break (BFs) and reduction fractions (RFs) are higher in protein content compared to the other flour fractions due to aleurone and bran contamination [21]. Other than the protein content, the relative gluten content and composition also play a crucial role in determining the functionality of flour. Hermans et al. [22] showed that the sub-aleurone has a significantly higher relative gluten content than the inner endosperm without significant differences in the HMW-GS protein proportion. Hermans et al. (Contribution 7) studied how the protein gradient within the starchy endosperm is reflected in the gluten content and composition of mill fractions, using three cultivars grown without and with N-fertilization (300 kg N ha^{-1}). The increasing protein content in successive break fractions (BFs) was shown to reflect the protein gradient within the starchy endosperm. They found an increase in protein content, also in the reduction fractions (RFs), that could be ascribable to increased levels of aleurone and bran contamination. The BFs were used to investigate the magnitude of the gradient in gluten content within the starchy endosperm and assess the effect of cultivar and N-fertilization level on it. They showed that, in all cultivars, the gluten content increased from the inner to outer endosperm to a greater extent without N-fertilization than with it. Furthermore, no consistent gradient for the gluten composition was observed when N-fertilization was applied. This study gives information for the milling industry to manage flour streams of different raw materials and optimize the process.

Secondary processing is used to transform grain/flour into edible products. This process can be studied both to identify the optimal experimental conditions to improve the performance of the end products and to evaluate the effect of the process as such on some qualitative characteristics. These two aspects have been investigated on some puffed cereals and pasta, respectively.

Puffed cereals are ready-to-eat cereal grain formulations suitable for human consumption without requiring further cooking. Non-optimal storage conditions of temperature and moisture could cause microbial spoilage and deterioration affecting their shelf life. Previous studies examined the moisture sorption and plasticity characteristics of biscuits [23] and wafers [24] and the relationship between moisture sorption and the texture of different baked products, such as corn cakes [25], by exploring their glass transition points as a critical factor involved in food deterioration and shelf life. Takahashi and Fujii (Contribution 8) analyzed the rheological and processing properties of puffed brown rice, barley, adlay, and amaranth by assessing alterations in the breaking behavior, monomolecular adsorption, moisture content, and glass transition points to evaluate the effects of moisture absorption on their rupture properties. All puffed cereals were stored under different temperatures and humidity conditions. They identified the ideal conditions for ensuring the desired crispy texture by maintaining a humidity content of <8% and a storage temperature of <40 °C.

In the case of pasta products, color plays an important role in attracting consumers and the pasta market. Troccoli et al. (Contribution 9) analyzed the traits involved in color expression, such as carotenoid pigments, yellow and brown indices, and oxidative enzymes such as lipoxygenase, responsible for pigment loss that occurs during pasta-making [26,27], and peroxidase and polyphenoloxidase, which influence brown hue, negatively affecting yellowness [28]. This study was carried out in semolina and pasta produced from eighteen durum wheat genotypes grown in eight environments. After processing, a decrease in the yellowness, measured both as the carotenoid pigment content and the yellow index, was observed, with a slight increase in the brown index. A multiple regression analysis performed on semolina traits demonstrated that the semolina yellow index was the main predictor for pasta color, while enzyme activities did not emerge as having a determining role. As a significant effect of genotype, environment, and their interaction was demonstrated on all traits investigated, to identify the genotype/environment that better combine all traits that positively (carotenoid pigment content and yellow index) or negatively (oxidative enzyme activities and brown index) influence color expression, the High-Performance Index tool, developed by Troccoli et al. [29], was applied. This information is very useful to predict final pasta color and perform suitable durum wheat breeding programs for its improvement.

With growing demand for functional, natural, and low-calorie ingredients, extensive research is being carried out to identify novel components that can be useful in the enrichment of formulations for various cereal-based products. Functional ingredients may also be derived from by-products or food processing waste, with the aim of sustainably using natural resources, providing additional economic benefits to food businesses fitting into a circular economy context.

Among the different by-products, mango peel is interesting as it is rich in dietary fiber, polyphenols, proteins and carotenoids, making it a promising source of functional ingredients in the production of various processed foods, such as fermented beverages, jellies, chips, pasta, biscuits, bread, and snacks [30]. Aviles et al. (Contribution 10) developed nutritious, high-quality snacks made from oats, dehydrated mango pulp, and mango peel flour. The authors prepared ten different mixtures, each with unique combinations and proportions of the three components. The optimal formulation able to produce the best snack quality consisted of 44.38% oats, 5.36% mango peel, and 29.24% mango pulp. Furthermore, they found that the low water activity (Aw) values enhanced quality stability at room temperature, ensuring a longer shelf life. These snacks were found to be microbiologically safe, easy to handle, visually appealing, and capable of maintaining their shape during market distribution.

Another example of by-product use is represented by rice bran [31]. Due to its unique fatty acid composition and functional components, the derived oil is recommended by the World Health Organization as one of the three major healthy edible oils, along with corn and sesame oils. It is rich in tocopherol, γ-glutamate, phytosterol, and other unsaponifiable substances, which have remarkable antioxidant, anti-inflammatory, cancer-preventive, hypoglycemic, lipid-lowering, and neuroprotective effects, becoming more beneficial to human health [32]. Rice bran is produced in large quantities annually. However, due to its low cost and high susceptibility to spoilage (3–5 days), it is mainly used to feed animals, while only 2–5% is used as a raw material for the production of rice bran oil for high production costs [32]. Huang et al. (Contribution 11) investigated the nutritional value and safety of rice bran oil. Furthermore, they have considered the stability of raw rice bran materials and the extraction and refining processes of rice bran oil and discussed food applications and sub-health regulations. The authors stated that a delayed stabilization treatment of rice bran seriously affects the overall quality of rice bran oil. Compared with traditional solvent extraction, new extraction technologies have been reported that have improved the yield and nutritional value of rice bran oil, but most of them are still in the basic research phase due to a lack of economical and applicable supporting production equipment. Therefore, despite the functional value and the health benefits of rice bran oil, its extraction is difficult to transfer to industry, thus limiting its production and consumption.

The articles published in this Special Issue range from the characterization and valorization of genetic material, from a food-chain perspective, to obtaining nutritionally superior cereal-based products suitable for target consumers in a sustainability perspective.

Author Contributions: Conceptualization, G.M.B. and D.B.M.F.; writing—original draft preparation, G.M.B. and D.B.M.F.; writing—review and editing, G.M.B. and D.B.M.F. All authors have read and agreed to the published version of the manuscript.

Acknowledgments: As Guest Editors, we would like to thank all the authors whose manuscripts have been published in this Special Issue, thus contributing to its success.

Conflicts of Interest: The authors declare no conflicts of interest.

List of Contributions:

1. Gazza, L.; Menga, V.; Taddei, F.; Nocente, F.; Galassi, E.; Natale, C.; Lanzanova, C.; Paone, S.; Fares, C. Nutritional Traits, Pasting Properties and Antioxidant Profile of Selected Genotypes of Sorghum, Oat and Maize Eligible for Gluten-Free Products. *Foods* **2024**, *13*, 990. https://doi.org/10.3390/foods13070990.
2. Pontieri, P.; Troisi, J.; Calcagnile, M.; Aramouni, F.; Tilley, M.; Smolensky, D.; Guida, M.; Del Giudice, F.; Merciai, A.; Samoylenko, I.; et al. Nutritional Composition, Fatty Acid Content, and Mineral Content of Nine Sorghum (*Sorghum bicolor*) Inbred Varieties. *Foods* **2024**, *13*, 3634. https://doi.org/10.3390/foods13223634.
3. Kayisoglu, C.; Altikardes, E.; Guzel, N.; Uzel, S. Germination: A Powerful Way to Improve the Nutritional, Functional, and Molecular Properties of White- and Red-Colored Sorghum Grains. *Foods* **2024**, *13*, 662. https://doi.org/10.3390/foods13050662.
4. Castro-Campos, F.G.; Esquivel-Fajardo, E.A.; Morales-Sánchez, E.; Rodríguez-García, M.E.; Barron-Garcia, O.Y.; Ramirez- Gutierrez, C.F.; Loarca-Pina, G.; Gaytán-Martínez, M. Resistant Starch Type 5 Formation by High Amylopectin Starch–Lipid Interaction. *Foods* **2024**, *13*, 3888. https://doi.org/10.3390/foods13233888.
5. Dragičević, V.; Simić, M.; Kandić Raftery, V.; Vukadinović, J.; Dodevska, M.; Đurović, S.; Brankov, M. Screening of Nutritionally Important Components in Standard and Ancient Cereals. *Foods* **2024**, *13*, 4116. https://doi.org/10.3390/foods13244116.
6. Teobaldi, A.G.; Carrillo Parra, E.J.; Barrera, G.N.; Ribotta, P.D. The Properties of Damaged Starch Granules: The Relationship Between Granule Structure and Water–Starch Polymer Interactions. *Foods* **2025**, *14*, 21. https://doi.org/10.3390/foods14010021.
7. Hermans, W.; Busschaert, J.; De Bondt, Y.; Langenaeken, N.A.; Courtin, C.M. The Wheat Starchy Endosperm Protein Gradient as a Function of Cultivar and N-fertilization Is Reflected in Mill Stream Protein Content and Composition. *Foods* **2023**, *12*, 4192. https://doi.org/10.3390/foods12234192.
8. Takahashi, A.; Fujii, K. Mechanical and Thermal Properties and Moisture Sorption of Puffed Cereals Made from Brown Rice, Barley, Adlay, and Amaranth. *Foods* **2025**, *14*, 189. https://doi.org/10.3390/foods14020189.
9. Troccoli, A.; Ficco, D.B.M.; Platani, C.; D'Egidio, M.G.; Borrelli, G.M. Prediction of Pasta Colour Considering Traits Involved in Colour Expression of Durum Wheat Semolina. *Foods* **2025**, *14*, 392. https://doi.org/10.3390/foods14030392.
10. Aviles-Rivera, Y.A.; Valdez-Torres, J.B.; Campos-Sauceda, J.P.; Heredia, J.B.; Hinojosa-Gómez, J.; Muy-Rangel, M.D. Designing of an Oat-Mango Molded Snack with Feasible Nutritional and Nutraceutical Properties. *Foods* **2024**, *13*, 3402. https://doi.org/10.3390/foods13213402
11. Huang, W.; Liu, B.; Shi, D.; Cheng, A.; Chen, G.; Liu, F.; Dong, J.; Lan, J.; Hong, B.; Zhang, S.; Ren, C. Research Progress on the Quality, Extraction Technology, Food Application, and Physiological Function of Rice Bran Oil. *Foods* **2024**, *13*, 3262. https://doi.org/10.3390/foods13203262.

References

1. Ficco, D.B.M.; Petroni, K.; Mistura, L.; D'Addezio, L. Polyphenols in Cereals: State of the Art of Available Information and Its Potential Use in Epidemiological Studies. *Nutrients* **2024**, *16*, 2155. [CrossRef] [PubMed]
2. Tortora, R.; Capone, P.; De Stefano, G.; Imperatore, N.; Gerbino, N.; Donetto, S.; Monaco, V.; Caporaso, N.; Rispo, A. Metabolic syndrome in patients with coeliac disease on a gluten-free diet. *Alim. Pharmacol. Therap.* **2015**, *41*, 352–359. [CrossRef]
3. Kulamarva, A.G.; Venkatesh, R.S.; Raghavan, V.G.S. Nutritional and Rheological Properties of Sorghum. *Int. J. Food Prop.* **2009**, *12*, 55–69. [CrossRef]
4. Stefoska-Needham, A. Sorghum and health: An overview of potential protective health effects. *J. Food Sci.* **2024**, *89*, A30–A41. [CrossRef] [PubMed]
5. Buerstmayr, H.; Krenn, N.; Stephan, U.; Grausgruber, H.; Zechner, E. Agronomic performance and quality of oat (*Avena sativa* L.) genotypes of worldwide origin produced under Central European growing conditions. *Field Crops Res.* **2007**, *101*, 343–351. [CrossRef]
6. Zilić, S.; Serpen, A.; Akıllıoğlu, G.; Gökmen, V.; Vančetović, J. Phenolic compounds, carotenoids, anthocyanins and antioxidant capacity of colored maize (*Zea mays* L.) kernel. *J. Agric. Food Chem.* **2012**, *60*, 1224–1231. [CrossRef]
7. Li, C.; Oh, S.-G.; Lee, D.-H.; Baik, H.-W.; Chung, H.-J. Effect of Germination on the Structures and Physicochemical Properties of Starches from Brown Rice, Oat, Sorghum, and Millet. *Int. J. Biol. Macromol.* **2017**, *105*, 931–939. [CrossRef]
8. Liu, S.; Wang, W.; Lu, H.; Shu, Q.; Zhang, Y.; Chen, Q. New Perspectives on Physiological, Biochemical and Bioactive Components during Germination of Edible Seeds: A Review. *Trends Food Sci. Technol.* **2022**, *123*, 187–197. [CrossRef]
9. Farooq, M.A.; Yu, J. Recent Advances in Physical Processing Techniques to Enhance the Resistant Starch Content in Foods: A Review. *Foods* **2024**, *13*, 2770. [CrossRef]
10. Hu, H.; Jiang, H.; Sang, S.; McClements, D.J.; Jiang, L.; Wen, J.; Jin, Z.; Qiu, C. Research Advances in Origin, Applications, and Interactions of Resistant Starch: Utilization for Creation of Healthier Functional Food Products. *Trends Food Sci. Technol.* **2024**, *148*, 104519. [CrossRef]
11. Castro-Campos, F.G.; Morales-Sánchez, E.; Cabrera-Ramírez, Á.H.; Martinez, M.M.; Rodríguez-García, M.E.; Gaytán-Martínez, M. High Amylose Starch Thermally Processed by Ohmic Heating: Electrical, Thermal, and Microstructural Characterization. *Innov. Food Sci. Emerg. Technol.* **2023**, *87*, 103417. [CrossRef]
12. Sun, S.; Hong, Y.; Gu, Z.; Cheng, L.; Li, Z.; Li, C. An Investigation into the Structure and Digestibility of Starch-Oleic Acid Complexes Prepared under Various Complexing Temperatures. *Int. J. Biol. Macromol.* **2019**, *138*, 966–974. [CrossRef]
13. Garutti, M.; Nevola, G.; Mazzeo, R.; Cucciniello, L.; Totaro, F.; Bertuzzi, C.A.; Caccialanza, R.; Pedrazzoli, P.; Puglisi, F. The Impact of Cereal Grain Composition on the Health and Disease Outcomes. *Front. Nutr.* **2022**, *9*, 888974. [CrossRef]
14. Basile, G.; De Maio, A.C.; Catalano, A.; Ceramella, J.; Iacopetta, D.; Bonofiglio, D.; Saturnino, C.; Sinicropi, M.S. Ancient Wheat as Promising Nutraceuticals for the Prevention of Chronic and Degenerative Diseases. *Curr. Med. Chem.* **2023**, *30*, 3384–3403. [CrossRef] [PubMed]
15. Majzoobi, M.; Jafarzadeh, S.; Teimouri, S.; Ghasemlou, M.; Hadidi, M.; Brennan, C.S. The Role of Ancient Grains in Alleviating Hunger and Malnutrition. *Foods* **2023**, *12*, 2213. [CrossRef]
16. Ficco, D.B.M.; Borrelli, G.M.; Miedico, O.; Giovanniello, V.; Tarallo, M.; Pompa, C.; De Vita, P.; Chiaravalle, A.E. Effects of grain debranning on bioactive compounds, antioxidant capacity and essential and toxic trace elements in purple durum wheats. *LWT—Food Sci. Technol.* **2020**, *118*, 108734. [CrossRef]
17. Yu, J.; Wang, S.; Wang, J.; Li, C.; Xin, Q.; Huang, W.; Zhang, Y.; He, Z.; Wang, S. Effect of laboratory milling on properties of starches isolated from different flour millstreams of hard and soft wheat. *Food Chem.* **2015**, *172*, 504–514. [CrossRef]
18. Liu, C.; Li, L.; Hong, J.; Zheng, X.; Bian, K.; Sun, Y.; Zhang, J. Effect of mechanically damaged starch on wheat flour, noodle and steamed bread making quality. *Int. J. Food Sci. Technol.* **2014**, *49*, 253–260. [CrossRef]
19. Shewry, P.R.; Wan, Y.; Hawkesford, M.J.; Tosi, P. Spatial distribution of functional components in the starchy endosperm of wheat grains. *J. Cereal Sci.* **2020**, *91*, 102869. [CrossRef]
20. Hermans, W.; Mutlu, S.; Michalski, A.; Langenaeken, N.A.; Courtin, C.M. The Contribution of Sub-Aleurone Cells to Wheat Endosperm Protein Content and Gradient Is Dependent on Cultivar and N-Fertilization Level. *J. Agric. Food Chem.* **2021**, *69*, 6444–6454. [CrossRef]
21. Jacobs, P.J.; Hemdane, S.; Claes, S.; Mulders, L.; Langenaeken, N.A.; Dewettinck, K.; Courtin, C.M. Wheat bran-associated subaleurone and endosperm proteins and their impact on bran-rich bread-making. *J. Cereal Sci.* **2018**, *81*, 99–107. [CrossRef]
22. Hermans, W.; Geisslitz, S.; De Bondt, Y.; Langenaeken, N.A.; Scherf, K.A.; Courtin, C.M. Nano LC-MS/MS protein analysis on laser-microdissected wheat endosperm tissues: A comparison between aleurone, sub-aleurone and inner endosperm. *Food Chem.* **2024**, *437*, 137735. [CrossRef]
23. Sampaio, M.R.; Marcos, K.S.; Moraes, I.C.F.; Perez, V.H. Moisture adsorption behavior of biscuits formulated using wheat, oatmeal and passion fruit flour. *J. Food Process. Preserv.* **2009**, *33*, 105–113. [CrossRef]

24. Martinez-Navarrete, N.; Moraga, G.; Talens, P.; Chiralt, A. Water sorption and the plasticization effect in wafers. *Int. J. Food Sci. Technol.* **2004**, *39*, 555–562. [CrossRef]
25. Higo, A.; Wada, Y.; Sato, Y. Relationship between physical properties and water layers in commercial confectionary products during moisture sorption processing. *J. Jpn. Soc. Food Sci. Technol.* **2012**, *59*, 572–582. [CrossRef]
26. Borrelli, G.M.; Troccoli, A.; Di Fonzo, N.; Fares, C. Durum wheat lipoxygenase activity and other quality parameters that affect pasta color. *Cereal Chem.* **1999**, *76*, 335–340. [CrossRef]
27. Carrera, A.; Echenique, V.; Zhang, W.; Helguera, M.; Manthey, F.; Schrager, A.; Picca, A.; Cervigni, G.; Dubcovsky, J. A deletion at the *Lpx-B1* locus is associated with low lipoxygenase activity and improved pasta color in durum wheat (*Triticum turgidum* ssp. *durum*). *J. Cereal Sci.* **2007**, *45*, 67–77. [CrossRef]
28. Cabas-Luhmann, P.A.; Manthey, F.A.; Elias, E.M. Variations of colour, polyphenol oxidase and peroxidase activities during the production of low temperature dried pasta in various durum wheat genotypes. *Int. J. Food Sci. Technol.* **2021**, *56*, 4700–4709. [CrossRef]
29. Troccoli, A.; De Leonardis, A.M.; Platani, C.; Borrelli, G.M. High Performance Index as a tool to identify the best combination of pearled fractions and durum wheat genotypes for semolina and pasta colour improvement. *Int. J. Food Sci. Technol.* **2021**, *56*, 4799–4806. [CrossRef]
30. Kumar, A.G.; Singh, P.G.; Beer, K.; Pongener, P.; Ravi, C.S.; Singh, S.; Verma, A.; Singh, A.; Thakur, M.; Tripathy, S.; et al. A review on valorization of different byproducts of mango (*Mangifera indica* L.) for functional food and human health. *Food Biosci.* **2022**, *48*, 101783. [CrossRef]
31. Tan, B.L.; Norhaizan, M.E.; Chan, L.C. Rice Bran: From Waste to Nutritious Food Ingredients. *Nutrients* **2023**, *15*, 2503. [CrossRef] [PubMed]
32. Punia, S.; Kumar, M.; Sandhu, K.S.; Whiteside, W.S. Rice-bran oil: An emerging source of functional oil. *J. Food Process. Preserv.* **2021**, *45*, e15318. [CrossRef]

foods

MDPI

Article

Nutritional Traits, Pasting Properties and Antioxidant Profile of Selected Genotypes of Sorghum, Oat and Maize Eligible for Gluten-Free Products

Laura Gazza [1], Valeria Menga [2], Federica Taddei [1], Francesca Nocente [1], Elena Galassi [1], Chiara Natale [1], Chiara Lanzanova [3], Silvana Paone [2] and Clara Fares [2,*]

[1] CREA-IT Consiglio Per la Ricerca in Agricoltura e L'analisi Dell'economia Agraria Centro di Ricerca Ingegneria e Trasformazioni Agroalimentari, Via Manziana, 30, 00189 Roma, Italy; laura.gazza@crea.gov.it (L.G.); federica.taddei@crea.gov.it (F.T.); francesca.nocente@crea.gov.it (F.N.); elena.galassi@crea.gov.it (E.G.); chiara.natale@crea.gov.it (C.N.)

[2] CREA-CI Consiglio Per la Ricerca in Agricoltura e L'analisi Dell'economia Agraria Centro di Ricerca Cerealicoltura e Colture Industriali, S.S. 673, km 25.200, 71122 Foggia, Italy; valeria.menga@crea.gov.it (V.M.); silvana.paone@crea.gov.it (S.P.)

[3] CREA-CI Consiglio Per la Ricerca in Agricoltura e L'analisi Dell'economia Agraria Centro di Ricerca Cerealicoltura e Colture Industriali, Via Stezzano, 24, 24126 Bergamo, Italy; chiara.lanzanova@crea.gov.it

* Correspondence: clara.fares@crea.gov.it; Tel.: +39-0881-742972; Fax: +39-0881-713150

Abstract: The technological and nutritional traits of food-grade sorghum hybrids, hulled/naked oat varieties and maize genotypes of different colors were studied for novel and healthier gluten-free foods. Oat genotypes showed the highest protein content, followed by maize and sorghum. The total starch and the total dietary fiber content were quite similar among the three species. Great variation was found in the amylose content, and the highest was in sorghum (27.12%), followed by oat 16.71% and maize 10.59%. Regarding the pasting profile, the rank of Peak Viscosity was sorghum (742.8 Brabender Unit, BU), followed by maize (729.3 BU) and oat (685.9 BU). Oat and sorghum genotypes had similar average breakdown (407.7 and 419.9 BU, respectively) and setback (690.7 and 682.1 BU, respectively), whereas maize showed lower values for both parameters (384.1 BU and 616.2 BU, respectively). The total antioxidant capacity, only in maize, significantly correlated with total flavonoid, phenolic and proanthocyanidin contents, indicating that all the measured compounds contributed to antioxidant capacity. The study indicated the importance of sounding out the nutritional and technological characteristics of gluten-free cereals in order to select suitable cultivars to be processed in different gluten-free foods with better and healthier quality.

Keywords: gluten-free; pasting properties; antioxidant activity; pigmented maize; food grade sorghum; oat

1. Introduction

In the last 50 years, an increasing trend in the incidence of coeliac disease (CD) has been observed, but at present, a lifelong, strict gluten-free diet is the only effective treatment. However, lifelong gluten exclusion from the diet may have a negative impact on the quality of life of CD patients [1]. Furthermore, nutritional complications are frequent, as gluten-free food is often rich in lipids and poor in vitamins, antioxidants and fibers [2]. There is an increasing interest in the search for gluten-free cereal species with good baking performance suitable for the diet of CD patients [3], but insufficient information is available to identify genotypes for obtaining gluten-free pasta of good acceptability and technological quality [4]. The rheological properties of doughs could be used as a tool to select appropriate raw materials and investigate their behavior under different transformation processes in terms of suitable product formulation. Among the gluten-free cereal species, sorghum, oat and maize represent valuable raw materials for the development of products with improved nutritional value.

Sorghum (*Sorghum bicolor* L.) is the fifth cereal crop worldwide, and it is considered among the species that could ensure global food security in the climate change scenario. Besides its easy adaptability to a wide range of growing conditions, being a C4 plant, sorghum has higher photosynthetic nitrogen- and water-use efficiencies than C3 plants, hence good tolerance to drought, which is expected to increase in the future. Sorghum, like other cereals, is an excellent source of starch and proteins. It represents the major staple food crop for a fair share of the populations living in the semi-arid tropics [5], whereas in Western countries, it is mainly intended for feed and ethanol production. In recent years, the interest in sorghum for human foods has increased due to its nutritional characteristics, such as the presence of bioactive compounds (especially flavonoids), oligo-elements (mainly magnesium and manganese) and fatty acid profile (omega 6 and 3). Moreover, the low glycaemic index and the absence of gluten make it beneficial to diabetics and safe for people affected by celiac disease or gluten intolerances [6]. Around the world, it is consumed as porridges, bread, cookies, tortillas, extruded products and beverages [7], and in recent years it has been processed into noodles and pasta [8,9].

Oat (*Avena sativa* L.) is the seventh cereal crop worldwide, but with a general decreasing trend [10]; in particular, in Western Europe, the area dedicated to oat has halved in the last 30 years. Oat is rich in substances beneficial to the human body, vitamin E (tocols), phytic acid, phenolic compounds and avenanthramides are the most abundant antioxidants, but flavonoids and sterols are also present. These compounds are concentrated in the outer layers of the seed [11]. The introduction of oats in the formulation of GF products is very recent, and the AIC (Associazione Italiana Celiachia) [12] has fully included it in the gluten-free cereals. In addition to the antioxidants listed above, the richness of beta-glucans and insoluble fiber represent functional compounds that, in the formulation of GF products, can make a significant contribution to improving the health quality of these foods. Several studies have indicated that the prevalence of celiac disease in patients with type 1 diabetes mellitus (T1DM) is between 3 and 10%, which is more than ten times the prevalence in the general population [13]. Both diseases result from a complex interaction between genetic susceptibility and environmental exposure, being associated with the major histocompatibility complex class II antigen DQ2 and sharing non-HLA loci [14]. These insights make the need to develop low glycemic index foods more compelling.

Maize (*Zea mays* L.) is one of the most important crops cultivated worldwide due to its huge versatility and multiple uses such as food, forage and industrial purposes. In Latin America, Asia and Africa, it is used for the preparation of traditional foods [15], but in recent years, in Western countries, the use of this crop for gluten-free foods has increased due to the increase in consumption of GF foods. Maize is a good source of starch, proteins and lipids, and it also contains several bioactive compounds that are important for human health [16]. The different types of maize are characterized by a grain color varying from white to yellow, pink, red, blue or black. Pigmented maize contains many secondary metabolites, such as phenolic compounds, anthocyanins, carotenoids and tocols [17,18]. The principal distinction between blue and purple maize varieties is the different localization in the distribution of anthocyanins. In purple maize varieties, these flavonoids are localized most abundantly in the pericarp of the kernel, while blue varieties typically produce anthocyanins in the aleurone [19]. Because the content of proteins and antioxidants in traditional GF flours is low, the use of nutrient and dietary fiber-rich ingredients has been proposed to improve the nutritional quality of such foods [20]. Therefore, the aim of this study was to analyze the technological, biochemical, and nutritional characteristics of twenty-six genotypes of food-grade sorghum, oat and maize in order to find eligible genotypes for developing novel and healthier gluten-free products.

2. Materials and Methods

2.1. Plant Material

Seeds and flours of food grade/zootechnical use hybrids of sorghum, genotypes of naked/hulled seed of oat with both clear and colored glumes, and white/pigmented maize genotypes, were analyzed.

Oat, maize and sorghum genotypes were derived from seed collections of CREA Research Centre for Cereal and Industrial Crops, Foggia and Bergamo and Research Centre for Engineering and Agro-food Processing of Rome, respectively. In particular, were analyzed: (i) one sorghum genotype for zootechnical use (cultivar Aralba, RV-Venturoli, Bologna, Italy), four food-grade hybrids PSE7431, PSEAG4E44, DMS (Padana Sementi Elette, Padova, Italy), Diamond (APSOV Sementi, Voghera, PV, Italy), and three Bolivian food-grade hybrids SW6129, SW6143W, SW6237W (SemWest Semillas, LaPaz Bolivia); (ii) six naked oat genotypes (Abel, Konradin, Kripton, Kynom, Lexic and Terra) and three hulled genotypes (Corneil, Genziana and Nigra); (iii) two maize Bolivian landraces, Hualtaco (white) and Kully (purple), collected in the frame of International Cooperation Project (P.S.G.O. km0 Bolivia), four Italian landraces VA116W and VA522W (white) VA 572 (yellow), VA1268 (red), and three pigmented genotypes VA116 (purple), VA522 (blue) and VA572 (black), the latter three obtained by crossing VA116W, VA522W, VA572 with germplasm of Bolivian origin "Kully", Mexican "Azul" and Spanish "Millo Corvo".

2.2. Chemical and Physical Characterization

All samples were milled to whole meal flour using a laboratory mill (Retsch ZM 200 Haan, Dusseldorf, Germany) at 12,000 rpm and 0.5 or 1.0-mm sieve, depending on the requirements of each analysis. All analyses were performed in triplicate. The sample moisture was measured using a thermobalance (Sartorius MA 40, Goettingen, Germany) at 120 °C just before the chemical analyses in order to express all data on a dry weight basis (dw). Protein content was measured according to the ICC 105/2 method [21]. Factors of 6.25 for sorghum and maize and 5.83 for oats were used to convert nitrogen to protein [22]. Total and resistant starch (TS and RS) content were determined by enzymatic method using the Megazyme (Bray, Dublin, Ireland) kits K-TSTA and K-RSTAR according to AOAC methods 996.11 [23] and 2002.02 [24], respectively. Amylose content was determined using the Megazyme Amylose/Amylopectin assay kit K-AMYL. The content of total dietary fiber (TDF) was measured using an enzymatic kit for fiber determination (Bioquant, Merck, Darmstadt, Germany) according to the AOAC Official Method 991.42 [25]. Protein, TS, RS, TDF and Amylose content were expressed as percentage w/w on dry weight basis. The methods ISO 520:2010 [26] and ISO 7971-1:2009 [27] were used to determine the thousand kernel weight (TKW) and test weight (TW). The β-glucan content was determined by enzymatic method according to AOAC Method 995.16 [28].

2.3. Total Flavonoid, Phenolic, Proanthocyanidin Contents and Antioxidant Capacities

Phenolic compounds were extracted according to Beta et al. [29], with minor modifications. The samples (1 g), ground with a 1.0 mm– sieve, were extracted using 8 mL methanol acidified with 1N HCl (80:20; v/v) for 30 min in an ultrasonic bath (BRANSONIC 2200, Branson, Danbury, NY, USA). The mixtures were centrifuged at 1000× g for 15 min and were used for the determination of the total phenolic content (TPC), total proanthocyanidin content (TPAC), total flavonoid content (TFC) and total antioxidant capacity (TAC) by ABTS assay.

Determination of total phenolic content was performed according to the procedure described by Singleton and Rossi [30]. Ferulic acid was used as the standard, and the data were expressed in mg ferulic acid equivalents (FA) per g of dry matter. A calibration curve was prepared by using increasing concentrations of ferulic acid (Sigma-Aldrich, Milano, Italy).

The total proanthocyanidin content (TPAC) was determined according to the modified vanillin assay [31] and expressed in mg catechin equivalents (CE) per g of dry matter.

The total flavonoid content (TFC) was determined according to Eberhardt et al. [32], using catechin as standard. Data were expressed in mg catechin equivalents (CE) per g of dry matter.

The measure of total antioxidant capacity (TAC) was determined according to Re et al. [33], with some modifications. ABTS (2,2-azino-bis-[3-ethylbenzothiazoline 6-sulphonic acid]) was dissolved in water at a 7 mM concentration. ABTS radical cation was produced, allowing the reaction of the ABTS stock solution with potassium persulfate (2.45 mM, final concentration) for 16 h in the dark and at room temperature before use. ABTS radical cation solution was diluted with ethanol to an absorbance of 0.70 ± 0.02 at 734 nm. The antioxidant capacity was expressed as Trolox equivalent antioxidant capacity (TEAC) in mmol of Trolox equivalents (TE)/kg sample on a dry matter basis.

2.4. *Micro-Visco Amylograph Analysis*

Pasting properties of the finely ground cereal samples (<0.5 mm) were measured using a micro visco-amylograph (Brabender OHG, Duisburg, Germany) according to Marengo et al. [34] with slight modifications. Fifteen grams (on a 14% moisture basis) of oat, maize or sorghum were suspended in 100 mL of distilled water and heated in the visco-amylograph by using the following time-temperature profile: heating from 30 up to 95 °C at a rate of 3 °C/min, holding at 95 °C for 20 min, cooling to 30 °C at a rate of 3 °C/min under constant stirring (250 rpm). The torque measuring range was 300 cmg. The viscosity was expressed in Brabender units (BU). The parameters resulting from this analysis were: pasting temperature (PT), defined as the minimum temperature for cooking the flour; peak viscosity (PV), defined as the highest viscosity reached during heating; peak temperature reached at PV, breakdown (BD), which represents the difference between the viscosity at the peak and at the minimum, setback, defined as the difference between the viscosity at the end of cooling period and the minimum viscosity at the start of cooling period.

2.5. *Statistical Analyses*

Replicated results were expressed as mean ± standard deviation. A one-way analysis of variance was performed with the MSTATC program (Michigan State University, East Lansing, MI, USA), followed by the Duncan multiple range test for a post-hoc comparison of means, applied to assess significant differences ($p \leq 0.05$) for each considered parameter. Correlation analysis between the parameters was performed by using Past software (4.03 free version) through Pearson's r coefficient ($p \leq 0.05$).

3. Results and Discussion

3.1. *Physical and Nutritional Traits*

3.1.1. Thousand Kernel Weight

Thousand kernel weight (TKW, Table 1) is an important technological parameter, playing a large role in flour yield at milling [35]. Little variability was found amongst sorghum hybrids, with a mean value of 23.26 g; the minimum and the maximum values were observed in PSE7431 (21.30 g) and DMS (25.62 g), respectively. These values were similar to those reported by Galassi et al. [36] and fell in the range found by Aruna et al. [37], who analyzed 60 sorghum genotypes amongst germplasm, cultivated and parental lines of the hybrids. Oat genotypes exhibited TKW values ranging from 23.70 g (Genziana, hulled genotype) to 12.55 g (Kripton, naked genotype); the average TWK was 16.49 g and 21.78 g for naked and hulled genotypes, respectively. This variability was partially ascribable to the presence of husks in hulled genotypes but also to the minus kernel weight, typically of naked genotypes, as ascertained also by Buerstmayr et al. [38] in a large collection of 120 oat genotypes. Maize genotypes showed great variability in TKW values, ranging from a minimum of 246.0 g (VA116W) to a maximum of 741.5 g (Hualtaco), with an average of 375.7 g.

Table 1. Technological and nutritional traits of the twenty-six genotypes of sorghum, oat and maize.

	SAMPLES	TKW (g)		Protein (%)		TDF (%)		β-Glucan (%)		RS (%)		TS (%)		Amylose (%)	
		Mean	sd	Mean	sd	Mean	sd	Mean	sd	Mean	sd	Mean	sd	Mean	sd
SORGHUM	Aralba	24.33 ^{ab}	1.82	10.62 ^{c}	0.19	9.40 ^{c}	0.55	n.d.		0.48 ^{e}	0.05	69.49 ^{c}	0.35	23.90 ^{b}	0.87
	Diamond	23.76 ^{ac}	0.20	12.68 ^{a}	0.04	10.89 ^{c}	0.21	n.d.		0.40 ^{e}	0.01	57.77 ^{f}	0.98	26.51 ^{b}	3.28
	DMS	25.62 ^{a}	0.20	11.56 ^{b}	0.03	11.41 ^{b}	0.22	n.d.		0.44 ^{e}	0.00	73.04 ^{ab}	0.97	26.56 ^{b}	3.18
	PSEAG4E44	22.00 ^{bc}	0.50	9.44 ^{ef}	0.18	12.21 ^{a}	0.18	n.d.		0.51 ^{e}	0.02	65.57 ^{e}	1.49	54.40 ^{a}	2.84
	PSE7431	21.30 ^{c}	2.30	9.80 ^{de}	0.11	11.75 ^{ab}	0.06	n.d.		0.66 ^{d}	0.05	67.02 ^{de}	1.11	16.83 ^{c}	1.21
	SW6129	22.93 ^{bc}	0.70	9.20 ^{f}	0.36	8.52 ^{f}	0.33	n.d.		2.32 ^{a}	0.10	69.19 ^{cd}	1.18	27.37 ^{b}	0.92
	SW6143W	22.7 0 ^{bc}	1.50	10.91 ^{c}	0.46	10.22 ^{d}	0.40	n.d.		1.48 ^{c}	0.09	72.22 ^{b}	1.38	15.64 ^{c}	1.74
	SW6237W	23.47 ^{abc}	1.72	9.90 ^{d}	0.39	8.23 ^{f}	0.66	n.d.		2.02 ^{b}	0.12	74.88 ^{a}	0.12	25.75 ^{b}	4.79
	Means ± SE	23.26 ± 1.72		10.51 ± 1.15		10.33 ± 1.44				1.04 ± 0.75		68.65 ± 5.18		27.12 ± 11.68	
OAT	Abel	15.60 ^{f}	0.42	13.42 ^{e}	0.06	9.84 ^{ce}	0.33	3.97 ^{b}	0.10	0.51 ^{bc}	0.00	51.65 ^{d}	0.25	17.11 ^{c}	0.44
	Corneil	19.75 ^{c}	0.07	17.00 ^{a}	0.00	10.11 ^{cd}	1.21	4.63 ^{a}	0.16	0.52 ^{b}	0.00	53.58 ^{cd}	1.24	16.64 ^{c}	0.23
	Genziana	23.70 ^{a}	0.14	13.18 ^{f}	0.03	10.03 ^{ce}	0.03	3.07 ^{cd}	0.59	0.51 ^{bc}	0.00	60.67 ^{a}	0.83	15.37 ^{c}	1.70
	Konradin	20.10 ^{c}	0.28	13.65 ^{d}	0.02	9.75 ^{ce}	0.39	3.07 ^{cd}	0.25	0.48 ^{de}	0.00	61.38 ^{a}	1.76	16.45 ^{c}	1.72
	Kripton	12.55 ^{g}	0.07	13.55 ^{de}	0.02	11.38 ^{b}	0.01	3.01 ^{cd}	0.21	0.53 ^{b}	0.01	59.96 ^{ab}	1.02	28.98 ^{b}	1.01
	Kynom	18.55 ^{d}	0.07	13.71 ^{d}	0.03	8.89 ^{e}	0.34	2.52 ^{d}	0.11	0.47 ^{e}	0.00	56.57 ^{bc}	3.66	8.88 ^{d}	0.59
	Lexic	16.45 ^{e}	0.07	13.40 ^{e}	0.06	8.91 ^{de}	0.96	3.18 ^{c}	0.04	0.55 ^{a}	0.01	56.94 ^{bc}	0.97	7.08 ^{d}	1.13
	Nigra	21.90 ^{b}	0.14	14.17 ^{c}	0.13	10.79 ^{bc}	0.44	2.76 ^{cd}	0.13	0.50 ^{cd}	0.01	58.91 ^{ab}	1.34	32.78 ^{a}	5.01
	Terra	15.70 ^{f}	0.14	14.52 ^{b}	0.17	14.25 ^{a}	0.20	4.80 ^{a}	0.23	0.48 ^{de}	0.00	55.16 ^{c}	1.08	7.12 ^{d}	1.43
	Means ± SE	18.26 ± 3.40		14.07 ± 1.14		10.44 ± 1.64		3.45 ±0.82		0.50 ± 0.03		57.20 ± 3.44		16.71 ± 8.91	
MAIZE	Hualtaco	741.50 ^{a}	2.1	6.93 ^{i}	0.13	8.19 ^{b}	0.88	n.d.		0.08 ^{b}	0.01	70.91^{a}	0.01	11.27 ^{b}	2.58
	Kully	322.00 ^{d}	1.4	7.98 ^{h}	0.14	9.77 ^{a}	0.34	n.d.		0.21 ^{a}	0.07	70.67 ^{a}	1.12	10.72 ^{b}	2.95
	VA 116 W	246.00 ^{h}	1.4	10.99 ^{e}	0.06	10.90 ^{a}	0.39	n.d.		0.15 ^{ab}	0.01	64.34 ^{c}	0.69	21.64 ^{a}	0.89
	VA 116 W Purple	305.50 ^{e}	2.1	13.35 ^{a}	0.06	10.16 ^{a}	0.33	n.d.		0.24 ^{a}	0.07	66.65 ^{bc}	2.22	8.26 ^{b}	0.32
	VA 1268 Red	421.50 ^{b}	2.1	11.87 ^{d}	0.12	10.60 ^{a}	0.66	n.d.		0.18 ^{a}	0.01	67.00 ^{bc}	0.83	6.33 ^{b}	0.68
	VA 522 W	355.50 ^{c}	2.1	13.01 ^{b}	0.10	9.43 ^{ab}	0.42	n.d.		0.17 ^{ab}	0.06	66.74 ^{bc}	0.04	10.35 ^{b}	3.46
	VA 522 W Blue	422.50 ^{b}	2.1	10.01 ^{f}	0.08	8.12 ^{b}	1.19	n.d.		0.23 ^{a}	0.02	71.09 ^{a}	2.17	10.30 ^{b}	1.66
	VA 572	286.50 ^{f}	2.1	12.34 ^{c}	0.05	10.70 ^{a}	0.64	n.d.		0.17 ^{ab}	0.01	69.18 ^{ab}	2.31	8.61 ^{b}	1.08
	VA 572 Black	280.00 ^{g}	1.4	9.38 ^{g}	0.10	10.32 ^{a}	0.09	n.d.		0.24 ^{a}	0.02	71.64 ^{a}	2.62	7.80 ^{b}	1.81
	Means ± SE	375.70 ± 145.6		10.70 ± 2.19		9.80 ± 1.11				0.18 ± 0.06		68.69 ± 2.77		10.59 ± 4.55	

Results presented in the table are expressed as the mean value and standard deviation (SD) for three replications. Different letters in the same column indicate significant differences at $p \leq 0.05$. sd = standard deviation; SE = standard error of mean; TKW = Thousand kernel weight; TDF = Total Dietary Fibre; RS = Resistant Starch; TS = Total Starch; n.d. = not determined.

3.1.2. Proteins

Besides their nutritional properties, proteins are important for the processing aptitude of cereals as they contribute, with the amylose and amylopectin chains, to giving a better texture to gluten-free products [39,40]. Among the three species analyzed, oat genotypes showed, on average, the highest protein content (14.07%), ranging from 13.18% to 17.00% observed in Genziana and Corneil, respectively (Table 1). Maize exhibited a mean value of 10.70% and ranged from 6.93% to 13.35% in Hualtaco and VA116 purple, respectively. Sorghum hybrids presented a mean value of 10.51% and ranged from 9.20% to 12.68% in PSE7431 and Diamond, respectively. It is also noteworthy that maize hybrids revealed huger variability when compared to both sorghum hybrids and oat genotypes for this parameter.

3.1.3. Total Dietary Fiber and β-Glucans

High dietary fiber intake is associated with health benefits for the intestine, and fiber content should be taken into the right consideration in gluten-free cereals alternative to the more diffused refined rice (average TDF content reported in literature is 0.5%) for the manufacturing of gluten-free foods, since a long-term food choice of individuals following a strict gluten-free diet often resolved in not sufficient fiber intake, anyway generally under the recommended daily allowance (RDA = 25 g/die for an adult) [41]. On average, total dietary fiber content was quite similar amongst the three cereal species (Table 1), but a fair variability was observed, especially in sorghum hybrids and oat genotypes. Moreover, the high content of TDF in our samples of maize could be ascribable to the milling process adopted in our study, i.e., whole grain, and it was comparable with the study performed on whole grain maize by Blandino et al. (2017), who found about 11% TDF in their genotypes [42]. It should be noted that in oats, a certain amount of β-glucan also contributes to the TDF content. Indeed, β-glucan is a component of the soluble fiber, which plays a major role in a number of the putative health benefits attributed to oat and barley products, mainly the reduction of blood glucose and -cholesterol [43]. In addition to their health-promoting properties, β-glucans are used as thickeners, emulsifiers and fat substitutes in foods, which is of crucial importance mainly in the manufacturing of GF products. The β-glucan mean content was 3.45%, and the highest significant content was measured in Terra (4.80% naked genotype) and Corneil (4.63% hulled genotype). In maize and sorghum, the level of β-glucan was under the limit of the detection of the method.

3.1.4. Starch Characterization

Maize and sorghum hybrids had similar total starch contents (TS), 68.69% on average, about 10 percentage points more than oat genotypes (Table 1); in particular, the highest TS content (74.88%) was detected in the SW6237W sorghum hybrid, whereas the lowest was recorded in oat cultivar Abel (51.65%). The high total starch content of sorghum and maize hybrids makes these cereals a valuable alternative raw material to produce highly nutritious gluten-free foods or for the extraction of starch and derivatives for industrial purposes.

Because of the health benefits related to foods with increased resistant starch (RS) and decreased glycemic index, there is a growing interest in developing foods with higher resistant starch contents, even if a very high amount of resistant starch could affect the digestibility of foods and intestinal functions [44]. RS content in all the gluten-free cereals analyzed was found below the quantification limit of the method (2%). Nevertheless, sorghum exhibited the highest average content (1.04%), followed by oat (0.50%) and maize, with very low values (0.18%). The highest content of RS was observed in the Bolivian SW sorghum samples, with an average of almost 2% (Table 1).

The same trend was observed for the amylose content, the highest values being reported in sorghum hybrids, on average 27.12%, followed by oat (16.71%) and maize (10.59%), confirming the positive correlation between RS and amylose content, as previously observed by Biselli et al. [45] in rice.

3.1.5. Pasting Properties

Starch functionality in foods is determined by amylose and amylopectin and the physical organization of these macromolecules into the granular structure [46]. Amylopectin, with its multiple branched chains of (1–4)-α-glucans interlinked by (1–6)-α-linkages, is the major component of starch, with the unbranched amylose accounting for the minor fraction. These two components have different solubility, with amylopectin being more soluble than amylose [47]. The micro-visco amylograph analysis mimics the cooking process and, thus, allows the measurement of the pasting properties of a flour-water mixture during continuous heating and mixing for a given period. The pasting profile provides a very important indication of the structural changes affecting the starch granules during gelatinization and retrogradation.

Sorghum had the highest mean value of the pasting temperature (PT) (Table 2) (74.1 °C), followed by maize (71.5 °C) and oat (60.1 °C); consequently, oat flours required less energy to cook than the other two species. Generally, lower PT relates to a lower degree of swelling of starch granules and lower starch content [48]. Observing the rank obtained, the lowest value of oat is found as this species showed the lowest starch content (TS), counterbalanced with the highest protein content (Table 1). The rank of the tree species of the mean value for the peak viscosity was the same as observed for PT (Table 2). The highest mean value was recorded in sorghum at 742.8 (BU), followed by maize at 729.3 (BU) and oat at 685.9 (BU). Among sorghum genotypes, a great variability was found, and Diamond, SW6129 and SW6237W showed the highest significant values, more than 800 BU, while PSE7431 showed the lowest value (512.5 BU). High peak viscosity is recognized as the marker of starch gelatinization intensity [49]. Therefore, SW6129 and SW6237W hybrids that have the highest PV and setback could be appreciable for food products that request stable thickening after heating and cooling, like in sauces production. On the contrary, PSEAG4E44, with low PV and setback, is more suitable for baking goods because the gel is not too hard and less inclined to water separation, thus having a longer shelf life (i.e., less staling). The mean value of viscosity of the breakdown was 419.9 BU, and six genotypes showed no statistical difference; conversely, the highest value was found in PSEAG4E44 (505 BU), and the lowest was in PSE7431 (350.5 BU). The breakdown provides information about the rigidity or fragility of swollen starch granules, and it is an index of paste stability during cooking. Accordingly, out of 7 sorghum genotypes analyzed, six had similar starch resistance to mechanical and thermal stresses, and only PSEAG4E44 had different behavior, confirming good pasting properties.

Table 2. Micro Visco Analyzer parameters of the twenty-six genotypes of sorghum, oat and maize.

	SAMPLES	Pasting Temperature (°C)		Peak Viscosity (BU)		Breakdown (BU)		Setback (BU)	
		Mean	sd	Mean	sd	Mean	sd	Mean	sd
SORGHUM	Aralba	76.50 ^{a}	0.57	754.50 ^{ab}	130.8	421.50 ^{b}	115.3	671.50 ^{cd}	67.2
	Diamond	76.45 ^{a}	0.49	822.50 ^{a}	40.3	435.50 ^{b}	70.0	687.50 ^{bc}	34.6
	DMS	76.90 ^{a}	0.00	766.00 ^{ab}	69.3	431.50 ^{b}	71.4	646.50 ^{d}	61.5
	PSEAG4E44	65.20 ^{d}	0.14	697.50 ^{b}	136.5	505.00 ^{a}	121.6	611.50 ^{e}	43.1
	PSE7431	73.50 ^{c}	0.28	512.50 ^{c}	98.3	350.50 ^{c}	72.8	699.50 ^{abc}	64.3
	SW6129	74.85 ^{b}	0.07	808.00 ^{a}	90.5	406.00 ^{b}	99.0	706.50 ^{abc}	27.6
	SW6143W	74.75 ^{b}	0.35	768.00 ^{ab}	91.9	407.00 ^{b}	84.9	725.00 ^{a}	31.1
	SW6237W	74.85 ^{b}	0.92	813.00 ^{a}	80.6	402.50 ^{b}	84.1	708.50 ^{ab}	44.5
	Means ± SE	74.13 ± 3.67		742.80 ± 120.8		419.90 ± 79.0		682.10 ± 50.9	
OAT	Abel	61.56 ^{ab}	0.76	595.00 ^{cd}	7.1	404.00 ^{bc}	5.7	710.00 ^{ab}	7.1
	Corneil	58.65 ^{bc}	1.20	617.00 ^{cd}	8.5	423.00 ^{b}	24.0	705.50 ^{ab}	43.1
	Genziana	50.65 ^{d}	3.75	932.50 ^{a}	46.0	548.00 ^{a}	32.5	671.50 ^{ab}	60.1
	Konradin	62.25 ^{ab}	0.21	591.50 ^{cd}	87.0	353.00 ^{bc}	60.8	637.00 ^{b}	53.7
	Kripton	66.15 ^{a}	0.35	752.00 ^{b}	53.7	429.00 ^{b}	17.0	588.50 ^{b}	48.8
	Kynom	65.45 ^{a}	1.91	683.00 ^{bc}	33.9	372.00 ^{bc}	41.0	646.50 ^{ab}	40.3
	Lexic	58.55 ^{bc}	1.20	532.50 ^{d}	37.5	333.50 ^{c}	48.8	802.00 ^{a}	137.2
	Nigra	54.45 ^{cd}	1.77	782.50 ^{b}	0.7	431.00 ^{b}	8.5	656.50 ^{ab}	46.0
	Terra	62.90 ^{ab}	5.66	687.50 ^{bc}	60.1	375.50 ^{bc}	20.5	798.50 ^{a}	34.6
	Means ± SE	60.07 ± 5.24		685.90 ± 123.4		407.70 ± 65.8		690.70 ± 83.1	
MAIZE	Hualtaco	67.15 ^{d}	1.20	1204.5 ^{a}	101.1	708.00 ^{a}	75.0	439.50 ^{b}	245.4
	Kully	67.25 ^{d}	1.48	736.5 ^{bd}	115.3	498.00 ^{b}	70.7	588.50 ^{ab}	4.9
	VA 116 W	72.45 ^{b}	0.49	784.5 ^{bc}	34.6	411.00 ^{bc}	32.5	683.50 ^{ab}	36.1
	VA 116 W Purple	74.55 ^{a}	0.21	666.5 ^{ce}	30.4	342.00 ^{ce}	21.2	684.00 ^{ab}	125.9
	VA 1268 Red	73.50 ^{ab}	0.42	735.5 ^{bd}	57.3	369.00 ^{cd}	52.3	646.00 ^{ab}	38.2
	VA 522 W	73.45 ^{ab}	0.35	633.5 ^{de}	72.8	268.50 ^{df}	61.5	670.50 ^{ab}	88.4
	VA 522 W Blue	70.65 ^{c}	0.49	837.0 ^{b}	59.4	420.00 ^{bc}	8.5	647.00 ^{ab}	38.2
	VA 572	74.50 ^{a}	0.28	553.5 ^{e}	57.3	203.00 ^{f}	25.5	715.00 ^{a}	29.7
	VA 572 Black	69.75 ^{c}	1.06	412.5 ^{f}	12.0	237.00 ^{ef}	15.6	472.00 ^{ab}	11.3
	Means ± SE	71.47 ± 2.88		729.3 ± 218.4		384.10 ± 153.2		616.20 ± 119.3	

Results presented in the table are expressed as the mean value and standard deviation (SD) for three replications. Different letters in the same column indicate significant differences at $p \leq 0.05$. sd = standard deviation; SE = standard error of mean; BU = Brabender Unit.

In oat, high variability was found for PV and Genziana and Nigra (hulled genotypes) had the highest value (932.5 and 782.5 BU, respectively). Among the naked genotypes, Kripton, Kynom and Terra recorded the best values (752.0, 683.0 and 687.5 BU, respectively). Despite the highest mean value of the breakdown (407.7 BU), a narrow variability among genotypes was found, and the highest value was in Genziana (548 BU). Setback indicated that the highest value was for Terra and Lexic (798.5 and 802.0 BU, respectively). Accordingly, Genziana could be used for sauce because of its high hot viscosity and cold viscosity. Terra and Lexic could be interesting for GF pasta as the high value of setback. Conversely, Kripton, with a lower retrogradation degree, could be more apt for baking goods. In maize, great variability among genotypes was observed for pasting temperature and the highest significant value was found for the genotype VA 572 (74.5 °C) and the lowest for Kully and Hualtaco (67.25 and 67.15 °C, respectively). Regarding the PV, high variability was found, and the best values were found for Hualtaco, which showed the highest PV with respect to all the genotypes of the three species (1204.5 BU). This behavior indicates that Hualtaco, a white genotype of Andean origin, requires a short time for cooking and reaches a high viscosity with respect to all other genotypes of this study, but with less stable starch (high BD) and thus, not suitable for food preparation. The maize mean value of the breakdown was 384.1 BU and was the lowest in comparison to oat and sorghum. Generally, a low value of BD is associated with great stability of starch granules and a decrease in the rate of collapsing of starch granules. In this view, VA572 black could be interesting because of its low breakdown and setback and because it is suitable for baking goods.

3.1.6. Antioxidant Profile

Amongst the species analyzed, sorghum hybrids showed the lowest total flavonoid content (TFC), with an average value of 0.44 mgCE/g (Table 3). The highest and the lowest content occurred in PSE7431 and SW6129 (Bolivian genotype), respectively, with an average value of 0.85 mgCE/g and 0.31 mgCE/g, respectively. Statistical analysis showed that TFC observed in Aralba, Diamond, DMS and SW6237W were very similar. Regarding the total polyphenol content (TPC), the average value observed in sorghum was 2.03 mgFA/g. It is still found that PSE7431 was the genotype with the highest polyphenol content, with an average value of 2.59 mgFA/g. The lowest TPC content was observed in DMS, with an average value of 1.67 mgFA/g. The proanthocyanidins (TPAC) are oligomers and polymers of flavan-3-ol monomer units, and they have potent antioxidant activities. The role of these compounds is to protect against oxidative stress that is induced by free radicals by increasing the capacity of the cell for the absorbing of such oxygen radicals, which prevents the generation of free radicals, stimulates the detoxification enzymes, decreases lipid peroxidation, and reduces cell proliferation [50]. For this parameter, the greatest variation was observed, and five sorghum genotypes showed low TPAC levels that ranged between 0.01 mgCE/g in DMS and 0.53 mgCE/g in SW6237W. Aralba and SW6143W showed values ranging between 1.81 mgCE/g and 3.19 mgCE/g, respectively, whereas PSE7431 reached the highest content (17.07 mgCE/g). The average TPAC content of sorghum was the highest (2.93 mgCE/g) with respect to oat (1.06 mgCE/g) and maize (0.55 mgCE/g). Moreover, sorghum hybrids showed the highest TAC compared to oats (three times higher) and maize hybrids (about 23% higher), with an average content of 35.20 mmol TEAC/kg. The highest and lowest antioxidant capacity occurred in PSEAG4E44 and PSE7431, respectively, with an average value of 44.38 and 26.89 mmolTEAC/kg, respectively.

Table 3. Total flavonoid, phenolic, proanthocyanidin contents and antioxidant capacities of the twenty-six genotypes of sorghum, oat and maize.

	SAMPLES	TFC (mgCE/g)		TPC (mgFA/g)		TPAC (mgCE/g)		TAC (ABTS) (mmol TE/Kg)	
		Mean	sd	Mean	sd	Mean	sd	Mean	sd
SORGHUM	Aralba	0.35^{cd}	0.01	1.87^{cd}	0.07	1.81^{c}	0.01	31.66^{cd}	0.90
	Diamond	0.33^{cd}	0.00	1.77^{de}	0.04	0.11^{f}	0.03	30.57^{d}	0.01
	DMS	0.34^{cd}	0.03	1.67^{e}	0.02	0.01^{f}	0.01	34.68^{c}	1.16
	PSEAG4E44	0.45^{bc}	0.01	2.26^{b}	0.04	0.67^{d}	0.11	44.38^{a}	2.71
	PSE7431	0.85^{a}	0.02	2.59^{a}	0.17	17.07^{a}	0.02	26.89^{e}	2.07
	SW6129	0.31^{d}	0.04	1.70^{de}	0.04	0.08^{f}	0.03	43.77^{a}	0.13
	SW6143W	0.54^{b}	0.17	2.41^{b}	0.08	3.19^{b}	0.02	31.53^{cd}	0.39
	SW6237W	0.38^{cd}	0.04	1.96^{c}	0.01	0.53^{e}	0.02	38.15^{b}	1.03
	Means ± SE	0.44 ± 0.18		2.03 ± 0.33		2.93 ± 5.56		35.20 ± 6.13	
OAT	Abel	1.50^{a}	0.00	1.77^{a}	0.06	0.99^{bc}	0.04	11.94^{b}	0.32
	Corneil	0.97^{fg}	0.12	1.62^{b}	0.04	1.28^{a}	0.07	13.01^{a}	0.09
	Genziana	1.22^{cd}	0.08	1.39^{cd}	0.02	0.86^{c}	0.11	12.19^{b}	0.04
	Konradin	1.30^{bc}	0.04	1.46^{c}	0.05	1.17^{ab}	0.08	11.65^{b}	0.37
	Kripton	1.11^{de}	0.00	1.42^{c}	0.05	1.01^{bc}	0.01	11.69^{b}	0.43
	Kynom	0.94^{g}	0.08	1.31^{d}	0.03	1.04^{bc}	0.18	10.48^{c}	0.21
	Lexic	1.42^{ab}	0.04	1.57^{b}	0.00	1.03^{bc}	0.11	11.65^{b}	0.13
	Nigra	1.08^{ef}	0.04	1.31^{d}	0.03	1.04^{bc}	0.05	11.65^{b}	0.23
	Terra	1.36^{b}	0.04	1.56^{b}	0.05	1.11^{ab}	0.00	11.78^{b}	0.12
	Means ± SE	1.21 ± 0.20		1.49 ± 0.15		1.06 ± 0.13		11.78 ± 0.66	
MAIZE	Hualtaco	1.05^{g}	0.16	1.29^{f}	0.01	0.18^{d}	0.01	13.32^{de}	0.68
	Kully	29.43^{a}	0.28	9.63^{a}	0.18	2.40^{a}	0.19	89.16^{a}	7.90
	VA 116 W	1.19^{g}	0.12	1.54^{f}	0.03	0.06^{d}	0.00	12.72^{e}	0.51
	VA 116 W Purple	13.28^{b}	0.20	3.83^{b}	0.22	0.61^{c}	0.04	40.83^{b}	0.06
	VA 1268 Red	3.79^{d}	0.04	2.22^{d}	0.03	0.77^{b}	0.06	19.80^{c}	0.22
	VA 522 W	1.70^{f}	0.04	1.37^{f}	0.11	0.14^{d}	0.02	10.80^{e}	0.20
	VA 522 W Blue	2.35^{e}	0.16	1.85^{e}	0.04	0.12^{d}	0.02	19.14^{cd}	2.13
	VA 572	1.28^{g}	0.00	1.45^{f}	0.04	0.08^{d}	0.00	10.38^{e}	0.16
	VA 572 Black	9.55^{c}	0.12	3.27^{c}	0.12	0.59^{c}	0.05	41.79^{b}	1.92
	Means ± SE	7.07 ± 9.14		2.94 ± 2.59		0.55 ± 0.72		28.66 ± 25.04	

Results presented in the table are expressed as the mean value and standard deviation (SD) for three replications. Different letters in the same column indicate significant differences at $p \leq 0.05$. sd = standard deviation; SE = standard error of mean; TFC = total flavonoid content; TPC = total phenolic content; TPAC = total proanthocyanidin content; TAC = total antioxidant activity; TE = Trolox equivalent; FA = ferulic acid equivalent; CE = catechin equivalent.

The average value of TFC differed between hulled and naked oats, being 1.27 mgCE/g and 1.08 mgCE/g, respectively, while the mean content was 1.21 mgCE/g, varying from the highest and lowest content in Abel (1.50 mgCE/g) and Kynom (0.94 mgCE/g). Our results were higher with respect to Meenu et al. [51], who found in a large collection of naked and husked oat germplasm a range that varied from 0.14 mgCE/g to 0.46 mgCE/g in naked oat, and from 0.32 mgCE/g to 0.12 mgCE/g in hulled oat. These differences in TFC could be attributable to the great interaction with the environment, as asserted by Menga et al. [52]. The average value observed for TPC was 1.49 mgFA/g, and little, but no significant differences were observed between the mean content of naked (1.51 mgFA/g) and husked (1.44 mgFA/g) genotypes. Abel was the genotype with the highest polyphenol content (1.77 mgFA/g), and Kynom and Nigra with the lowest (1.31 mgFA/g). Similarly, also in Meenu et al. [51] work, no differences were found in the TPC content between naked and husked oats. It is noteworthy that TPC measured by the Folin–Ciocalteu method may not give a full picture of the quantity and, often, of the quality of phenolic compounds present in the extracts. Scarce variability was observed in the TPAC, and the average value observed was 1.06 mgCE/g, ranging between Corneil with 1.28 mgCE/g and Genziana with 0.86 mgCE/g. No differences were observed between the mean content of naked (1.06 mgCE/g) and hulled (1.06 mgCE/g) genotypes. The highest and lowest antioxidant capacity occurred in Corneil and Kynom (13.01 mmol TE/kg and 10.48 mmol TE/kg, respectively), and the average value observed was 11.78 mmol TE/kg. The mean content of naked and hulled was 11.53 mmol TE/kg and 12.28 mmol TE/kg, respectively. The antioxidant activity measured in the samples was higher than those reported by Meenu et al. [51], but they agreed to find higher activity in the hulled genotypes compared to the naked ones. In fact, large quantities of phenolic acids and avenanthramides, which have high antioxidant capacity, were concentrated in hulls, as reported by Varga et al. [53]. In this study that investigated the scavenging capacity of hulls and groats separately, an increased antioxidant activity of hulls compared to groats was demonstrated. Oat accessions showed interesting values of total antioxidant capacity mainly due to the presence of avenanthramides. Avenanthramides are a group of phenolic compounds consisting of anthranilic acid or hydroxy anthranilic acid, and hydroxyl cinnamic acid. They are powerful antioxidants and have also been shown to have antioxidative and anti-inflammatory effects in vivo, and they are unique to oats.

Pigmented maize contains secondary metabolites, mainly represented by phenolics and carotenoids [54]. Among the polyphenols, flavonoids are recognized as a source of antioxidative, antihypertensive and anti-inflammatory activities [55]. Flavonoids are also responsible for the pericarp color, which varies from cream to blue and encompasses crimson and violet. The mean TFC in maize (7.07 mgCE/g) was about six times with respect to oat (1.21 mgCE/g) and about sixteen times with respect to sorghum (0.44 mgCE/g). The genotype with the highest flavonoid content was Kully (purple), with a value of 29.43 mgCE/g. On the contrary, Hualtaco (white) was the genotype with the lowest flavonoid content, with an average value of 1.05 mgCE/g. No significant differences for TFC were observed between the white (Hualtaco and VA116w) and yellow genotypes (VA572) (Table 3). Moreover, the range observed for those genotypes was the same as previously reported by Suriano et al. [56] of about 1.16 mgCE/g, but higher than Rodriguez-Salinas et al. [57] in a collection of native maize genotypes from Northeast Mexico. Considering the introduction of the new pigmented genotypes, VA116 purple, with an average content of 13.28 mgCE/g, showed the greatest increase (of about eleven times) in TFC compared to the parental genotype VA116w (white). Indeed, purple-colored maize genotypes (Kully and VA116w purple) had higher phenolic compounds than light- or white-colored genotypes. Regarding the TPC, a large and significant difference was observed, with an average value of 2.94 mgFA/g. Kully, the native purple Andean genotype, showed the highest content with an average value of 9.63 mgFA/g, while Hualtaco, the native white Andean genotype, had the lowest content (1.29 mgFA/g). No significant differences for TPC were observed among the white (Hualtaco, VA116w, VA522w) and the yellow genotypes (VA572). A similar range was found by Žilić et al. [54], indicating that the genotypes with purple pericarp had the highest content; on the contrary, the white and yellow genotypes that have no anthocyanin showed the lowest content. These results are consistent with our study. Moreover, the positive effect of introgression of colored genotype in Italian maize lines (VA116 purple, VA522 blue and VA572 black) has led to an increase in the TPC compared to the parents of about 60%, 25% and 55%, respectively. Great variation was observed in the TPAC (average value of 0.55 mgCE/g), ranging between Kully (2.40 mgCE/g) and VA116w (0.06 mgCE/g). No significant differences were observed among the white-type genotypes (Hualtaco, VA116w, VA522w) and the yellow genotype (VA572). Our results were in accordance with the study of Suriano et al. [56] and Rodriguez-Salinas et al. [57]. The TAC measured by ABTS assay had an average content of 28.66 mmolTE/kg. Compared to sorghum, which has an average value of 35.20 mmolTE/kg but with rather homogeneous values between genotypes, greater variability was highlighted in maize. In this regard, the highest TAC was detected in Kully and VA116, the purple genotypes, with values of

89.16 and 40.83 mmolTE/kg, respectively. The lowest activity was observed in VA572 (10.38 mmolTE/kg), which was not statistically different from VA116w and VA522w, the white and yellow genotypes. Also, for this parameter, the introgression of colored genotype in Italian maize lines led to an increase in TAC compared to the parents (69%, 43%, and 75%, respectively). The two Bolivian genotypes Kully (purple) and Hualtaco (white) showed opposite antioxidant traits, with Kully having the highest levels of TFC, TPC, TPAC and TAC while Hualtaco had the lowest values. Therefore, in accordance with the literature, our data confirm that a darker-colored maize kernel had more antioxidant activity [54].

3.1.7. Correlations among the Different Investigated Traits

For a better understanding of the pasting properties, it must be highlighted that they are related to the amylose content. A high amylose content limits starch swelling by forming complexes with lipids, which in turn results in a lower peak viscosity and setback [58]. In sorghum, this feature occurs, and a negative correlation was found between amylose and PT ($p \leq 0.01$ r = −0786). This phenomenon is explained by the low solubility of the amylose molecule compared to amylopectin, which in turn produces the effect of lowering the temperature required for gelatinization. Accordingly, the PT provides an indication of the minimum temperature required to cook flour [59]. The same was not so clearly observed for oat genotypes because the high content of beta-glucan, acting as a hydrocolloid, contributes to increasing the viscosity [60]. Starch gelatinization was important for its capacity to form a matrix in which the gas bubbles are entrapped, producing a crucial impact in the gluten-free formulation of bakery products. According to Abdel-Aal [61], the addition of gel-forming starches (for example, pregelatinized starches) is a useful means to provide gas blockage as a stabilizing agent. Moreover, a negative and significant correlation was detected with a setback for oat and sorghum ($p \leq 0.05$ r = −0.573 and −0.608, respectively), as previously indicated by Acquistucci et al. [62]. Thus, flours with high amylose content develop low viscosity after heating and a low re-organization rate after the cooling period, providing a low final viscosity. Retrogradation is linked to the increase in the consistency or hardness of the starch [63], and in baking, this process causes the undesirable staling process of bread during storage [64]. In GF pasta production, instead, all flours that are more susceptible to gelatinization (PV) and retrogradation (setback) are preferable [62].

Sorghum showed a significant and negative correlation between fiber content and peak temperature, PV and setback ($p \leq 0.05$ r = −0.614, −0.556 and −0.525, respectively). This result is consistent because the fiber interferes with the gelatinization of the starch, causing a reduction in the parameters related to viscosity [65]. For oats, however, a positive correlation with the peak temperature is observed ($p \leq 0.05$ r = 0.493), probably due to the specific composition of the fiber, namely high beta-glucan content.

In maize, few significant correlations were found with resistant starch (negative correlation with PV, $p \leq 0.05$ r = −0.553), total starch (negative correlation with PT and setback, $p \leq 0.05$ r = −0.560 and −0.550, respectively) and none with amylose. Instead, several significant correlations between fiber and protein were found. Peak temperature and pasting temperature positively correlated with fiber ($p \leq 0.05$, r = 0.534 and 0.525, respectively) and proteins ($p \leq 0.01$ r = 0.885 and 0.960, respectively). The positive correlation between proteins and fiber with pasting temperature could be explained by the effect produced by these latter in delaying the gelation of amylose. Conversely, PV and BD negatively correlated with proteins and fiber ($p \leq 0.05$, r = −0.520 and −0.721 for proteins and r = −0.553 and −0.547 for fiber, respectively), as previously observed by Sandhu and Singh [59], because they form complex with amylose that in turn caused the decrease of viscosity (PV) and BD.

Regarding the total antioxidant capacity in maize, significant correlations were observed with TPC, TFC, and TPAC ($p \leq 0.01$, r = 0.982, 0.987 and 0.984, respectively) (On the contrary, in sorghum, only negative and significant correlations were found between TAC and TPAC ($p \leq 0.05$ r = −0.580), whereas in oat, a significant correlation was observed only with TPC ($p = 0.05$, r = 0.482). These features suggest that in maize, the analyzed compounds contributed the most to TAC, whereas in the other two species, other bioactive compounds impacted the TAC levels. Moreover, a fair but significant correlation with β-glucan in oats ($p \leq 0.01$, r = 0.601) suggested that the antioxidant capacity correlated with this soluble fiber component, located in the outermost part of the kernel [66].

4. Conclusions

According to the results of this research, we can assume that all three species possess technological indexes suitable for food processing. Moreover, among oats and maize, there are interesting nutritional components that could be desirable in food formulation for improving health. Our results clearly indicated that based on the type of food to be formulated, some identified genotypes present adequate characteristics for the formulation of GF pasta or bakery products. In this regard, among oat genotypes, Terra and Lexic could be interesting for GF pasta with a high value of Setback; conversely, Kripton showing a lower retrogradation degree could be more apt for baking goods. Among sorghum genotypes, SW6129 and SW6237W, which have the highest PV and Setback, could be appreciable for food products that request stable thickening after heating and cooling, like in sauces production, while PSEAG4E44 with low PV and setback, is more suitable for baking goods. For maize, VA572 black could be interesting for its low BD and setback and is suitable for baking goods, while Kully and VA116Wpurple are the most interesting for TAC. Therefore, the adequate blending of whole grain flours from the different GF cereals presented in this study could allow the manufacturing of products of good technological and healthier nutritional quality. Furthermore, the great challenge of the nutritional and sensorial aspects of GF products must be carefully evaluated. Finally, three different GXE trials are ongoing in Northern, Central and Southern Italian Regions, using the genotypes abovementioned of each species, selected for the most promising traits in terms of technological and nutritional performance. These multi-environmental trails will allow us to identify the effect of the area and the year of cultivation on the variability of the nutritional and pasting properties of the three cereal species.

Author Contributions: Conceptualization, L.G. and C.F.; methodology, V.M., F.T., F.N., E.G., C.N., C.L. and S.P.; validation, V.M., F.T., F.N., E.G., C.N., C.L., S.P., L.G. and C.F.; formal analysis, F.T. and E.G.; investigation, V.M., F.T., F.N., E.G., C.N., C.L. and S.P.; resources, C.F. and L.G.; data curation, C.F. and L.G.; writing—original draft preparation, C.F. and L.G.; writing—review and editing, V.M., F.T., F.N., E.G., C.N., C.L., S.P., L.G. and C.F.; visualization, E.G. and F.T.; supervision, C.F. and L.G.; project administration, C.F. and L.G.; funding acquisition, C.F. All authors have read and agreed to the published version of the manuscript.

Funding: This research was funded by MASAF, project: RETI 2020. CUP: 75H21000290001.

Institutional Review Board Statement: Not applicable.

Informed Consent Statement: Not applicable.

Data Availability Statement: The original contributions presented in the study are included in the article, further inquiries can be directed to the corresponding author.

Acknowledgments: Pigmented maize materials were developed in the frame of International Cooperation Project P.S.G.O. Km0 Bolivia Piccoli semi grandi opportunità, agro ecologia campesina famigliare e filiere Km 0 in Bolivia (2018–2021) funded by AICS-Agenzia Italiana per la Cooperazione allo sviluppo, AID 011.457.

Conflicts of Interest: The authors declare no conflicts of interest.

References

1. See, J.A.; Kaukinen, K.; Makharia, G.K.; Gibson, P.R.; Murray, J.A. Practical insights into gluten-free diets. *Nat. Rev. Gastroenterol. Hepatol.* **2015**, *12*, 580–591. [CrossRef]
2. Tortora, R.; Capone, P.; De Stefano, G.; Imperatore, N.; Gerbino, N.; Donetto, S.; Monaco, V.; Caporaso, N.; Rispo, A. Metabolic syndrome in patients with coeliac disease on a gluten-free diet. *Aliment. Pharmacol. Ther.* **2015**, *41*, 352–359. [CrossRef] [PubMed]

3. Picascia, A.; Camarca, M.; Malamisura, R.; Mandile, M.; Galatola, D. In celiac disease patients the in vivo challenge with the diploid Triticum monococcum elicits a reduced immune response compared to hexaploid wheat. *Mol. Nutr. Food Res.* **2020**, *64*, 1901032. [CrossRef] [PubMed]
4. Bongianino, N.F.; Steffolani, M.E.; Morales, C.D.; Biasutti, C.A.; León, A.E. Technological and Sensory Quality of Gluten-Free Pasta Made from Flint Maize Cultivars. *Foods* **2023**, *12*, 2780. [CrossRef]
5. Kulamarva, A.G.; Venkatesh, R.S.; Raghavan, V.G.S. Nutritional and Rheological Properties of Sorghum. *Int. J. Food Prop.* **2009**, *12*, 55–69. [CrossRef]
6. Del Giudice, F.; Massardo, D.R.; Pontieri, P.; Maddaluno, L.; De Vita, P.; Fares, C.; Ciacci, C.; Del Giudice, L. Development of a sorghum chain in the Italian Campania Region: From the field to the celiac patient's table. *J. Plant Interact.* **2008**, *3*, 49–55. [CrossRef]
7. Aguiar, E.V.; Santos, F.G.; Queiroz, V.A.V.; Capriles, V.D. A Decade of Evidence of Sorghum Potential in the Development of Novel Food Products: Insights from a Bibliometric Analysis. *Foods* **2023**, *12*, 3790. [CrossRef]
8. Xu, F.; Dube, N.M.; Han; Zhao, R.; Wang, Y.; Chen, J. The Effect of Zimbabwean Tannin-Free Sorghum Flour Substitution on Fine Dried Noodles Quality Characteristics. *J. Cereal Sci.* **2021**, *102*, 103320. [CrossRef]
9. Galassi, E.; Gazza, L.; Nocente, F.; Kouagang Tchakoutio, P.; Natale, C.; Taddei, F. Valorization of two African typical crops, Sorghum and Cassava, by the production of different dry pasta formulations. *Plants* **2023**, *12*, 2867. [CrossRef] [PubMed]
10. Bvenura, C.; Kambizi, L. Chapter 5—Future Grain Crops. In *Future Foods*; Rajeev, B., Ed.; Academic Press: Cambridge, MA, USA, 2022.
11. Peterson, D.M. Oat Antioxidants. *J. Cereal Sci.* **2001**, *33*, 115–129. [CrossRef]
12. Associazione Italiana Celiachia (AIC). Available online: https://www.celiachia.it/dieta-senza-glutine/gestire-gluten-free-diet/cereals-allowed/ (accessed on 13 February 2024).
13. Elfström, P.; Sundström, J.; Ludvigsson, J.F. Systematic review with meta-analysis: Associations between coeliac disease and type 1 diabetes. *Aliment. Pharmacol. Ther.* **2014**, *40*, 1123–1132. [CrossRef]
14. Smyth, D.J.; Plagnol, V.; Walker, N.M.; Cooper, J.D.; Downes, K.; Yang, J.H.; Howson, J.M.; Stevens, H.; McManus, R.; Wijmenga, C.; et al. Shared and distinct genetic variants in type 1 diabetes and celiac disease. *N. Engl. Med.* **2008**, *359*, 2767–2777. [CrossRef]
15. Rooney, L.W.; Serna Saldívar, S.O. Food use of whole corn and dry-milled fractions. In *Corn: Chemistry and Technology*, 2nd ed.; White, P.J., Johnson, L.A., Eds.; AACC: St. Paul, MN, USA, 2003; pp. 495–535.
16. Nuss, E.T.; Tanumihardjo, S.A. Maize: A paramount staple crop in the context of global nutrition. *Food Sci. Food Saf.* **2010**, *4*, 417–436. [CrossRef] [PubMed]
17. Hu, Q.P.; Xu, J.G. Profiles of carotenoids, anthocyanins, phenolics and antioxidant activity of selected color waxy corn grains during maturation. *J. Agr. Food Chem.* **2011**, *59*, 2026–2033. [CrossRef]
18. Zillić, S.; Serpen, A.; Akillioĝlu, G.; Vanĉetović, V. Phenolic compounds, carotenoids, anthocyanins and antioxidant capacity of colored maize (*Zea mays* L.) kernel. *J. Agric. Food Chem.* **2012**, *60*, 1224–1231. [CrossRef]
19. Li, Q.; Somavat, P.; Singh, V.; Chatham, L.; Mejia, E.G. A comparative study of anthocyanins distribution in purple and red corn coproducts from three conventional fractionation process. *Food Chem.* **2017**, *231*, 332–339. [CrossRef] [PubMed]
20. Menga, V.; Amato, M.; Phillips, T.D.; Angelino, D.; Morreale, F.; Fares, C. Gluten-free pasta incorporating chia (*Salvia hispanica* L.) as thickening agent: An approach to naturally improve the nutritional profile and the in vitro carbohydrate digestibility. *Food Chem.* **2017**, *221*, 1954–1961. [CrossRef]
21. International Association for Cereal Science and Technology. *ICC Standard Methods (Methods No. 105/2)*; ICC: Vienna, Austria, 2003.
22. Mariotti, F.; Tomé, D.; Mirand, P.P. Converting nitrogen into protein—Beyond 6.25 and Jones' factors. *Crit. Rev. Food Sci. Nutr.* **2008**, *48*, 177–184. [CrossRef]
23. Association of Official Analytical Chemists. *Official Methods of Analysis 996.11*, 18th ed.; AOAC: Gaithersburg, MD, USA, 2005.
24. Association of Official Analytical Chemists. *Official Methods of Analysis 2002.02, Resistant Starch in Starch and Plant Materials*; AOAC: Gaithersburg, MD, USA, 2002.
25. Association of Official Analytical Chemists. *Official Methods of Analysis 991*, 16th ed.; Cunniff, P., Ed.; AOAC: Gaithersburg, MD, USA, 1995; p. 42.
26. *Method 520: 2010*; Cereals and Pulses-Determination of the Mass of 1000 Grains. ISO: Geneva, Switzerland, 2010; p. 10.
27. *Method 7971-1: 2009*; Determination of Bulk Density, Called Mass per Hectolitre-Part 1: Reference Method. ISO: Geneva, Switzerland, 2009; p. 8.
28. McCleary, B.V.; Mugford, D.C.; Camire, M.C.; Gibson, T.S.; Harrigan, K.; Janning, M.; Meuser, F.; Williams, P. Determination of ß-glucan in barley and oats by streamlined enzymatic method: Summary of collaborative study. *J. AOAC Int.* **1997**, *80*, 580–583. [CrossRef]
29. Beta, T.; Nam, S.; Dexter, J.E.; Sapirstein, H.D. Phenolic content and antioxidant activity of pearled wheat and roller-milled fractions. *Cereal Chem.* **2005**, *82*, 390–393. [CrossRef]
30. Singleton, V.L.; Rossi, J.A. Colorimetry of Total Phenolics with Phosphomolybdic-Phosphotungstic Acid Reagents. *Am. J. Enol. Vitic.* **1965**, *16*, 144–158. [CrossRef]
31. Sun, B.; da-Silva, J.M.R.; Spranger, I. Critical Factors of Vanillin Assay for Catechins and Proanthocyanidins. *J. Agric. Food Chem.* **1998**, *46*, 4267–4274. [CrossRef]
32. Eberhardt, M.V.; Lee, C.Y.; Liu, R.H. Antioxidant activity of fresh apples. *Nature* **2000**, *405*, 903–904. [CrossRef]

33. Re, R.; Pellegrini, N.; Proteggente, A.; Pannala, A.; Yang, M.; Rice-Evans, C. Antioxidant activity applying an improved ABTS radical cation decolorization assay. *Free Radic. Biol. Med.* **1999**, *26*, 1231–1237. [CrossRef]
34. Marengo, M.; Barbiroli, A.; Bonomi, F.; Casiraghi, M.C.; Marti, A.; Pagani, M.A.; Manful, J.; Graham-Acquaah, S.; Ragg, E.; Fessas, D.; et al. Macromolecular Traits in the African Rice *Oryza glaberrima* and in Glaberrima/Sativa Crosses, and Their Relevance to Processing. *J. Food Sci.* **2017**, *82*, 2298–2305. [CrossRef]
35. Wang, K.; Fu, B.X. Inter-Relationships between Test Weight, Thousand Kernel Weight, Kernel Size Distribution and Their Effects on Durum Wheat Milling, Semolina Composition and Pasta Processing Quality. *Foods* **2020**, *9*, 1308. [CrossRef]
36. Galassi, E.; Taddei, F.; Ciccoritti, R.; Nocente, F.; Gazza, L. Biochemical and technological characterization of two C4 gluten-free cereals: *Sorghum bicolor* and *Eragrostis tef*. *Cereal Chem.* **2020**, *97*, 65–73. [CrossRef]
37. Aruna, O.A.; Oluwaseyi, I.O.; Babatunde, S.E.; Adeniyi, O. Effects of Different Rates of Poultry Manure and Split Applications of Urea Fertilizer on Soil Chemical Properties, Growth, and Yield of Maize. *Sci. World J.* **2020**, *2020*, 4610515.
38. Buerstmayr, H.; Krenn, N.; Stephan, U.; Grausgruber, H.; Zechner, E. Agronomic performance and quality of oat (*Avena sativa* L.) genotypes of worldwide origin produced under Central European growing conditions. *Field Crops Res.* **2007**, *101*, 343–351. [CrossRef]
39. Korus, J.; Chmielewska, A.; Witczak, M.; Ziobro, R.; Juszczak, L. Rapeseed protein as a novel ingredient of gluten-free bread. *Eur. Food Res. Technol.* **2021**, *247*, 2015–2025. [CrossRef]
40. Ziobro, R.; Juszczak, L.; Witczak, M.; Korus, J. Non-gluten proteins as structure forming agents in gluten free bread. *J. Food Sci. Technol.* **2016**, *53*, 571–580. [CrossRef]
41. Lee, A.R.; Ng, D.L.; Dave, E.; Ciaccio, E.J.; Green, P.H. The effect of substituting alternative grains in the diet on the nutritional profile of the gluten-free diet. *J. Hum. Nutr. Diet.* **2009**, *22*, 359–363. [CrossRef]
42. Blandino, M.; Alfieri, M.; Giordano, D.; Vanara, F.; Redaelli, R. Distribution of bioactive compounds in maize fractions obtained in two different types of large-scale milling processes. *J. Cereal Sci.* **2017**, *77*, 251–258. [CrossRef]
43. Wang, Q.; Ellis, P.R. Oat β-glucan: Physico-chemical characteristics in relation to its blood-glucose and cholesterol-lowering properties. *Br. J. Nutr.* **2014**, *112* (Suppl. S2), S4–S13. [CrossRef]
44. Bojarczuk, A.; Skąpska, S.; Khaneghah, A.M.; Marszałek, K. Health benefits of resistant starch: A review of the literature. *J. Funct. Foods* **2022**, *93*, 105094. [CrossRef]
45. Biselli, C.; Volante, A.; Desiderio, F.; Tondelli, A.; Gianinetti, A.; Finocchiaro, F.; Taddei, F.; Gazza, L.; Sgrulletta, D.; Cattivelli, L.; et al. GWAS for Starch-Related Parameters in Japonica Rice (*Oryza sativa* L.). *Plants* **2019**, *8*, 292. [CrossRef]
46. Annison, G.; Topping, D.L. Nutritional role of resistant starch: Chemical structure vs physiological function. *Ann. Rev. Nutr.* **1994**, *14*, 297–320. [CrossRef] [PubMed]
47. Cozzolino, D. The use of the rapid visco analyser (RVA) in breeding and selection of cereals. *J. Cereal Sci.* **2016**, *70*, 282–290. [CrossRef]
48. Batey, I.L. Interpretation of RVA curves. In *The RVA Handbook*; Crosbie, G.B., Ross, A.S., Eds.; AACC International: St Paul, MN, USA, 2007; pp. 19–31.
49. Alfieri, M.; Bresciani, A.; Zanoletti, M.; Pagani, M.A.; Marti, A.; Redaelli, R. Physical, chemical and pasting features of maize Italian inbred lines. *Eur. Food Res. Tech* **2020**, *246*, 2205–2214. [CrossRef]
50. Wang, L.S.; Stoner, G.D. Anthocyanins and their role in cancer prevention. *Cancer Lett.* **2008**, *269*, 281–290. [CrossRef]
51. Meenu, M.; Cozzolino, D.; Xu, B. Non-destructive prediction of total phenolics and antioxidants in hulled and naked oat genotypes with near-infrared spectroscopy. *J. Food Meas. Charact.* **2023**, *17*, 4893–4904. [CrossRef]
52. Menga, V.; Fares, C.; Troccoli, A.; Cattivelli, L.; Baiano, A. Effects of genotype, location and baking on the phenolic content and some antioxidant properties of cereal species. *Int. J. Food Sci. Tech* **2009**, *45*, 7–16. [CrossRef]
53. Varga, M.; Jójárt, R.; Fónad, P.; Mihály, R.; Palágyi, A. Phenolic composition and antioxidant activity of colored oat. *Food Chem.* **2018**, *268*, 153–161. [CrossRef]
54. Žilić, S.; Serpen, A.; Akıllıoğlu, G.; Janković, M.; Gökmen, V. Distributions of phenolic compounds, yellow pigments and oxidative enzymes in wheat grains and their relation to antioxidant capacity of bran and debranned flour. *J. Cereal Sci.* **2012**, *56*, 652–658. [CrossRef]
55. Zielińska, D.; Turemko, M.; Kwiatkowski, J.; Zieliński, H. Evaluation of Flavonoid Contents and Antioxidant Capacity of the Aerial Parts of Common and Tartary Buckwheat Plants. *Molecules* **2012**, *17*, 9668–9682. [CrossRef] [PubMed]
56. Suriano, S.; Balconi, C.; Valoti, P.; Redaelli, R. Comparison of total polyphenol, profile anthocyanins, color analysis, carotenoids and tocols in pigmented maize. *Food Sci. Tech* **2021**, *144*, 111257. [CrossRef]
57. Rodriguez-Salinas, P.A.; Zavala-Garcia, F.; Urias-Orona, V.; Muy-Rangel, D.; Heredia, J.B.; Nino-Medina, G. Chromatic, Nutritional and Nutraceutical Properties of Pigmented Native Maize (*Zea mays* L.) Genotypes from the Northeast of Mexico. *Arab. J. Sci. Eng.* **2019**, *45*, 95–112. [CrossRef]
58. Sang, Y.; Scott, B.; Seib, P.A.; Pedersen, J.; Shi, Y.C. Structure and functional properties of Sorghum starch differing in amylose content. *J. Agric. Food Chem.* **2008**, *56*, 6680–6685. [CrossRef] [PubMed]
59. Sandhu, K.S.; Singh Malhi, N. Some properties of corn grains and their flours I: Physicochemical, functional and chapatti-making properties of flours. *Food Chem.* **2007**, *01*, 938–946. [CrossRef]
60. Karp, S.; Wyrwisz, J.; Kurek, M.A. The impact of different levels of oat β-glucan and water on gluten-free cake rheology and physicochemical characterisation. *J. Food Sci. Technol.* **2020**, *57*, 3628–3638. [CrossRef]

61. Abdel-Aal, E.S.M. Functionality of starches and hydrocolloids in gluten-free foods. In *Gluten-Free Food Science Technology*; Gallagher, E., Ed.; Wiley-Blackwell: Hoboken, NJ, USA, 2009; pp. 200–224.
62. Acquistucci, R.; Francisci, R.; Bucci, R.; Ritota, M.; Mazzini, F. Nutritional and physicochemical characterization of Italian rice flours starches. *Food Sci. Tech Res.* **2009**, *15*, 507–518. [CrossRef]
63. Shrestha, A.K.; Halley, P.J. Starch modification to develop novel starch-biopolymer blends: State of art and perspectives. In *Starch Polymers: From Genetic Engineering to Green Applications*, 1st ed.; Halley, P.J., Avérous, L.R., Eds.; Elsevier: New York, NY, USA, 2014.
64. Gray, J.; Bemiller, J. Bread staling: Molecular basis and control. *Compr. Rev. Food Sci. Food Saf.* **2003**, *2*, 1–21. [CrossRef] [PubMed]
65. Laguna, L.; Sanz, T.; Sahi, S.; Fiszman, S.M. Role of fibre morphology in some quality features of fibre-enriched biscuits. *Int. J. Food Prop.* **2014**, *17*, 163–178. [CrossRef]
66. Emmons, C.L.; Peterson, D.M.; Paul, G.L. Antioxidant capacity of oat (*Avena sativa* L.) extracts. 2. In vitro antioxidant activity and contents of phenolic and tocol antioxidants. *J. Agric. Food Chem.* **1999**, *47*, 4894–4898. [CrossRef]

foods

MDPI

Article

Nutritional Composition, Fatty Acid Content, and Mineral Content of Nine Sorghum (*Sorghum bicolor*) Inbred Varieties

Paola Pontieri [1,*], Jacopo Troisi [2], Matteo Calcagnile [3], Fadi Aramouni [4], Michael Tilley [4], Dmitriy Smolensky [4], Marco Guida [5], Fabio Del Giudice [1], Antonio Merciai [1], Iryna Samoylenko [1], Alberto L. Chessa [6], Mariarosaria Aletta [7], Pietro Alifano [3] and Luigi Del Giudice [1]

1 Sezione di Igiene, Istituto di Bioscienze e BioRisorse-UOS Napoli-CNR c/o Dipartimento di Biologia, 80134 Naples, Italy; felicita24@libero.it (F.D.G.); antonio.merciai7@gmail.com (A.M.); irynak@hotmail.it (I.S.); luigi.delgiudice@ibbr.cnr.it (L.D.G.)
2 Department of Medicine and Surgery, Theoreosrl, Spin off of the University of Salerno, Via Degli Ulivi, 3, 84090 Montecorvino Pugliano, Italy; troisi@theoreosrl.com
3 Dipartimento di Medicina Sperimentale, Università del Salento, 73100 Lecce, Italy; matteo.calcagnile@unisalento.it (M.C.); pietro.alifano@unisalento.it (P.A.)
4 Center for Grain and Animal Health Research (CGAHR), Agricultural Research Service (ARS), United States Department of Agriculture (USDA), Manhattan, KS 66502, USA; fadi.aramouni@usda.gov (F.A.); michael.tilley@usda.gov (M.T.); dmitriy.smolensky@usda.gov (D.S.)
5 Department of Biology, University of Naples Federico II, 80126 Naples, Italy; marco.guida@unina.it
6 Sorghum Breeder, Fernando Camino N_ 15 P4 F, 29016 Málaga, Spain; alberto.chessa@outlook.com
7 DCSRSI SPR BIBLIOTECA, 80131 Naples, Italy; mariarosaria.aletta@igb.cnr.it
* Correspondence: paola.pontieri@ibbr.cnr.it

Abstract: *Sorghum* is a self-pollinating species belonging to the Poaceae family characterized by a resistance to drought higher than that of corn. Sorghum (*Sorghum bicolor* L. Moench) has been grown for centuries as a food crop in tropical areas where it has an increasing importance, particularly as a cereal option for people with celiac disease. Over the past fifty years, food-grade varieties and hybrid seeds with white pericarp have been developed, particularly in the United States, to maximize sorghum food quality. Nutrient composition, including moisture, protein, carbohydrates, dietary fiber, fat content, fatty acid composition, and mineral content, was determined for nine inbred varieties with a stabilized food-grade sorghum genotype selected in the USA and grown under typical Mediterranean conditions. Differences in these nutritional components were observed among the varieties considered. Notable differences were found for monounsaturated and polyunsaturated fats, while saturated fatty acids were similar in all varieties. Oleic, linoleic, and palmitic acids were the most abundant fatty acids in all nine lines. Differences were also noted in mineral content, particularly for K, Mg, Al, Mn, Fe, Cu, Zn, and Ba. Enzyme-linked immunosorbent assays (ELISAs) demonstrated the absence of gliadin-like peptides in all the sorghum varieties analyzed, confirming, thus, that these analyzed varieties are safe for consumption by celiac patients. Knowledge of the nutritional values of sorghum lines is relevant for breeding programs devoted to sorghum nutritional content and for beneficial properties to human health.

Keywords: sorghum pure lines; nutritional value; fatty acids; mineral elements

1. Introduction

Sorghum (*Sorghum bicolor* L. Moench) is a self-pollinating crop mainly used as a staple food in Africa and Asia; sorghum seeds are often used as raw materials for alcoholic beverages, sweets, and as a source of glucose [1–3]. Sorghum is the fifth leading crop in the world after wheat, maize, rice, and barley [4]. The United States is the world's largest producer and exporter of sorghum, generating roughly 20% of world sorghum production and nearly 80% of sorghum exports [2,3,5]. In several developing countries, sorghum has traditionally been used in food products and to prepare various health foods [3,6]. The crop is considered a source of safe food for celiac patients showing an immune reaction to gluten proteins found in several *Triticum* and closely related cereals [7]. Molecular evidence has directly demonstrated the absence of toxic gliadin-like peptides in sorghum, which can be considered safe for people with celiac disease [8]. This is why the potential role of sorghum in human health and disease prevention has gained increased attention in the past decade [1–3,9,10].

Recently, there has been a growing interest in sorghum as a food ingredient in Western societies: the use of sorghum in human food products has increased and expanded since early efforts in the United States to develop hybrids with white seeds (often called "food-grade" sorghum) for the production of gluten-free food [11]. Moreover, new technologies have been developed, aimed at enhancing the nutritional and functional values of sorghum proteins in industrial-scale processes [12]. Breeding programs of public and private institutions have released improved varieties adapted to semi-arid and tropical environments, including those that meet specific food and industrial requirements [11,13]. To meet the market demand, numerous collections of sorghum seeds have arisen all over the world, particularly in Ethiopia, China, and the United States, as well as at the International Crops Research Institute for the Semi-Arid Tropics (ICRISAT) [14]. With increasing world population and decreasing water supplies due to climate change, sorghum represents an important crop for future human use.

Thus, genetic improvement of sorghum through the crossing of pure lines is an important research area for increasing yield, disease resistance, environmental stress tolerance, and other desirable traits of sorghum. Moreover, the availability of pure lines with genetic diversity is essential for the effectiveness of crossing. The greater the genetic diversity among pure sorghum lines, the greater the chances of obtaining hybrids with improved characteristics. Generally, sorghum breeding programs aim to acquire know-how and technologies that allow for the cultivation, storage, and milling of new white sorghum hybrids for human consumption in an economical manner, making them appealing to farmers and end-users for gluten-free diets for celiac patients and healthy diets for everyone based on sorghum flour [3].

This study aimed to characterize nine inbred sorghum lines developed in the United States and grown in Mediterranean environments from a nutritional standpoint. The specific goals were to evaluate levels of lipids, fiber, total protein, carbohydrates, and minerals. To that end, the study of the nutritional composition of sorghum inbreds would help breeders select hybrids with superior quality and nutritional values. Furthermore, one of the aims of the study was to identify varieties with superior nutritional attributes and to demonstrate that all varieties were safe to produce foods suitable for consumption by celiac patients.

The aim of this study is to evaluate the nutritional properties and functional food traits of nine inbred sorghum varieties with a focus on their potential role in promoting health. We hypothesize that these varieties, especially those grown in the Mediterranean area, can serve as valuable sources of essential nutrients, such as proteins, carbohydrates, unsaturated fatty acids, and minerals. These varieties are anticipated to meet rising consumer demand for nutritious, gluten-free cereals and may support breeding programs focused on the development of high-quality, food-grade sorghum lines.

2. Materials and Methods

2.1. Sorghum Varieties

The sorghum inbreds and their respective seed sources used in this study are listed in Table 1. In 2021, an open field cultivation of sorghum was carried out in San Bartolomeo in Galdo (BN) located in the Fortore area of the Campania region, southern Italy (41°25′ N, 15°01′ E and 597 m.a.s.l.). The nine sorghum varieties were planted on April 2021, and the grain harvest took place on 28 October 2021. Monthly rainfall and maximum and minimum temperatures of the 2021 growing season recorded in the aforementioned area are demonstrated in Table 2, while soil physical–chemical properties of the aforementioned experimental area are shown in Table 3. The milling was carried out starting 1 month after the harvest of the sorghum grains which were stored in a dry environment at 16 °C.

Table 1. List of Sorghum inbreds grown in Italy.

Inbreds	Type
PL-1 = TX436	Normal food grade
PL-2 = 05MN5113	Normal food grade
PL-3 = 05MN5115	Normal food grade
PL-4 = Macia	Normal food grade
Tw = B.TXARG-1	Waxy food grade
N2 = SURENO	Normal food grade
N3 = DORADO	Normal food grade
N4 = R.TX436	Normal food grade
N5 = SEPON82	Normal food grade

The source of the seeds was Purdue University (M. Tunistra, Indiana, USA). These varieties are not currently available for commercial use.

Table 2. Monthly rainfall and maximum and minimum temperatures of the 2021 growing season recorded at San Bartolomeo in Galdo, Italy.

Month	T. Min (°C)	T. Max (°C)	T. Mean (°C)	Rainfall (mm)
April	7.8	14.4	11.1	78
May	9.8	18.6	14.2	30.8
June	14.1	27.8	20.9	29
July	18.8	28.5	23.6	15.2
August	17.4	29.4	23.4	29
September	14.7	23.4	19.0	24.6
October	10.8	13.4	12.1	1000
	Mean	Mean	Mean	Total
	13.3	22.2	17.7	172.3

Table 3. Physico-chemical properties of the soil of the experimental field located in San Bartolomeo in Galdo (BN), Campania (Italy).

Scheme 0	0–60 cm Depth
Clay (%)	42.6
Silt (%)	18.8
Sand (%)	39.4
pH	8.4
Exchangeable Ca (g/kg)	119
Available P (mg/kg)	16
Exchangeable K (meq/100 g)	1.4
Exchangeable Mg (meq/100 g)	1.6
Total Ca carbonate (g/kg)	68
Total N (g kg^{-1})	0.8
CSC (meq/100 g)	28
Organic C (g kg^{-1})	2.5

2.2. *Flour Sample Preparation*

Roughly a kilogram of grain was pulverized into flour utilizing a two-roller mill from Chopin Moulin CD1 (Chopin S.A., Villeneuve la Garenne, France). Post-grinding, the samples underwent screening via a planetary sieve with a screen size measuring 120 μm^2, manufactured by Buhler AG (Uzwil, Switzerland).

2.3. *Moisture Content*

The moisture content of the flour samples was measured as previously described [15]. Initially, a ceramic capsule was meticulously weighed following thorough desiccation at 100 °C under vacuum (25 mm Hg) using an oven (ISCO mod. NSV9035, Milan, Italy) and subsequently cooled down to room temperature inside a silica gel dryer. Then, a precisely weighed portion of the flour samples (approximately 2 g) was introduced into the dried ceramic capsule. The sample was then subjected to the same temperature and pressure conditions for around 5 h until a constant weight was achieved. At this point, the humidity had been extracted from the sample. The moisture content was calculated by determining the weight loss.

The moisture content of sorghum samples was calculated using the following equation:

$$Moisture\ (\%) = \frac{Weight\ of\ fresh\ sample - Weight\ after\ drying}{Weight\ of\ fresh\ sample} \times 100$$

2.4. Ash

Approximately 3 g of sorghum samples were weighed and placed in an incineration dish to determine the ash content. The dish was then ignited at roughly 550 °C and subsequently allowed to cool down within a desiccator. Once the dish had reached room temperature, it was immediately weighed, in line with the methodology established by AOAC [16] (AOAC, 1923).

For ash content, the following equation was used:

$$Ash\ (\%) = \frac{Weight\ of\ ash}{Initial\ sample\ weight} \times 100$$

2.5. Protein Content

Sorghum flour samples (2 g each) were analyzed for their nitrogen content by the AOAC [17] (AOAC, 1920) Kjeldahl method using a Mineral Six Digester and an Auto Disteam semi-automatic distillation unit (International PBI, Milan, Italy). The total protein content was subsequently calculated using a conversion factor of 6.25.

2.6. Total Lipid Content

The total lipid content was determined as previously described [18]. Initially, roughly 3 g of grain was ground down to a fine powder with liquid nitrogen, utilizing a mortar and pestle, and subsequently lyophilized using the FTS-System Flex-DryTM instrument. The ground whole meal was extracted for 4 h in a Soxhlet apparatus with chloroform ($CHCl_3$). The resulting extracts were dried out with a rotary evaporator to obtain crude extracts, and the weight of the fat extracted was subsequently determined.

2.7. Gas Chromatography of Fatty Acids

To conduct the esterification of the fatty acids found within the crude extracts, and subsequent gas chromatographic analysis of the fatty acid methyl esters, the protocols established by Pontieri et al. [15,18] were followed. Initially, the solid sorghum fat was melted in an oven set to 50 °C. A single drop of fat was then transferred into a 1.5 mL vial, followed by the addition of 1 mL of hexane and 100 µL of 2 N KOH methanolic solution. The vial was then vortexed for a duration of 5 min before being left to stand undisturbed for an additional 5 min to allow for complete stratification of the hexanic portion, which contained the methyl ester of the fatty acids. Chromatographic separation was achieved via a GC-2010 (Shimadzu, Kyoto, Japan) that was fitted with a DB-Wax column (Phenomenex, Torrance, CA, USA), measuring 30 m in length, 0.25 mm in internal diameter, and with a 0.25 µm film thickness. The GC conditions utilized were as follows: carrier gas, helium; pressure, 75 kPa; injector temperature, 220 °C; FID temperature, 250 °C; and oven program, 170 °C for 8 min, 2 °C/min to 185 °C for 10 min, 1 °C/min to 190 °C for 12 min, and 10 °C/min to 240 °C for 5 min.

2.8. Carbohydrates

The quantity of carbohydrates present in the samples was calculated by deducting the values obtained for moisture, ash, protein, and fat content, as explained in the procedure described by Arienzo et al. [19].

2.9. Fiber Content

The AOAC method from 1995 [20] (AOAC, 1995) was employed to determine the fiber content. In this method, the sample was digested under acidic conditions using 0.255 N H_2SO_4, followed by alkaline digestion with 0.223 N NaOH in an automatic digestor (Velp Scientific mod. FIWE3, Usmate Velate, Monza e Brianza, Italy). The lost mass of the sample after incineration was considered as fiber.

2.10. Total Minerals Determination

The procedure described by Tenore et al. [21] and Pontieri et al. [22,23] was followed for the determination of the mineral elements of interest employing quadrupole inductively coupled plasma mass spectrometry (ICP-QMS) on an 820-MS instrument (Bruker Daltonics, Billerica, MA, USA). Operational parameters included the following: plasma flow rate: 18 L per minute; auxiliary flow rate: 1.8 L per minute; sheath gas flow rate: 0.14 L per minute; nebulizer flow rate: 0.98 L per minute; RF power: 1.40 kilowatts; pump rate: 4 revolutions per minute; stabilization delay: 20 s; and voltage settings for various extraction lenses and other components. Control of reactive contaminants was achieved through the use of high purity helium and hydrogen gases. The stability of plasma was maintained by employing high radio frequency power. All chemicals utilized were of the highest commercially available purity. Prior to utilization, all glassware and plastic containers underwent thorough cleaning with 10% ultra-pure grade nitric acid followed by rinsing with ultra-pure water. Calibration solutions were prepared from multi-elemental standard stock solutions with a concentration of 20.00 milligrams per liter. The calibration curves were generated using nine calibration solutions. Reagent blanks, consisting of ultra-pure water, were also analyzed to ensure the purity of reagents and laboratory equipment. The determination process involved the use of an internal standard mixture containing specific isotopes aspirated online alongside the sample and standard solution. Analysis included quantification of 17 isotopes using a calibration curve method, while others were quantified using a semi-quantitative approach based on mean response factors from adjacent elements.

For samples analysis, ash content was dissolved in a 5% HNO_3 solution containing ultra-pure water, and the solution was filtered using regenerated cellulose filters that were free from ash.

2.11. ELISA Assay

The RIDASCREEN® Gliadin standard test kit (Art. No R7001, R-Biopharm AG, Darmstadt, Germany) ELISA-based sandwich method was used to identify gliadins in grain flour samples according to both Valdés et al. [24] and the manufacturer's instructions. Commercial gliadin standard 16–18% N (Sigma Aldrich, Milan, Italy) was used as the control.

2.12. Statistical Analysis

With the exception of total lipids analyses, which were performed in triplicate, all analyses were performed in quintuples ($n = 5$) (technical replicates), and the results are presented as mean ± SD. Data distributions were evaluated by the Shapiro–Wilk test. As all data were not normally distributed, differences in means were investigated using the Mann–Whitney non-parametric U test. This test is utilized to ascertain whether significant differences exist between two independent groups. Analysis of variance (ANOVA) was used to evaluate whether the different values were statistically significant or not. This test is employed to assess whether differences between group means stem from genuine distinctions in the groups themselves or if they are merely attributable to random variability. Tukey's post-hoc test was used to identify which samples were different. This test was employed subsequent to conducting an ANOVA to discern which pairs of groups exhibit significant differences from each other. While ANOVA solely indicates the presence of overall differences between groups, Tukey's post-hoc test aids in identifying the specific pairs of groups that differ significantly. A false discovery rate (FDR)-adjusted p-value was used to handle multiple comparisons. PAST 4 was employed to generate non-metrical multidimensional (NM-MDS) scaling based on the Bray–Curtis index [25].

3. Results

3.1. Weather Conditions and Soil Characteristics

The amount of rainfall during the crop growing season varied from 78 to 1000 mm (April–October) with a mean of 172.3 mm, mainly concentrated during spring and early autumn. Climatic results were measured in the year 2021 using a meteorological station located near the experimental field. Maximum and minimum temperatures, as well as precipitation, were collected monthly (Table 2) [26]. The soil of the area is mainly clayey, deep, and with a good water retention capacity, as reported in Table 3. It is well known that grain composition can vary significantly, influenced by the genotype and the growth environment, such as temperature, soil characteristics, fertilizers, and other factors. An example is the work published by Rooney (2004) [27], where it was demonstrated that high-nitrogen fertilizer levels increase grain protein content and decrease the amount of total carbohydrates.

3.2. Chemical Composition

The chemical composition of nine food-grade sorghum inbreds developed in the United States and grown in southern Italy is shown in Table 4. Protein content was higher in PL-2 and N3 inbreds and relatively lower in N2. The fat content was slightly higher in PL-1 than in PL-2, N3, N4, and N5, while a lower fat content was observed in PL-3, PL-4, Tw, and N2. Variations were also observed in the total carbohydrate content where Tw, N2, N3, and N5 showed a higher content than the other inbreds. Finally, a higher fiber content was evident for N4, PL-2, and PL-3 inbreds compared to the others. Particularly significant was the variation interval shown by the fiber content.

Table 4. Nutritional composition of 9 food-grade sorghum inbred varieties grown under typical Mediterranean conditions.

Parameter	Moisture (%)	Ash (%)	Total Proteins (%)	Fats (%)	Total Carbohydrates (%)	Sugars (%)	Fibers (%)
PL-1	12.2 ± 0.9^{dg}	1.7 ± 0.1^{bcdeg}	11 ± 0.4^{fg}	2.93 ± 0.12^{cdefh}	68.74 ± 2.75	1.6 ± 0.2^{f}	3.43 ± 0.14^{bcdefghi}
PL-2	11.7 ± 0.6^{dg}	2 ± 0.2^{acdf}	12.2 ± 1.2^{cdfi}	2.61 ± 0.29	65.9 ± 5.27	1.7 ± 0.2^{efg}	5.59 ± 0.56^{acdefghi}
PL-3	12.3 ± 0.6^{dgi}	2.5 ± 0.3^{abfhi}	10.5 ± 0.4^{bfg}	2.3 ± 0.07^{afghi}	65.54 ± 5.24	1.5 ± 0^{df}	6.86 ± 0.69^{abdefgi}
PL-4	10.6 ± 0.7^{abc}	2.4 ± 0.2^{abfhi}	10.7 ± 0.5^{bfg}	2.45 ± 0.27^{a}	69.41 ± 6.25	1.8 ± 0.2^{cefgi}	4.44 ± 0.13^{abcefghi}
Tw	11.8 ± 0.8^{g}	2.3 ± 0.2^{afhi}	10.7 ± 0.6^{fg}	2.36 ± 0.09^{ah}	71.15 ± 6.4	1.4 ± 0.1^{bdh}	1.69 ± 0.08^{abcdfghi}
N2	11.2 ± 1.3	1.7 ± 0.1^{bcdeg}	9 ± 0.3^{abcdegh}	2.47 ± 0.1^{ac}	72.87 ± 8.74	1.3 ± 0.1^{abcdhi}	2.76 ± 0.3^{abcdeghi}
N3	10.5 ± 0.4^{abce}	2.2 ± 0.1^{afhi}	12.5 ± 0.4^{acdefhi}	2.63 ± 0.26^{c}	70.74 ± 6.37	1.4 ± 0.1^{bdh}	1.43 ± 0.17^{abcdefhi}
N4	11.4 ± 1	1.9 ± 0.2^{cdeg}	10.8 ± 0.6^{fg}	2.63 ± 0.11^{ace}	65.54 ± 1.97^{i}	1.6 ± 0.1^{efg}	7.73 ± 0.85^{abdefgi}
N5	10.9 ± 0.9^{c}	1.8 ± 0.1^{cdeg}	10 ± 1.2^{bg}	2.66 ± 0.27^{c}	73.47 ± 6.61^{h}	1.5 ± 0.1^{df}	1.17 ± 0.09^{abcdefgh}
Int.Var [1]	10.5–12.3	1.7–2.5	9.03–12.5	2.3–2.93	65.54–73.47	1.3–1.8	1.17–7.73

[1] Interval of variation. The letter a in superscript indicates a value significantly different compared to the PL-1 variety, b compared to PL-2 variety, c compared to PL-3 variety, d compared to PL-4 variety, e compared to Tw variety, f compared to N2 variety, g compared to N3 variety, h compared to N4 variety, i compared to N5 variety.

3.3. Fatty Acid Composition of Total Lipids

The percentages of total fatty acids, as well as aggregated as saturated, monounsaturated, and polyunsaturated fats of the nine inbreds, are shown in Table 5. A higher content of total monounsaturated and polyunsaturated fats was found in both PL-1 and N5, while N3 showed a polyunsaturated fat content comparable to that found in PL-1. The saturated fatty acid content was almost similar in all varieties except for the N5 inbred which showed a higher content than the others. Oleic, linoleic, and palmitic acids were the most abundant fatty acids in all nine samples. Erucic acid was absent in six out of the nine lines. According to the NM-MDS (Figure 1A), the varieties N2, N4, Tw, and PL4 exhibit similarity in terms of fatty acid content. Variety N3 is not far from this first cluster and is also considered similar. Varieties PL2 and PL3 share similarity, as do PL-3 and N5.

Table 5. Fatty acid content of 9 food-grade sorghum inbred varieties grown under typical Mediterranean conditions. The letter a in superscript indicates a value significantly different compared to the PL-1 variety, b compared to PL-2 variety, c compared to PL-3 variety, d compared to PL-4 variety, e compared to Tw variety, f compared to N2 variety, g compared to N3 variety, h compared to N4 variety, i compared to N5 variety.

Parameter	PL-1	PL-2	PL-3	PL-4	Tw	N2	N3	N4	N5	Int.Var [4]
Myristic C14:0	$0.082 \pm 0.003^{bcdefghi}$	0.054 ± 0.004^{ai}	0.05 ± 0.003^{af}	$0.0\,5 \pm 0.002^{af}$	0.05 ± 0.004^{af}	$0.057 \pm 0.003^{acdeghi}$	0.05 ± 0.003^{afhi}	0.049 ± 0.003^{af}	0.046 ± 0.004^{abf}	0.05–0.082
Palmitic C16:0	11.58 ± 0.58^{defghi}	11.84 ± 0.59^{defghi}	11.88 ± 0.59^{defghi}	13.47 ± 0.53^{abc}	13.46 ± 1.07^{abc}	14.22 ± 0.56^{abchi}	13.45 ± 0.94^{abc}	13.40 ± 0.40^{abcf}	13.26 ± 0.53^{abcf}	11.58–13.47
Palmitoleic C16:1 9 c	$0.291 \pm 0.015^{bdefghi}$	$0.215 \pm 0.009^{acdefghi}$	0.31 ± 0.009^{bdefghi}	0.34 ± 0.01^{abcefghi}	$0.600 \pm 0.018^{abcdghi}$	$0.567 \pm 0.028^{abcdghi}$	$0.369 \pm 0.011^{abcdefhi}$	$0.516 \pm 0.021^{abcdefg}$	$0.500 \pm 0.025^{abcdefg}$	0.215–0.600
Margaric C17:0	0.082 ± 0.007^{bcdegi}	$0.046 \pm 0.002^{acefghi}$	0.070 ± 0.004^{abd}	$0.050 \pm 0.004^{acefghi}$	0.070 ± 0.004^{abd}	0.073 ± 0.004^{bd}	0.069 ± 0.004^{abd}	0.074 ± 0.002^{bd}	0.07 ± 0.004^{abd}	0.046–0.082
Margaroleic C17:1 10 c	0.08 ± 0.00^{bcdfgh}	0.06 ± 0.00^{acdfh}	0.07 ± 0.00^{abdfgh}	0.071 ± 0.00^{abcfgh}	0.07 ± 0.01^{h}	0.062 ± 0.00^{abcdgh}	0.06 ± 0.00^{acdfh}	0.00 ± 0.00^{abcdefgi}	0.07 ± 0.01^{h}	0.00–0.08
Stearic C18:0	1.22 ± 0.10^{f}	1.33 ± 0.05^{cdefhi}	1.22 ± 0.08^{bf}	1.205 ± 0.07^{bf}	1.21 ± 0.08^{bf}	1.08 ± 0.03^{abcdeghi}	1.24 ± 0.07^{f}	1.24 ± 0.03^{bf}	1.18 ± 0.07^{bf}	1.08–1.33
Oleic C18:1 9c	35.92 ± 1.07^{cdefghi}	35.08 ± 2.80^{cdegi}	41.06 ± 2.46^{abdefgh}	30.65 ± 2.45^{abci}	30.39 ± 0.91^{abcgi}	31.97 ± 2.55^{acgi}	27.61 ± 2.20^{abcefhi}	32.33 ± 1.94^{acgi}	40.95 ± 2.04^{abdefgh}	27.61–41.06
Linoleic C18:2 9c12c	46.95 ± 1.40^{cdegi}	47.79 ± 1.91^{cdegi}	42.54 ± 2.12^{abdefgh}	50.76 ± 1.52^{abci}	50.98 ± 1.52^{abci}	49.38 ± 2.96^{ci}	54.28 ± 4.34^{abci}	49.59 ± 3.96^{ci}	40.66 ± 1.22^{abdefgh}	42.54–54.28
Linolenic C18:3 c9c12c15	2.55 ± 0.17^{bcdefghi}	2.14 ± 0.08^{acefgh}	1.49 ± 0.04^{abdehi}	2.05 ± 0.10^{acefgh}	1.83 ± 0.05^{abcdfgi}	1.45 ± 0.10^{abdehi}	1.57 ± 0.07^{abdei}	1.71 ± 0.13^{abcdfi}	2.01 ± 0.10^{acefgh}	1.45–2.55
Arachidic C20:0	0.14 ± 0.01^{bcegi}	0.16 ± 0.01^{af}	0.17 ± 0.01^{adfh}	0.15 ± 0.01^{c}	0.16 ± 0.01^{af}	0.14 ± 0.01^{bcegi}	0.16 ± 0.01^{af}	0.151 ± 0.01^{c}	0.16 ± 0.01^{af}	0.14–0.17
Eicosenoic C20:1 11c	0.220 ± 0.009^{bdgi}	$0.160 \pm 0.008^{acdefghi}$	0.220 ± 0.009^{bdgi}	$0.200 \pm 0.014^{abcfghi}$	0.210 ± 0.017^{bfgi}	0.238 ± 0.017^{bde}	0.250 ± 0.015^{abcde}	0.229 ± 0.011^{bdi}	0.250 ± 0.008^{abcdeh}	0.160–0.250
Behenic C22:0	$0.160 \pm 0.005^{bcdefghi}$	$0.077 \pm 0.002^{acdefghi}$	$0.130 \pm 0.007^{abdefghi}$	$0.110 \pm 0.009^{abcefghi}$	$0.040 \pm 0.003^{abcdfghi}$	$0.070 \pm 0.005^{abcdeghi}$	$0.070 \pm 0.003^{abcdehi}$	$0.060 \pm 0.002^{abcdefg}$	$0.060 \pm 0.002^{abcdefg}$	0.040–0.160
Lignoceric C24:0	$0.067 \pm 0.005^{cdefghi}$	0.060 ± 0.004^{cdefgh}	$0.040 \pm 0.001^{abdefghi}$	0.110 ± 0.004^{abcei}	0.12 ± 0.006^{abcdfi}	0.102 ± 0.008^{abcegi}	0.120 ± 0.01^{abcfi}	0.110 ± 0.009^{abci}	$0.060 \pm 0.002^{acdefgh}$	0.040–0.120
Erucic C22:1 13c	0.000 ± 0.000^{dgh}	0.000 ± 0.000^{dgh}	0.000 ± 0.000^{dgh}	$0.011 \pm 0.000^{abcefghi}$	0.000 ± 0.000^{dgh}	0.000 ± 0.000^{dgh}	$0.008 \pm 0.000^{abcdefhi}$	$0.003 \pm 0.000^{abcdefgi}$	00.000 ± 0.000^{dgh}	0.00–0.008
MSF [1]	1.07 ± 0.11^{defghi}	0.93 ± 0.07^{defgi}	0.96 ± 0.07^{defgi}	0.77 ± 0.07^{abci}	0.74 ± 0.04^{abcfhi}	0.81 ± 0.03^{abcegi}	0.74 ± 0.04^{abcfhi}	0.87 ± 0.07^{aegi}	1.53 ± 0.09^{abcdefgh}	0.74–1.53
PSF [2]	1.47 ± 0.15^{cef}	1.33 ± 0.15^{ci}	1.03 ± 0.07^{abdefghi}	1.32 ± 0.13^{cgi}	1.27 ± 0.05^{acgi}	1.27 ± 0.06^{acgi}	1.49 ± 0.06^{cef}	1.37 ± 0.12^{ci}	1.59 ± 0.14^{bcdefh}	1.03–1.59
SF [3]	0.39 ± 0.01^{bcdei}	0.35 ± 0.02^{acghi}	0.31 ± 0.02^{abdfghi}	0.37 ± 0.01^{aci}	0.35 ± 0.03^{agi}	0.39 ± 0.04^{ci}	0.4 ± 0.03^{bcei}	0.39 ± 0.03^{ci}	0.54 ± 0.04^{abcdefgh}	0.31–0.54

[1] Monounsaturated fats; [2] polyunsaturated fats; [3] saturated fats, [4] interval of variation.

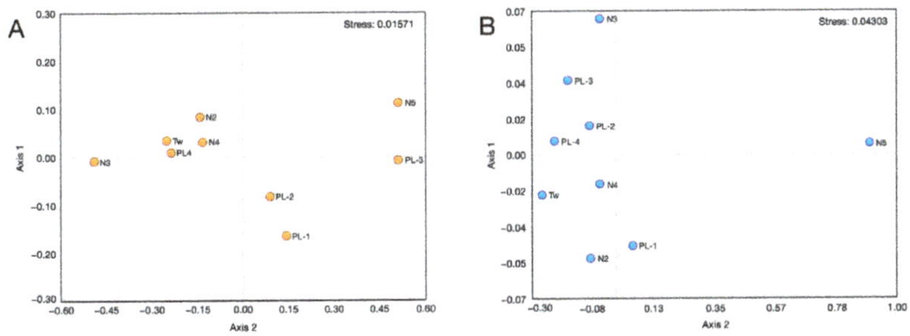

Figure 1. Non-metrical multidimensional scaling (NM-MDS) based on fatty acids content (**A**) and macronutrients, micronutrients, and trace elements (**B**).

3.4. Mineral Content

The results of the macro-elements, micro-elements, and trace elements of the nine inbreds of sorghum are reported in Table 6, Table 7, and Table 8, respectively. The content of macro-elements followed the sequence K > Mg > Ca > Na in the nine samples analyzed. The content of micro-elements followed the sequence Fe > Zn > Al > Mn >Cr >Cu> Ba > Ni > Pb > Mo > Ag > V > Sn > Co > As > Se >Be > Tl, while the content of trace elements followed the sequence Cd > Hg > U > Sb. Variations in the content of the elements were found among the nine inbreds of sorghum analyzed. On average, K and Mg were, respectively, the most abundant macro-elements, while Fe, Zn, Al, and Mn were the most abundant micro-elements and Cd, Hg, and U the most abundant trace elements (with some exceptions). Furthermore, the potassium content of the nine inbred samples averaged about 60 times higher than that of sodium. Notably, the K:Na ratio of each of the PL-1, N2, N3, and N4 inbreds was higher, while the K:Na ratio of the Tw inbred was lower than that of the other sorghum inbreds analyzed. The K:Na ratio was higher than the recommended ratio 5.0 [26] for human diet. The high K:Na ratio suggests that sorghum inbreds may be suitable for improving health problems due to sodium retention. In fact, diets with a higher K:Na ratio are recommended for particular patients [28].

Table 6. Nutritionally essential macro-element content of 9 food-grade sorghum inbred varieties grown under typical Mediterranean conditions (mg kg^{-1}). The letter a in superscript indicates a value significantly different compared to the PL-1 variety, b compared to PL-2 variety, c compared to PL-3 variety, d compared to PL-4 variety, e compared to Tw variety, f compared to N2 variety, g compared to N3 variety, h compared to N4 variety, i compared to N5 variety.

Parameter	Na	Mg	K	Ca
PL-1	0.14 ± 0.01^{bcdefghi}	6.16 ± 0.37^{bcdg}	13.66 ± 0.82^{g}	0.51 ± 0.03^{bcdefghi}
PL-2	0.24 ± 0.01^{acdefgh}	7.06 ± 0.42^{aefghi}	14.04 ± 0.7^{g}	0.82 ± 0.04^{adefghi}
PL-3	0.32 ± 0.01^{abdefghi}	7.32 ± 0.37^{aefghi}	14.25 ± 1^{g}	0.83 ± 0.04^{adefghi}
PL-4	0.29 ± 0.01^{abcefghi}	6.99 ± 0.21^{aefghi}	14.09 ± 0.56^{g}	0.96 ± 0.06^{abcefhi}
Tw	0.35 ± 0.01^{abcdfghi}	6.05 ± 0.24^{bcdg}	13.04 ± 0.91^{gh}	1.73 ± 0.12^{abcdfghi}
N2	0.19 ± 0.01^{abcdegi}	5.88 ± 0.29^{bcdg}	14.38 ± 1.01^{eg}	0.72 ± 0.02^{abcdeghi}
N3	0.21 ± 0.01^{abcdefi}	8.2 ± 0.25^{abcdefhi}	$16.38 \pm 0.82^{abcdefhi}$	0.99 ± 0.03^{abcefhi}
N4	0.2 ± 0.01^{abcdei}	5.91 ± 0.35^{bcdg}	14.54 ± 0.87^{eg}	0.68 ± 0.02^{abcdefgi}
N5	0.25 ± 0.01^{acdefgh}	6.3 ± 0.44^{bcdg}	14.14 ± 0.42^{g}	0.63 ± 0.03^{abcdefgh}
Int.Var [1]	0.14–0.35	5.88–8.2	13.04–16.38	0.51–1.73

[1] Interval of variation.

Table 7. Nutritionally essential micro-element content of 9 food-grade sorghum inbred varieties grown under typical Mediterranean conditions (mg kg^{-1}).

Parameter	PL-1	PL-2	PL-3	PL-4	Tw	N2	N3	N4	N5	Int.Var [1]
Be	<0.01	0.02 ±0.00	0.03 ±0.00	0.04 ±0.00	0.05 ±0.00	0.02 ±0.00	0.02 ±0.00	0.02 ±0.00	0.02 ±0.00	<0.01–0.05
Al	13.91 ± 0.83 ^{bcdefghi}	65.47 ± 3.93 ^{acdefghi}	107.15 ± 3.21 ^{abdefghi}	131.88 ± 5.28 ^{abcefghi}	178.44 ± 8.92 ^{abcdfghi}	54.42 ± 3.27 ^{abcdeghi}	42.62 ± 1.7 ^{abcdefhi}	56.77 ± 3.97 ^{abcdefgi}	25 ± 1.75 ^{abcdefgh}	13.91–178.44
V	<0.01 ^{bcdef}	<0.01 ^{acdefghi}	0.15 ± 0.01 ^{abdefghi}	0.22 ± 0.01 ^{abcefghi}	0.35 ± 0.02 ^{abcdfghi}	0.08 ± 0.00 ^{abcdeghi}	0.04 ± 0.00 ^{bcdef}	<0.01 ^{bcdef}	<0.01 ^{bcdef}	<0.01–0.35
Cr	5.47 ± 0.38 ^{bcdefghi}	17.79 ± 0.53 ^{acdefghi}	3.67 ± 0.18 ^{abefghi}	3.55 ± 0.18 ^{abefghi}	4.18 ± 0.21 ^{abcdfghi}	4.69 ± 0.14 ^{abcdeghi}	3.07 ± 0.09 ^{abcdefhi}	6.91 ± 0.21 ^{abcdefgi}	5.71 ± 0.34 ^{bcdefgh}	3.07–17.79
Mn	36.33 ± 2.18 ^{bcdeg}	41.78 ± 2.92 ^{acdgh}	55.12 ± 3.31 ^{abdefhi}	47.39 ± 3.32 ^{abcfghi}	46.53 ± 3.26 ^{acfghi}	40.09 ± 2.81 ^{cdegh}	56.79 ± 1.7 ^{abdefhi}	34.95 ± 1.75 ^{bcdefg}	37.54 ± 2.63 ^{cdeg}	34.95–56.79
Fe	113.17 ± 6.79 ^{cdefgh}	117.2 ± 8.2 ^{cdefh}	154.25 ± 4.63 ^{abdefghi}	182.04 ± 5.46 ^{abcfghi}	193.66 ± 13.56 ^{abcfghi}	135.67 ± 5.43 ^{abcdei}	128.12 ± 6.41 ^{acdei}	132.68 ± 9.29 ^{abcdei}	113.42 ± 6.81 ^{cdefgh}	113.17–193.66
Co	0.06 ± 0.00 ^{bcdefhi}	0.07 ± 0.00 ^{adefghi}	0.07 ± 0.00 ^{adefghi}	0.08 ± 0.00 ^{abceghi}	0.11 ± 0.01 ^{abcdfgh}	0.08 ± 0.00 ^{abceghi}	0.06 ± 0.00 ^{bcdefhi}	0.05 ± 0.00 ^{abcdefgi}	0.12 ± 0.01 ^{abcdfgh}	0.05–0.12
Ni	1.89 ± 0.06 ^{bcdefghi}	1.45 ± 0.1 ^{acdfi}	0.95 ± 0.03 ^{abdefghi}	1.32 ± 0.04 ^{abcefi}	1.61 ± 0.1 ^{acdfghi}	2.37 ± 0.09 ^{abcdeghi}	1.40 ± 0.07 ^{acefi}	1.40 ± 0.06 ^{acefi}	3.9 ± 0.23 ^{abcdefgh}	0.95–2.37
Cu	11.05 ± 0.66 ^{bcdefghi}	16.15 ± 0.97 ^{adefghi}	16.39 ± 0.98 ^{adefghi}	13.77 ± 0.41 ^{abcfghi}	13.67 ± 0.68 ^{abcfghi}	9.26 ± 0.46 ^{abcdegi}	24.13 ± 0.72 ^{abcdefhi}	9.46 ± 0.38 ^{abcdegi}	33.79 ± 2.37 ^{abcdefgh}	9.26–33.79
Zn	97.85 ± 3.91 ^{bcdghi}	107.93 ± 5.4 ^{aefgi}	107.76 ± 6.47 ^{aefgi}	107.68 ± 3.23 ^{aefgi}	96.59 ± 3.86 ^{bcdghi}	93.9 ± 6.57 ^{bcdghi}	130.44 ± 6.52 ^{abcdefhi}	113.06 ± 4.52 ^{aefgi}	578.52 ± 28.93 ^{abcdefgh}	93.9–578.52
As	0.03 ± 0.00 ^{bcdefghi}	0.09 ± 0.00 ^{adefghi}	0.09 ± 0.00 ^{adefghi}	0.10 ± 0.00 ^{abcfgi}	0.10 ± 0.00 ^{abcfgi}	0.05 ± 0.00 ^{abcdeh}	0.05 ± 0.00 ^{abcdeh}	0.11 ± 0.01 ^{abcfgi}	0.05 ± 0.00 ^{abcdeh}	0.03–0.11
Se	0.02 ± 0.00 ^{bcdefghi}	0.01 ± 0.00 ^{acdefghi}	0.03 ± 0.00 ^{abegi}	0.03 ± 0.00 ^{abegi}	0.04 ± 0.00 ^{abcdfgh}	0.03 ± 0.00 ^{abegi}	0.05 ± 0.00 ^{abcdefhi}	0.03 ± 0.00 ^{abegi}	0.04 ± 0.00 ^{abcdfgh}	0.01–0.05
Mo	0.26 ± 0.01 ^{bcdefghi}	0.39 ± 0.02 ^{acdfghi}	0.34 ± 0.02 ^{abdefghi}	0.60 ± 0.02 ^{abcefghi}	0.42 ± 0.02 ^{acdfhi}	0.53 ± 0.03 ^{abcdeghi}	0.43 ± 0.02 ^{abcdfhi}	0.77 ± 0.02 ^{abcdefgi}	0.70 ± 0.03 ^{abcdefgh}	0.26–0.77
Ag	0.10 ± 0.01 ^{i}	0.10 ± 0.01 ^{i}	0.10 ± 0.01 ^{i}	0.10 ± 0.01 ^{i}	0.10 ± 0.01 ^{i}	0.10 ± 0.01 ^{i}	0.10 ± 0.01 ^{i}	0.10 ± 0.01 ^{i}	0.45 ± 0.01 ^{abcdefgh}	0.1–0.45
Sn	0.08 ± 0.01 ^{cdeghi}	0.07 ± 0.00 ^{cdefghi}	0.10 ± 0.00 ^{abhi}	0.10 ± 0.01 ^{abhi}	0.10 ± 0.01 ^{abhi}	0.09 ± 0.01 ^{bhi}	0.10 ± 0 ^{abh}	0.13 ± 0.01 ^{abcdefgi}	0.05 ± 0.00 ^{abcdefgh}	0.05–0.14
Ba	1.21 ± 0.08 ^{bcdefghi}	2.48 ± 0.07 ^{acdefghi}	3.86 ± 0.27 ^{abdefghi}	5.39 ± 0.27 ^{abcefghi}	6.3 ± 0.44 ^{abcdfghi}	1.97 ± 0.14 ^{abcdei}	1.9 ± 0.11 ^{abcdeh}	2.12 ± 0.13 ^{abcdegi}	1.73 ± 0.1 ^{abcdefh}	1.21–6.3
Tl	<0.01	<0.01	<0.01	<0.01	<0.01	<0.01	<0.01	<0.01	<0.01	<0.01
Pb	0.26 ± 0.01 ^{abcdefghi}	0.39 ± 0.02 ^{acdefghi}	0.63 ± 0.03 ^{abdefhi}	0.76 ± 0.05 ^{abcfghi}	0.81 ± 0.05 ^{abcfghi}	1.05 ± 0.05 ^{abcdeghi}	0.65 ± 0.05 ^{abdefhi}	0.54 ± 0.03 ^{abcdefgi}	0.96 ± 0.03 ^{abcdefgh}	0.26–1.05

[1] Interval of variation. The letter a in superscript indicates a value significantly different compared to the PL-1 variety, b compared to PL-2 variety, c compared to PL-3 variety, d compared to PL-4 variety, e compared to Tw variety, f compared to N2 variety, g compared to N3 variety, h compared to N4 variety, i compared to N5 variety.

Table 8. Nutritionally essential trace element content of 9 food-grade sorghum inbred varieties grown under typical Mediterranean conditions (μg kg^{-1}).

Parameter	U	Sb	Hg	Cd
PL-1	10.01 ± 0.70 ^{bcdefghi}	3.0 ± 0.2 ^{bcdefhi}	46.5 ± 1.8	112.5 ± 3.3 ^{bcdefghi}
PL-2	25.27 ± 1.52 ^{acdefghi}	8.0 ± 0.5 ^{acdefghi}	47.0 ± 3.2	25.0 ± 0.7 ^{acdefghi}
PL-3	38.26 ± 1.15 ^{abdefghi}	5.5 ± 0.3 ^{abdefgi}	47.0 ± 2.8	27.0 ± 1.3 ^{abdefghi}
PL-4	49.41 ± 2.47 ^{abcfghi}	13.5 ± 0.5 ^{abcefghi}	49.0 ± 3.4	15.5 ± 0.6 ^{abcefghi}
Tw	54.11 ± 3.79 ^{abcfghi}	6.0 ± 0.1 ^{abcdfghi}	46.5 ± 2.3	13.0 ± 0.6 ^{abcdfghi}
N2	15.09 ± 1.06 ^{abcdegi}	2.5 ± 0.1 ^{abcdeghi}	46.0 ± 2.3	17.0 ± 0.5 ^{abcdeghi}
N3	33.6 ± 2.02 ^{abcdefhi}	3.0 ± 0.1 ^{bcdefhi}	47.1 ± 1.8	48.0 ± 2.8 ^{abcdefhi}
N4	15.28 ± 0.76 ^{abcdegi}	5.5 ± 0.2 ^{abdefgi}	47.5 ± 1.43	42.0 ± 2.1 ^{abcdefgi}
N5	8.83 ± 0.44 ^{abcdefgh}	<0.1 ^{abcdefgh}	47.7 ± 2.82	379.5 ± 26.5 ^{abcdefgh}
Int.Var [1]	10.01–54.11	<0.1–13.5	46.0–49.0	13.0–379.5

[1] Interval of variation. The letter a in superscript indicates a value significantly different compared to the PL-1 variety, b compared to PL-2 variety, c compared to PL-3 variety, d compared to PL-4 variety, e compared to Tw variety, f compared to N2 variety, g compared to N3 variety, h compared to N4 variety, i compared to N5 variety.

According to the NM-MDS (Figure 1B), the N5 variety exhibits distinct characteristics in terms of macronutrients, micronutrients, and trace elements compared to all other samples. Despite variations in some features, such as Ni and Co content, the remaining varieties share common characteristics.

Our results for Mg, K, and Zn were lower than those reported recently by Jacimovic et al. [29]. On the contrary, our Fe levels were much higher (Table 9). Mineral content of grains is affected by factors such as grain variety, soil composition, and weather conditions. This might explain the differences in the values between the two studies.

Table 9. Mineral content of the sorghum grains was compared to Recommended Dietary Allowance/Adequate Intake (RDA/AI) for these minerals. Additionally, a comparison was made with the findings from Jaćimović et al. (2023) [30].

Mineral	mg/100 g Sorghum	US RDA/AI	Jaćimović et al., 2023 [30]
Mg	0.6–0.8	400 (adult males age < 50)	57.6–92.7
Fe	11.3–19.4	18 (adult males and females age > 50)	1.4–3.4
K	1.3–1.6	3400	96.4–232.3
Zn	9.4–57.8	11 (adult males)	1.3–2.5

3.5. Immunochemical Evidence for the Absence of Gluten in Sorghum Inbreds

The results of immunochemical measurement of gliadin concentration in the sorghum flour from all samples tested demonstrated that gluten levels in all sorghum inbreds were less than 5 ppm (the detectable limit is 5 ppm) (Table 10). Those values are well below the 20 ppm threshold that has been proposed to be safe for celiac patients [24].

Table 10. Gliadin content (as ppm) in flours by using sandwich R5 enzyme-linked immunosorbent assay (ELISA).

Sorghum Inbreds	Content (ppm) [2]
PL-1	<5
PL-2	<5
PL-3	<5
PL-4	<5
Tw	<5
N2	<5
N3	<5

Table 10. *Cont.*

Sorghum Inbreds	Content (ppm) [2]
N4	<5
N5	<5
Wheat gliadin standard [1]	56

[1] Gliadin standard from wheat (Sigma). [2] Mean values from 3 measurements.

4. Discussion

Depending on the region of cultivation, the type of sorghum and the purpose of its production vary greatly, and the primary focus of sorghum breeders around the world is to improve nutritional and quality traits, yield, maturity, and adaptability [11,13,27]. In recent years there has been a growing interest worldwide in both functional and nutraceutical foods, including sorghum, and research has focused on identifying the mechanisms associated with their preventive or therapeutic potential [10]. As a consequence, the focus of sorghum breeding is the development of hybrid varieties characterized by high nutritional value and quality and by high yield, high adaptability, early maturation, and resistance tolerance to disease, pests, and stress [11,27,30].

With the aim of obtaining higher-quality sorghum cultivars, genetic improvement programs should be directed towards the use of pure lines of sorghum whose nutritional properties are known, following a genetic strategy to predict hybrids with desired nutritional and qualitative parameters [31,32]. Sorghum exhibits high genetic diversity, with many thousands of accessions and landraces collected and developed worldwide, particularly in collections in the USA, Ethiopia, China, and at the International Crops Research Institute for the Semi-Arid Tropics (ICRISAT) [14]. It has been demonstrated that grain composition can vary significantly due to genetic and environmental factors [15,18,22,27,33]. Inbred varieties and hybrids exhibit significant variations in yield potential, adaptability, and grain-quality characteristics; thus, cultivar selection is one of the most important considerations in crop management [11]. The genetic diversity of sorghum provides an opportunity to enhance the crop at the genetic level in interaction with the most suitable growth environment.

Currently, around the world, a wide range of genetic diversity of sorghum is available. However, inbred lines for the development of hybrids with improved nutritional and functional properties have not yet been fully evaluated [34]. Due to the growing interest in sorghum as a functional food and nutraceutical for human health [35], we have analyzed the nutritional composition, fatty acid content, and mineral content of nine food-grade sorghum lines grown under Mediterranean conditions, in an area with predominantly clayey soil, deep, and with good water retention capacity, that is, a suitable environment for the selection of inbred lines with high agronomic value and high nutritional quality.

Among the inbreds analyzed, those with the best characteristics could be used for hybrid development and germplasm improvement.

The composition profiles of the nine inbreds had some differences in both protein and carbohydrate percentages. As described in the Results section, PL-2 and N3, having a higher content of protein and fat, could be used in inbred breeding. Additionally, the higher fiber content of N4, PL-3, PL-4, PL-1, and particularly PL-2 may have health benefits beyond those conferred by the high protein content.

The amounts of total saturated fat were similar in the genotypes considered, but much lower than the values of total monounsaturated and polyunsaturated fats. Inbred varieties PL-1, PL-2, and N5 had higher amounts of both monounsaturated and polyunsaturated fatty acids, so these three lines may have a slight nutritional advantage. Linoleic, oleic, and palmitic acids were predominant over other fatty acids in all inbreds. Unsaturated fatty acids are important for human nutrition, as they are main components of biological membranes and play a role in modulating membrane fluidity. Furthermore, unsaturated fatty acids, unlike those saturated, do not have a cholesterogenic property and reduce the risk of thrombosis. Due to these characteristics, unsaturated fatty acids are strongly recommended to reduce the risk of atherosclerosis [10].

The content of each macro-element measured shows that the primary minerals are K, followed by Mg, a finding consistent with the available literature [36].

Furthermore, while the concentrations of the macro-elements were similar among all inbreds, both Ca and Na were higher in the Tw and lower in the PL-1 lines than in other inbreds. These data support previous studies indicating that the mineral content of sorghum is influenced by both genetic and environmental factors [33]. With regards to the macro-element content, we found a K:Na ratio higher than that recommended in the human diet for varieties of sorghum [26]. A higher K:Na ratio may improve bone health, reduce muscle loss, and moderate other chronic diseases such as hypertension [28].

Furthermore, it has been suggested that the K:Na ratio is an indicator of the quality of the diet of pregnant women and that the Na/K ratio in community settings is a potential population-based approach to address hypertension [37,38]. Prioritizing potassium-rich foods and limiting sodium intake can improve overall health outcomes, especially in reducing risks associated with chronic conditions.

It can also be noted from our data that the magnesium content was higher than typically found in corn (on average, 0.47 g kg^{-1}) and wheat flour (on average, 0.25 g kg^{-1}) [39]. Due to their high magnesium content, the inbreds considered in this study can be a good source of magnesium. Magnesium is an important macro-element necessary for the function of several enzyme systems [36].

In the nine inbreds, differences were noted in the concentration of micro-elements that could be due to the sorghum genotype, to soil conditions, and to the plant maturity state at harvest [23,40]. Fe, the most abundant micro-element, is an essential micro-element in human nutrition, and its deficiency is a serious public health threat worldwide. The expanding production of sorghum for human use in the United States and in the Mediterranean countries [3] is also motivated by the high levels of Fe in this crop, and by the beneficial nutritional effects of this micro-element. The results reported in the present study show a high content of Zn in all inbreds considered. This finding is notable because Zn deficiencies are a public health concern worldwide. It is important to underline that, as regards the trace elements, their concentration in all nine inbreds analyzed in this study did not exceed the maximum allowed by Regulation (CE) n. 41/2009. The nine inbred varieties of sorghum used in this study each have their own stabilized genetic heritage, which differs from one another, and they were all grown in the same typical Mediterranean environment. The integrity of the results, which report differences in nutritional composition, fatty acids, and mineral content, is ensured by previous studies that have demonstrated that the composition of sorghum grain can vary significantly due to genetic and environmental growth factors [15,18,22,27,33].

Finally, in this study, the nine inbreds we have characterized were also tested for gliadin concentration to confirm previous reports on the safety of sorghum for people with celiac disease [8].

The PL-2 and N3 genotypes have a high content of proteins, carbohydrates, fibers, unsaturated fatty acids, and minerals and appear to be the most suitable for the selection of hybrids with high nutritional quality. Substantial research has been conducted with the aim of developing the cultivation of sorghum lines in the Mediterranean area for use in the production of human food products [15,18,33].

The aim of this research was achieved, mainly comparing the nutritional composition of inbred sorghum varieties in order to assess the variation in nutritional properties and identify varieties with improved nutritional characteristics, thus providing greater potential health benefits for consumers. Furthermore, this research contributes to the body of knowledge on the nutritional composition of sorghum, particularly for sorghum varieties grown outside of the main sorghum breeding regions.

5. Conclusions

Today, particularly in developed countries, there is a significant demand for functional and nutraceutical healthy food. Adopting a diet based on food-grade white sorghum can promote disease prevention, leading to substantial cost savings for national health services. Food-grade sorghum is increasingly vital in the developed world, particularly as a cereal option for individuals with celiac disease. When developing food-grade sorghum varieties, several key grain traits, including grain quality, are considered. The functional food and feed traits and characteristics of sorghum (e.g., pasting, temperature, taste, amylase activity, shelf-life, etc.) are not fully understood. However, studies on cereal crops, including sorghum, suggest that these traits may be influenced by both genetic characteristics and the growth environment.

Breeding programs of both public and private institutions have released improved varieties adapted to semi-arid and tropical environments, including those that meet specific food and industrial requirements. To meet market demand, numerous collections of sorghum seeds have emerged worldwide, particularly in Ethiopia, China, and the United States, as well as at the International Crops Research Institute for the Semi-Arid Tropics (ICRISAT). The present study supports the strategy of evaluating sorghum nutritional properties such as protein and carbohydrate contents, levels of unsaturated fatty acids, and minerals of nine inbred sorghum varieties developed in the USA and grown in the Mediterranean area. The current research provides valuable information on the nutrient composition of various inbred sorghum varieties and supports the burgeoning sorghum breeding programs focused on the unique health benefits of consuming whole grain sorghum. The findings of this study underscore the importance of developing sorghum inbred lines with superior quality traits for use in breeding programs worldwide. These programs aim to select new sorghum hybrids that align with the preferences and nutritional needs of end-consumers, thereby ensuring a safe and healthy diet.

Author Contributions: P.P.: conceptualization, formal analysis, writing—original draft preparation. J.T.: conceptualization, formal analysis. M.C.: writing—review and editing, visualization. F.A.: writing—original draft preparation, writing—review and editing. M.T.: formal analysis. D.S.: writing—original draft preparation, writing—review and editing. M.G.: writing—review and editing, visualization. F.D.G.: writing—review and editing. A.M.: resources (cultivation of the nine sorghum inbred varieties in the field), formal analysis. I.S.: resources (cultivation of the nine sorghum inbred varieties in the field), formal analysis. A.L.C.: resources (cultivation of the nine sorghum inbred varieties in the field). M.A.: writing—review and editing, visualization. P.A.: conceptualization. L.D.G.: conceptualization, writing—original draft preparation. All authors have read and agreed to the published version of the manuscript.

Funding: The research was supported by MIUR—PON03PE_00060_2, Decreto di Concessione (27/06/2014), Progetto 2 "Progettazione, sviluppo e produzione di cibi funzionali e/o arricchiti", (OR4: Farine funzionali e prodotti da forno per celiaci ed intolleranti al glutine) to P. Pontieri; CNR-DISBA project "NutrAge" (project nr. 7022); and CNR project FOE-2019 DBA.AD003.139. Part of this work was supported by USDA-ARS.

Informed Consent Statement: Not applicable.

Data Availability Statement: The original contributions presented in the study are included in the article. Further inquiries can be directed to the corresponding author/s.

Acknowledgments: The authors are grateful to Francesco Salamini for revision, helpful discussion, and critical reading of the manuscript. Thanks also to Mitch Tuinstra for a generous gift of the food-grade white sorghum inbreds. The technical assistance of both Federico Gomez Paloma and Concetta Porzio is acknowledged. Names are necessary to report factually on available data; however, the U.S. Department of Agriculture neither guarantees nor warrants the standard of the product and use of the name by the U.S. Department of Agriculture implies no approval of the product to the exclusion of others that may also be suitable.

Conflicts of Interest: The authors declare no conflicts of interest.

References

1. Anglani, C. Sorghum for human food: A review. *Plant Food Hum. Nutr.* **1998**, *52*, 85–89. [CrossRef]
2. Awika, J.M.; Rooney, L.W. Sorghum phytochemicals and their potential aspects on human health. *Phytochemistry* **2004**, *65*, 1199–1221. [CrossRef]
3. Pontieri, P.; Del Giudice, L. Sorghum: A novel and healthy food. In *The Encyclopedia of Food and Health*; Caballero, B., Finglas, P., Toldrà, F., Eds.; Academic Press: Oxford, UK, 2016; pp. 33–42.
4. FAO. FAOSTAT. World Crops Website. 2017. Available online: www.fao.org/faostat/en/#data/QC (accessed on 30 March 2019).
5. USDA (United States Department of Agriculture)—Foreign Agriculture Division. 2003 Data. Available online: https://fas.usda.gov/ (accessed on 30 March 2019).
6. Awika, J.M. Chapter 3—Sorghum: Its unique nutritional and health-promoting attributes. In *Gluten-Free Ancient Grains*; Taylor, J.R.N., Awika, J., Eds.; Woodhead Publishing: Sawston, UK, 2017; pp. 21–54.
7. Ciacci, C.; Maiuri, L.; Caporaso, N.; Bucci, C.; Del Giudice, L.; Massardo, D.R.; Pontieri, P.; Di Fonzo, N.; Bean, S.R.; Ioerger, B.; et al. Celiac disease: In vitro and in vivo safety and palatability of wheat-free sorghum food products. *Clin. Nutr.* **2007**, *26*, 799–805. [CrossRef] [PubMed]
8. Pontieri, P.; Mamone, G.; De Caro, S.; Tuinstra, M.R.; Roemer, E.; Okot, J.; De Vita, P.; Ficco, D.B.; Alifano, P.; Pignone, D.; et al. Sorghum, a healthy and gluten-free food for celiac patients as demonstrated by genome, biochemical and immunochemical analyses. *J. Agric. Food Chem.* **2013**, *61*, 2565–2571. [CrossRef] [PubMed]
9. Smolensky, D.; Rhodes, D.; McVey, D.S.; Fawver, Z.; Perumal, R.; Herald, T.; Noronha, L. High-polyphenol sorghum bran extract inhibits cancer cell growth through ROS induction, cell cycle arrest, and apoptosis. *J. Med. Food.* **2018**, *21*, 990–998. [CrossRef] [PubMed]
10. Stefoska-Needham, A.; Tapsell, L. Considerations for progressing a mainstream position for sorghum, a potentially sustainable cereal crop, for food product innovation pipelines. *Trends Food Sci. Tech.* **2020**, *97*, 249–253. [CrossRef]
11. Tuinstra, M.R. Food-grade sorghum varieties and production considerations: A review. *J. Plant Interact.* **2008**, *3*, 69–72. [CrossRef]
12. de Mesa-Stonestreet, N.J.; Alavi, S.; Bean, S.R. Sorghum proteins: The concentration, isolation, modification, and food applications of kafirins. *J. Food Sci.* **2010**, *75*, R90–R104. [CrossRef]
13. Prasad, V.R.; Govindaraj, M.; Djanaguiraman, M.; Djalovic, I.; Shailani, A.; Rawat, N.; Singla-Pareek, S.L.; Pareek, A.; Prasad, P.V. Drought and high temperature stress in sorghum: Physiological, genetic, and molecular insights and breeding approaches. *Int. J. Mol. Sci.* **2021**, *22*, 9826. [CrossRef]
14. Rosenow, D.T.; Dahlberg, J.A. Collection, conversion, and utilization of sorghum. In *Sorghum: Origin, History, Technology, and Production*; Smith, C.W., Frederiksen, R.A., Eds.; John Wiley& Sons: New York, NY, USA, 2000; pp. 309–328.
15. Pontieri, P.; Di Fiore, R.; Troisi, J.; Bean, S.R.; Roemer, E.; Okot, J.; Alifano, P.; Pignone, D.; Del Giudice, L.; Massardo, D.R. Chemical composition and fatty acid content of white food sorghums grown in different environments. *Maydica* **2011**, *56*, 1–7.
16. AOAC. Association of Analytical Chemists Official Method, 923.03. Ash of flour, direct method. *J. AOAC Inter.* **1923**, *7*, 132.
17. AOAC. *Association of Analytical Chemists Official Method, 920.87-1929. Protein (Total) in Flour*; AOAC: Rockville, MD, USA, 2001.
18. Pontieri, P.; Di Maro, A.; Tamburino, R.; De Stefano, M.; Tilley, M.; Bean, S.R.; Roemer, E.; De Vita, P.; Alifano, P.; Del Giudice, L.; et al. Chemical composition of selected food-grade sorghum varieties grown under typical mediterranean conditions. *Maydica* **2010**, *55*, 139–143.
19. Arienzo, M.; De Martino, A.; Capasso, R.; Di Maro, A.; Parente, A. Analysis of carbohydrates and amino acids in vegetable waste waters by ion chromatography. *Phytochem. Anal.* **2003**, *14*, 74–82. [CrossRef] [PubMed]
20. AOAC. Association of Analytical Chemists Official Method, 962.09. Fiber (crude) in animal feed and pet food. In *Official Methods of Analysis AOC International*, 16th ed.; AOAC: Rockville, MD, USA, 1995.
21. Tenore, G.C.; Troisi, J.; Di Fiore, R.; Basile, A.; Novellino, E. Chemical composition, antioxidant and antimicrobial properties of Rapa Catozza Napo¬letana (Brassica rapa L. var. rapa DC.) seed meal, a promising protein source of Campania region (southern Italy) horticultural germplasm. *J. Sci. Food Agric.* **2012**, *92*, 1716–1724. [CrossRef]
22. Pontieri, P.; Troisi, J.; Calcagnile, M.; Bean, S.R.; Tilley, M.; Aramouni, F.; Boffa, A.; Pepe, G.; Campiglia, P.; Del Giudice, F.; et al. Chemical composition, fatty acid and mineral content of food-grade white, red and black sorghum varieties grown in the mediterranean environment. *Foods* **2022**, *11*, 436. [CrossRef]
23. Pontieri, P.; Troisi, J.; Di Fiore, R.; Di Maro, A.; Bean, S.R.; Tuinstra, M.R.; Roemer, E.; Boffa, A.; Del Giudice, A.; Pizzolante, G.; et al. Mineral content in grains of seven food-grade sorghum hybrids grown in a Mediterranean environment. *Aust. J. Crop. Sci.* **2014**, *8*, 1550–1559.
24. Valdés, I.; García, E.; Llorente, M.; Méndez, E. Innovative approach to low-level gluten determination in foods using a novel sandwich enzyme-linked immunosorbent assay protocol. *Eur. J. Gastroenterol. Hepatol.* **2003**, *15*, 465–474. [CrossRef]
25. Hammer, Ø.; Harper, D.A. Past: Paleontological statistics software package for educaton and data anlysis. *Palaeontol. Electron.* **2001**, *4*, 1.
26. Szentmihályi, K.; Kéry, Á.; Then, M.; Lakatos, B.; Sándor, Z.; Vinkler, P. Potassium-sodium ratio for the characterization of medicinal plant extracts with diuretic activity. *Phytother. Res. Int. J. Devoted Pharmacol. Toxicol. Eval. Nat. Prod. Deriv.* **1998**, *12*, 163–166. [CrossRef]
27. Rooney, L.W. Sorghum improvement-integrating traditional and new tecnology to produce improved genotypes. *Adv. Agron.* **2004**, *83*, 37–109.

28. Arbeit, M.L.; Nicklas, T.A.; Berenson, G.S. Considerations of dietary sodium/potassium/energy ratios of selected foods. *J. Am. Coll. Nutr.* **1992**, *11*, 210–222. [CrossRef] [PubMed]
29. Jaćimović, S.; Kiprovski, B.; Ristivojević, P.; Dimić, D.; Nakarada, Đ.; Dojčinović, B.; Sikora, V.; Teslić, N.; Pantelić, N.Đ. Chemical composition, antioxidant potential, and nutritional evaluation of cultivated Sorghum Grains: A combined experimental, theoretical, and multivariate analysis. *Antioxidants* **2023**, *12*, 1485. [CrossRef] [PubMed]
30. Xin, Z.; Wang, M.; Hugo, E.; Cuevas, H.E.; Chen, J.; Harrison, M.N.; Ace Pugh, N.A.; Morris, G. Sorghum genetic, genomic, and breeding resources. *Planta* **2021**, *254*, 114. [CrossRef]
31. Kumar, A.A.; Reddy, B.V.S.; Sharma, H.C.; Hash, C.T.; Rao, P.S.; Ramaiah, B.; Reddy, O.S. Recent advances in sorghum genetic enhancement research at ICRISAT. *Am. J. Plant Sci.* **2011**, *2*, 589–600. [CrossRef]
32. Ribeiro, L.P.; Tardin, F.D.; de Menezes, C.B.; Baldoni, A.B.; Teodoro, P.E.; Bhering, L.L. Combining yield, earliness and plant height in a single genotype: A proposal for breeding in grain sorghum (*Sorghum bicolor* L.). *Rev. FCA UNCuyo.* **2021**, *53*, 11–21. [CrossRef]
33. Wu, G.; Johnson, S.K.; Bornman, J.F.; Bennett, S.J.; Clarke, M.W.; Singh, V.; Fang, Z. Growth temperature and genotype both play important roles in sorghum grain phenolic composition. *Sci. Rep.* **2016**, *6*, 21835. [CrossRef]
34. Wagaw, K.; Tadesse, T. Combining Ability and Heterosis of Sorghum (*Sorghum bicolor* L. Moench) Hybrids for Grain and Biomass Yield. *Am. J. Plant Sci.* **2020**, *11*, 2155–2171. [CrossRef]
35. Widowati, S.; Luna, P. Nutritional and Functional Properties of Sorghum (*Sorghum bicolor* (L.) Moench)-based Products and Potential Valorisation of Sorghum Bran. *IOP Conf. Ser. Earth Environ.* **2022**, *1024*, 012031. [CrossRef]
36. Afify, A.M.R.; El-Beltagi, H.S.; Abd El-Salam, S.M.; Omran, A.A. Effect of soaking, cooking, germination and fermentation processing on proximate analysis and mineral content of three white sorghum varieties (*Sorghum bicolor* L. Moench). *Not. Bot. Horti Agrobot. Cluj-Napoca* **2012**, *40*, 92–98.
37. Vulin, M.; Magušić, L.; Metzger, A.M.; Muller, A.; Drenjančević, I.; Jukić, I.; Šijanović, S.; Lukić, M.; Stanojević, L.; Davidović Cvetko, E.; et al. Sodium-to-Potassium Ratio as an Indicator of Diet Quality in Healthy Pregnant Women. *Nutrients* **2022**, *14*, 5052. [CrossRef]
38. Kogure, M.; Nakaya, N.; Hirata, T.; Tsuchiya, N.; Nakamura, T.; Narita, A.; Suto, Y.; Honma, Y.; Sasaki, H.; Miyagawa, K. Sodium/potassium ratio change was associated with blood pressure change: Possibility of population approach for sodium/potassium ratio reduction in health checkup. *Hypertens. Res.* **2021**, *44*, 225–231. [CrossRef] [PubMed]
39. Saxholt, E.; Christensen, A.T.; Møller, A.; Hartkopp, H.B.; Hess-Ygil, K.; Hels, O.H. *Danish Food Composition Databank, Revision 7*; Department of Nutrition, National Food Institute, Technical University of Denmark: Kongens Lyngby, Denmark, 2008.
40. Taleon, V.; Dykes, L.; Rooney, W.L.; Rooney, L.W. Effect of genotype and environment on flavonoid concentration and profile of black sorghum grains. *Cereal Sci.* **2012**, *56*, 470–475. [CrossRef]

foods

MDPI

Article

Germination: A Powerful Way to Improve the Nutritional, Functional, and Molecular Properties of White- and Red-Colored Sorghum Grains

Cagla Kayisoglu [1], Ebrar Altikardes [2], Nihal Guzel [2] and Secil Uzel [2,*]

[1] Scientific Technical Application and Research Center, Hitit University, 19030 Çorum, Türkiye; caglakayisoglu@hitit.edu.tr
[2] Department of Food Engineering, Hitit University, 19030 Çorum, Türkiye; ebrar6k@gmail.com (E.A.); nihalguzel@hitit.edu.tr (N.G.)
* Correspondence: seciltursoy@hitit.edu.tr; Tel.: +90-364-2274533/1220

Abstract: This study explored the effects of the germination of red and white sorghum grains (*Sorghum bicolor* [Moench (L.)]) for up to seven days on various properties of the grain. Germination enriched sorghum's nutritional and sensory qualities while mitigating existing anti-nutritional factors. The study employed Fourier-transformed infrared spectroscopy (FT-IR) and scanning electron microscopy techniques to support its findings. Germination increased protein and lipid content but decreased starch content. White sorghum grains showed elevated calcium and magnesium but decreased iron, potassium, and zinc. Red sorghum grains showed a consistent decrease in mineral content during germination. Germination also increased fiber and lignin values in both sorghum varieties. The results of the FT-IR analysis demonstrate that germination induced significant changes in the molecular structure of white sorghum samples after 24 h, whereas this transformation was observed in red sorghum samples at four days. Total phenolic content (TPC) in red sorghum ranged from 136.64 ± 3.76 mg GAE/100 g to 379.5 ± 6.92 mg GAE/100 g. After 72 h of germination, the germinated seeds showed a threefold increase in TPC when compared to ungerminated seeds. Similarly, the TPC of white sorghum significantly increased ($p < 0.05$) from 52.84 ± 3.31 mg GAE/100 g to 151.76 mg GAE/100 g. Overall, during the 7-day germination period, all parameters showed an increase, and the germination process positively impacted the functional properties that contributed to the health benefits of white and red sorghum samples.

Keywords: germination; sorghum bicolor; FT-IR; SEM; bioactive compounds

1. Introduction

Sorghum (*Sorghum bicolor* [Moench (L.)]) underwent domestication approximately 3000 to 5000 years ago and is currently recognized as the fifth most important crop worldwide in terms of carbohydrate content, following wheat, maize, rice, and barley. Sorghum exhibits remarkable tolerance to drought and serves as a staple food in various semi-arid, arid, and tropical regions globally [1–3]. Particularly in dry and semi-arid regions of Africa, Asia, and Latin America, sorghum has widespread use as both animal feed and human sustenance, including bread, snacks, and nonfermented and fermented beverages [1,3,4]. Sorghum is rich in energy, proteins, carbohydrates, vitamins, minerals (such as iron and potassium), and phenolic and flavonoid compounds, which offer potential health-promoting benefits such as anti-carcinogenic, antibacterial, and antioxidant properties, making it suitable for human consumption [3]. The proximate composition of sorghum varies across different varieties. Certain sorghum types, such as red, brown, and black varieties containing color pigments in the pericarp layers, have garnered attention due to their higher phenolic content (generally comprising phenolic acids, flavonoids, and condensed tannins) [5]. Despite its variations, sorghum typically contains significant amounts of macronutrients, including lipids (1.6–6%), proteins (7–15%), and carbohydrates (50–60%). A 100 g serving of sorghum, similar to maize and wheat, provides substantial amounts of resistant starch and can supply approximately 400 calories [3,6]. Furthermore, its lack of gluten proteins makes it a viable alternative for individuals with celiac disease or wheat allergies [3,7–9].

However, sorghum's nutritional components, particularly its proteins, are less digestible for humans and monogastric animals due to anti-nutritional factors like tannins and phytic acid. To enhance the nutritional value of sorghum and fully exploit its potential as food or animal feed, it is essential to mitigate these undesired components. Tannins and sorghum proteins interact to decrease protein and starch digestibility. This interaction is significant for human and animal nutrition, as sorghum proteins and tannins can form complexes that render proteins indigestible and inhibit digestive enzymes [8,10]. Tannins, including hydrolyzable (e.g., gallic tannins and ellagitannins), condensed (e.g., proanthocyanidins), and complex tannins, are water-soluble phenolic compounds. Condensed tannins can bind to and precipitate protein components, making them unavailable and indigestible. Moreover, tannins can bind not only to proteins but also to minerals, thus rendering them inaccessible, and they can inhibit hydrolytic enzymes such as trypsin, α-amylase, glucoamylase, and lipase. Another anti-nutritional factor in sorghum is phytate, which forms insoluble complexes with proteins and mineral cations, reducing the bioavailability of trace minerals and protein digestibility [1,3,7]. Therefore, various approaches such as fermentation, germination, and hydrolysis are employed to convert bound phenolic compounds into free forms and enhance the digestibility of protein and starch [11].

Germination represents a cost-effective and efficient technology that facilitates structural modifications and the synthesis of new compounds with heightened biological activity, improved nutritional value, and enhanced grain stability. Germination is considered a complex, selective, and efficient method for improving the nutritional and functional value and also reducing the anti-nutritional factors of grains and legumes. During germination, a large number of bound enzymes stored in dry plant seeds are activated, and macromolecular storage substances (starch, proteins, and fat) are hydrolyzed and transformed into more digestible forms. Meanwhile, anti-nutrient components such as tannins and phytic acid can be reduced thanks to hydrolysis triggered by germination. Germination leads to the accumulation of bioactive compounds that are beneficial to human health, such as γ-amino butyric acid (GABA), polyphenols, flavonoids, and vitamins [12].

Germinated sorghum and its flour are suitable for producing specialty foods and value-added products. In comparison to native sorghum, germinated sorghum is considered significantly healthier since the germination process reduces or eliminates anti-nutritional factors such as tannins, phytates, and protease inhibitors, thereby improving nutrient digestibility and vitamin and mineral availability and generating various bioactive compounds that promote health. The germination of sorghum leads to increased levels of albumin, globulin, free amino nitrogen, protease, and kafirin. Additionally, it enhances the riboflavin availability, nutrient digestibility, antioxidant activity, and sensory acceptability of sorghum-based products [3,7].

The present study aims to subject red and white sorghum varieties currently limited to animal feed use, especially in our country, Turkey, to the germination process. This endeavor seeks to expand their utilization in human nutrition and investigate the effects of germination at different time intervals on the functional composition of sorghum.

2. Materials and Methods

2.1. Materials and Reagents

The red and white sorghum (*Sorghum bicolor* L.) grains were provided by a commercial company (Ingro, Turkey). The sorghum grains were cleaned of impurities by sieving and kept at a refrigerator temperature (+4 °C) until the germination process. The germination process was carried out in two replications. All chemicals used in the study were of analytical purity.

2.2. Germination Process

Germination was accomplished as determined by [3] with some modifications. Ungerminated red and white sorghum grains were designated as Control Groups and labeled as C_R and C_W, respectively. Approximately 100 g of sorghum samples were rinsed three times with distilled water (1 min each). After that, the grains were soaked in distilled water at a ratio of 1:1 (sample to water) and soaked for 12 h at ambient temperature (23–25 °C). Thanks to soaking, the microbial load was decreased, and the grain moisture level was optimized between 35% and 40%, which is suitable for germination. Following the filtration of the soaking water, the Soaked Control Groups were labeled as SC_R and SC_W (Soaked Control Red and White Sorghum). Subsequently, the soaked samples underwent a germination process for a duration of 7 days, with sampling conducted every 24 h. During germination, the grains were moistened with water once a day. The samples taken from different intervals of germination were lyophilized. After lyophilization, the germinated and ungerminated (control) samples were milled in an analytical hammer mill (Brabender Laboratory Mill, (Duisburg, Germany) equipped with a sieve aperture of 500 µm. The samples were packed hermetically and finally stored at a refrigerator temperature (+4 °C) until further analysis. The germination process for each sorghum sample was performed in two replications.

2.3. Proximate Composition

The moisture content of the samples was determined using a halogen moisture analyzer (Shimadzu MOC63U, Kyoto, Japan) according to the measurement conditions—set as 5 g of the sample at 130 °C.

Samples (~500 g) were scanned using a near-infrared transmission spectroscopy (NIR) Perten Grain Analyzer (Model DA 7250, Perten Instruments, Springfield, IL, USA) to determine the protein, starch, ash, and lipid contents. Before scanning, the NIR device was calibrated and validated with the results from wet chemical testing of 50 samples of each trait scanned.

2.4. Determination of Mineral Element Composition Using Inductively Coupled Plasma Emission Spectrometry (ICP-OES)

Macroelements (Ca, K and Mg) and microelements (Fe and Zn) were determined by an Inductively Coupled Plasma-Optical Emission Spectrophotometer (ICP-OES, ICAP6500, Thermo Scientific, (Waltham, MA, USA) according to the method described by [13]. The ICP-OES operating conditions are given in Table 1. A closed microwave digestion unit (Berghof Instruments, Speedwave, Germany) equipped with Teflon vessels was used to mineralize 0.30 g of each sample, to which 5.0 mL of ultrapure 65% m/v nitric acid and 2 mL of hydrogen peroxide 35% m/v were added, in order to determine the dissolved metals and non-metals. The sample solution was filtered through a membrane filter (pore size 0.45 µm) and filled to an exact volume of 0.25 mL.

Table 1. ICP-OES operating conditions.

Operating Conditions	
Power	1150 W
Auxiliary Gas Flow	0.5 L/min
Coolant gas	12 L/min
Nebulizer gas	0.70 L/min

Calibration samples were prepared from multi-element standard solutions (100 mg mL^{-1}). Standard solutions (Chem-Lab, Zedelgem, Belgium) were diluted to concentrations in a range between 0.1 and 10 mg/L for microelements and between 1 and 40 mg/L for macroelements.

2.5. Crude Fiber Content

The Acid Detergent Fiber (ADF) and Neutral Detergent Fiber (NDF) composition of the samples were determined according to the modified method of [14].

2.6. Fourier Transform Infrared (FT-IR) Spectroscopy Characterization

Chemical changes in sorghum samples due to germination were evaluated using a nitrogen-cooled Attenuated Total Reflectance FTIR (Thermo Scientific Nicolet Nexus 470, Waltham, MA, USA) equipped with a mercury/cadmium/telluride detector at a resolution of 4 cm^{-1} [15]. The spectra of the samples were scanned in the range 4000–400 cm^{-1}. Data were processed using Omnic software (OMNIC 9.0).

2.7. Scanning Electron Microscopy (SEM)

The morphological properties of the germinated sorghum grains were determined using a scanning electron microscope (ANATEK, FEI/Quanta 450 FEG, CZECHIA). The samples that were coated with a thin film of gold were mounted on an aluminum stub plated with double-sided adhesive tape and examined at 10,000 kV of accelerating voltage. The magnification value, pressure, and scale were set as 1000×, 3.04×10^{-3} Pa, and 100 µm, respectively.

2.8. Color Measurement

The CIE color parameters of soaked, germinated, and control samples were measured using a Minolta CM-3600d (Minolta, Osaka, Japan) colorimeter as described by [16].

2.9. Determination of Total Phenolic Content

The total phenolic content (TPC) was measured by a Folin–Ciocalteu assay [17] with minor modification according to [18]. Sorghum extracts were obtained by finely grinding wholegrain sorghum in a laboratory mill (Brabender, Laboratory Mill SM4, Duisburg, Germany) and passing it through a 0.5 mm sieve. Then, 50 mg of grounded grain sample was mixed with 1.5 mL of extraction solvent of methanol, water, and hydrochloric acid solution (50:48.5:1.5, *v/v/v*). Extraction was carried out in an ultrasonic bath (Sonorex DigiPlus DL 255 H, Bandelin, Taufkirchen, Germany) at a 37 kHz frequency, 25 ± 2 °C, and 100% amplitude for 30 min. The extract was then centrifuged (Sigma 3K30, Taufkirchen, Germany) at 5000× *g* for 15 min. The extraction process was repeated, and the supernatants were combined until the last extraction volume was reached at 5 mL for white sorghum and 10 mL for red sorghum samples. The extraction procedure was performed in duplicate for all samples. A 500 µL of extract was mixed with 500 µL of Folin's reagent (0.2 N) and 1 mL of Na_2CO_3 solution (7.5%). The mixture was incubated at room temperature for 60 min in the dark. Absorbance measurements were taken at 720 nm (Shimadzu UV-1800, Kyoto, Japan), and results were expressed as mg Gallic Acid Equivalents (GAE)/100 g dry weight (R^2 = 0.9973).

2.10. Determination of Radical Scavenging Activity

The 2.2-diphenyl-1-picrylhydrazyl (DPPH) radical-scavenging activity assay was performed according to the method described by [19], with minor modifications. Then, 100 µL of the extract was mixed with the DPPH solution (0.1 mM) of 1900 µL that was allowed to stand for 30 min at room temperature in the dark. The absorbance was measured at 515 nm. The results were expressed as mg Ascorbic Acid Equivalents (AAE)/100 g dry weight (R^2 = 0.9988).

2.11. Statistical Analysis

All measurements (mean ± standard deviation) were performed in triplicate. Statistical analyses were performed using one-way ANOVA (SPSS'16) at a significance level of 0.05. Tukey's test was used to differentiate between mean values.

3. Results and Discussion

3.1. Germination Process and Proximate Composition

The germination process for a duration of 7 days (Figure 1) resulted in significant changes in the chemical composition of sorghum grains. The change in the proximate composition (moisture level, protein, ash, lipid, and total starch contents) of red and white sorghum grains is given in Table 2.

Figure 1. Images of germinated red and white sorghum grains.

Table 2. The proximate chemical composition of red and white sorghum grains.

Sample	Moisture	Ash	Protein	Lipid	Starch
C_W	11.31 ± 0.06 ^{m}	2.05 ± 0.05 ^{b}	12.58 ± 0.59 ^{g}	3.49 ± 0.05 ^{b}	53.71 ± 3.42 ^{j}
SC_W	2.95 ± 0.20 ^{cd}	2.43 ± 0.18 ^{c}	13.49 ± 0.12 ^{h}	4.52 ± 0.42 ^{ef}	46.76 ± 0.23 ^{h}
G_{W1}	2.34 ± 0.11 ^{b}	2.88 ± 0.05 ^{de}	12.14 ± 0.16 ^{fg}	4.49 ± 0.64 ^{ef}	44.9 ± 0.43 ^{g}
G_{W2}	3.49 ± 0.34 ^{ef}	3.15 ± 0.19 ^{ef}	11.88 ± 0.67 ^{f}	4.85 ± 0.17 ^{fg}	41.41 ± 1.49 ^{f}
G_{W3}	4.24 ± 0.31 ^{g}	3.23 ± 0.12 ^{f}	12.20 ± 0.32 ^{fg}	5.28 ± 0.15 ^{h}	40.59 ± 1.59 ^{f}
G_{W4}	6.37 ± 0.49 ^{i}	3.60 ± 0.14 ^{g}	10.82 ± 0.65 ^{e}	5.41 ± 0.21 ^{hi}	35.19 ± 0.38 ^{d}
G_{W5}	8.04 ± 0.33 ^{j}	4.43 ± 0.09 ^{i}	11.73 ± 0.61 ^{f}	5.79 ± 0.08 ^{ij}	27.80 ± 1.84 ^{b}
G_{W6}	8.04 ± 0.33 ^{j}	4.75 ± 0.1 ^{j}	14.35 ± 0.04 ^{i}	6.08 ± 0.28 ^{j}	26.65 ± 0.62 ^{b}
G_{W7}	8.85 ± 0.54 ^{k}	5.20 ± 0.08 ^{k}	15.02 ± 0.05 ^{j}	6.12 ± 0.25 ^{j}	22.59 ± 0.66 ^{a}
C_R	11.35 ± 0.057 ^{m}	1.25 ± 0.06 ^{a}	9.91 ± 0.57 ^{abcd}	2.26 ± 0.14 ^{a}	61.87 ± 0.04 ^{k}
SC_R	2.44 ± 0.31 ^{m}	3.10 ± 0.69 ^{ef}	10.13 ± 0.52 ^{bcd}	3.74 ± 0.56bc	49.06 ± 0.46 ^{i}
G_{R1}	1.90 ± 0.063 ^{a}	2.78 ± 0.09 ^{d}	10.30 ± 0.07 ^{cde}	3.42 ± 0.06 ^{b}	43.44 ± 0.57 ^{g}
G_{R2}	2.62 ± 0.075 ^{bc}	2.70 ± 0.18 ^{cd}	9.45 ± 0.28 ^{a}	3.64 ± 0.23 ^{bc}	45.04 ± 1.49 ^{g}
G_{R3}	3.11 ± 0.13 ^{de}	3.00 ± 0.08 ^{def}	9.87 ± 0.06 ^{abcd}	4.04 ± 0.31 ^{cd}	40.97 ± 0.32 ^{f}
G_{R4}	3.83 ± 0.11 ^{fg}	3.23 ± 0.05 ^{f}	9.74 ± 0.05 ^{abc}	4.3 ± 0.13 ^{de}	38.8 ± 0.30 ^{e}
G_{R5}	4.19 ± 0.29 ^{g}	3.28 ± 0.05 ^{f}	9.61 ± 0.22 ^{ab}	4.61 ± 0.07 ^{ef}	37.79 ± 0.24 ^{e}
G_{R6}	5.54 ± 0.12 ^{h}	3.30 ± 0.00 ^{f}	10.39 ± 0.24 ^{de}	5.14 ± 0.12 ^{gh}	35.42 ± 0.50 ^{d}
G_{R7}	5.97 ± 0.56 ^{i}	4.13 ± 0.09 ^{h}	12.28 ± 0.04 ^{fg}	5.33 ± 0.22 ^{h}	31.59 ± 0.35 ^{c}

Data (g/100 g DM) are expressed as mean values ± standard deviation. Means marked with the same letter are not statistically different from each other ($p < 0.05$) C_R: Control Red Sorghum, SC_R: Soak Control Red Sorghum, G_{R1}: Germination Sorghum Red 1 day, G_{R7}: Germination Sorghum Red 7 days; C_W: Control White Sorghum, SC_W: Soak Control White Sorghum, G_{W1}: Germination Sorghum White 1 days, G_{W7}: Germination Sorghum White 7 days.

In Table 2, it becomes apparent that white sorghum grains exhibit a more abundant composition regarding ash, protein, and lipid content. The protein content of both samples increased following a 7-day germination period, with a notable rise of 23% in red sorghum and 19% in white sorghum. Previous studies in the literature have indicated that this increase in protein content is attributed to the elevation of specific amino acids such as lysine, valine, phenylalanine, methionine, and tryptophan during germination [4]. Similar findings were reported by [1], who observed a significant increase in protein content after applying the germination process to corn in a study resembling ours. Furthermore, [20] conducted a study on quinoa germination (72 h) and found that protein content increased (control sample: 9.6 g/100 g, 72 h sample: 26 g/100 g), while lipid content decreased (control: 15.2 g/100 g, 72 h germinated sample: 7.6 g/100 g).

The increased crude protein content during germination can be explained by the production of enzymes by the developing seed, compositional changes resulting from the breakdown of other elements, and the synthesis of newly formed proteins. For instance, α-amylase enzymes break down starch granules, releasing packed proteins from the seed structure. Additionally, increased protease enzyme activity during germination leads to the breakdown of peptide components into amino acids, thereby increasing the protein content of germinated grains. Respiration during the germination process contributes to the biological synthesis of new amino acids, while losses in dry matter, particularly carbohydrates, also play a role in protein content enhancement [20,21].

Regarding other compositional changes, both samples exhibited an increase in ash, moisture, and lipid content during germination, while starch content decreased. The germ of grain seeds contains the majority of crude lipids, and during germination, enzymes hydrolyze triacylglycerol, resulting in the production of free fatty acids. These fatty acids undergo β-oxidation in the cytosol and mitochondria, generating vital energy to support seed development. Consequently, a decrease in crude fat content is expected during germination. However, in our study, an increase in lipid content was observed, which differs from the decrease reported in legume germination [20]. During the germination of sorghum, the increase in fat content can be attributed to various factors. Firstly, the mobilization of stored reserves within the seed, including lipids, proteins, and carbohydrates, provides energy for the growing seedling. As the seed undergoes metabolic processes, the breakdown of stored proteins and carbohydrates may lead to the synthesis and accumulation of fats. Additionally, the activation of lipid biosynthesis pathways during germination contributes to the production of new lipids required for cell division and expansion. Overall, the increase in fat content during sorghum germination reflects complex biochemical processes essential for seedling growth and development.

Moisture content decreased in both samples during germination, which can be attributed to the drying process that occurs after germination—an essential stage for completing the process. Additionally, extended storage periods contribute to moisture reduction, thereby slowing the growth of germs. The increase in ash content in both samples (red sorghum: 66%, white sorghum: 154%) aligns with the findings of previous studies [22,23]. It has been reported that the increase in ash during germination is due to a decrease in the amount of total soluble dry matter [22]. Discrepancies in results could be attributed to variations in varieties, geographic factors, treatment conditions, and analytical methods. Additionally, [1] found that proteins and carbohydrates in ash act as coenzymes in catalysis and are, therefore, highly consumed.

The amount of starch, crucial for sorghum digestibility, notably decreased during germination (red sorghum: 49%, white sorghum: 58%). The reduced starch content results from the increased activity of amylase and pullulanase enzymes, leading to the breakdown of starch molecules into maltose, maltotriose, and other oligosaccharides. Enzymes such as α-amylase, glucosidase, dextranase (produced in aleurone), and β-amylase (produced in endosperm) are activated during germination and contribute to starch hydrolysis [20,24]. Our study findings are consistent with those reported by [20,22,24].

3.2. Mineral Substance Composition with ICP-OES

White- and red-colored sorghum samples were analyzed in terms of macro- (calcium, potassium and magnesium) and microelement (iron and zinc) amounts following a 7-day germination period, and the results are given in Figure 2.

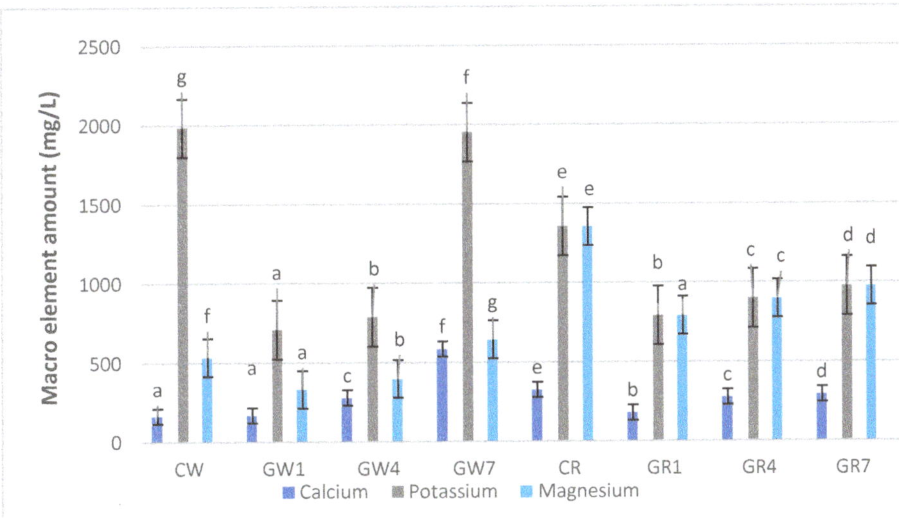

(a)

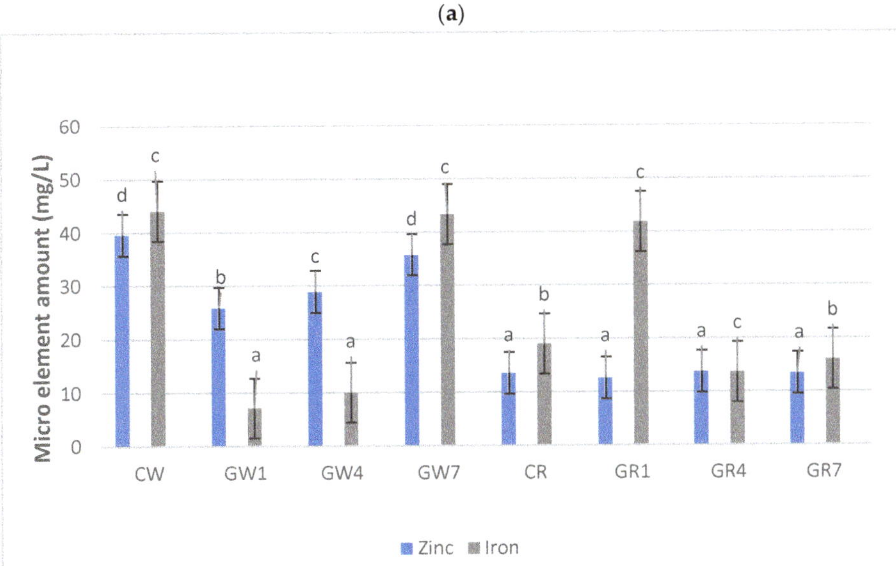

(b)

Figure 2. Effect of germination process on macro- (**a**) and micro- (**b**) element distribution of white and red sorghum samples. Data (g/100 g DM) are expressed as mean values ± standard deviation. Means marked with the same letter are not statistically different from each other ($p < 0.05$) C_W: Control White Sorghum, G_{W1}: Germination Sorghum White 1 days, G_{W4}: Germination Sorghum White 4 days, G_{W7}: Germination Sorghum White 7 days, C_R: Control Red Sorghum, G_{R1}: Germination Sorghum Red 1 day, G_{R4}: Germination Sorghum Red 4 days, G_{R7}: Germination Sorghum Red 7 days.

In Figure 2, it is evident that white-colored sorghum samples exhibit a more abundant profile concerning both macro- and micro-element content compared to their red-colored counterparts. Notably, the first day of the germination process resulted in a decrease in the levels of macro- and microelements in both sorghum samples. When considering the germination process as a whole, an overall increase, particularly in the levels of macroelements such as Ca, K, and Mg, was observed as the germination duration extended.

In the case of white sorghum grains subjected to a 7-day germination process, the calcium content increased from 163.23 mg/L to 584.43 mg/L, and the magnesium content increased from 534.3 mg/L to 643.98 mg/L. However, there was a reduction of 2.3%, 1.6%, and 9.7%, respectively, in the levels of iron, potassium, and zinc compared to the initial raw material. Conversely, red-colored sorghum samples displayed a consistent decrease in elemental content during germination. The calcium content increased from 326.21 mg/L to 292.53 mg/L, the iron content increased from 18.99 mg/L to 15.99 mg/L, the potassium content decreased from 1356.14 mg/L to 975.59 mg/L, and the magnesium content decreased from 500.19 mg/L. The zinc content exhibited a minimal decrease from 13.58 mg/L to 13.46 mg/L in the raw material.

These findings align with other studies in the literature. For instance, a study by [1] on germinated pea flour for 72 h indicated an 11.12% increase in calcium content and an 8.97% decrease in zinc content, which is consistent with our results. In another study, [1] reported that the germination of corn for 24 h led to a 79% increase in iron content and an 80% increase in zinc content. Furthermore, a study by [25] on germinated rice (normal brown, Heukjinju, and Keunnujami) indicated that calcium content increased and potassium content decreased. However, in the case of brown rice, all mineral substance values (calcium, magnesium, potassium, phosphorus) increased with germination. In a study conducted by [26], which investigated the effects of various treatments (soaking, cooking, germination, and fermentation) on sorghum composition, it was reported that the germination process had a diminishing effect on mineral element levels in different sorghum varieties. In the same study, it was noted that the soaking process also contributed to this reduction. Additionally, in a study where two different sorghum cultivars were soaked for 22 h at room temperature and subsequently germinated for 36 and 48 h, it was reported that the mineral content (iron, zinc, and calcium) of both cultivars increased significantly with prolonged germination. This increase was attributed to the loss of solute content in the washing and soaking water [27].

The variation in results among different studies can be attributed to factors such as the phytate content of sorghum, phytase activation, the degree of mineral binding, or the interaction of these factors [28], as indicated in the existing literature data.

3.3. Crude Fiber Composition

The Acid Detergent Fiber (ADF), Neutral Detergent Fiber (NDF) and Acid Detergent Lignin (ADL) contents of germinated red- and white-colored sorghum samples are given in Table 3.

The total plant cell wall matrix, comprising major components such as NDF (Neutral Detergent Fiber), cellulose, hemicellulose, and lignin, as delineated in Table 3, plays a crucial role in the structural composition of sorghum grains. The ADF (Acid Detergent Fiber) encompasses lignocellulosic materials composed of lignin and cellulose, while ADL (Acid Detergent Lignin) denotes crude lignin.

The germination process caused an increase in NDF, ADF, and ADL values in both red and white sorghum samples compared to the control samples. This increase, although not consistently linear with respect to the progression of germination time, was found to be approximately 84% and 12%, respectively, for red and white sorghum in terms of NDF values; around 26% and 260% for ADF values; and 3% and 171% for ADL values, respectively, on the seventh day of germination compared to the control samples.

Table 3. The effect of germination process on dietary fiber composition of sorghum samples.

Sample	Neutral Detergent Fiber (NDF, %)	Acid Detergent Fiber (ADF, %)	Acid Detergent Lignin (ADL, %)
C_W	25.08 ± 1.23 [d]	2.73 ± 0.40 [bc]	0.43 ± 0.07 [abcd]
SC_W	25.51 ± 0.69 [d]	0.97 ± 0.58 [ab]	0.26 ± 0.10 [a]
G_{W1}	15.98 ± 0.95 [bc]	0.74 ± 0.54 [a]	0.40 ± 0.08 [abcd]
G_{W2}	11.10 ± 0.91 [ab]	1.20 ± 0.32 [ab]	0.33 ± 0.13 [ab]
G_{W3}	11.30 ± 2.07 [ab]	0.98 ± 0.33 [ab]	0.35 ± 0.12 [abc]
G_{W4}	24.02 ± 5.28 [d]	3.09 ± 0.70 [c]	0.67 ± 0.01 [abcdef]
G_{W5}	30.69 ± 6.57 [e]	5.82 ± 0.50 [d]	1.02 ± 0.31 [ef]
G_{W6}	28.62 ± 0.35 [de]	6.06 ± 0.34 [de]	0.93 ± 0.21 [ef]
G_{W7}	29.05 ± 0.61 [de]	8.81 ± 1.93 [fg]	1.33 ± 0.43 [f]
C_R	9.30 ± 0.14 [a]	9.30 ± 0.14 [fg]	1.06 ± 0.05 [ef]
SC_R	7.72 ± 1.09 [a]	7.72 ± 1.08 [efg]	0.94 ± 0.00 [ef]
G_{R1}	7.84 ± 1.02 [a]	7.83 ± 1.02 [fg]	0.95 ± 0.08 [ef]
G_{R2}	7.17 ± 0.19 [a]	7.16 ± 0.19 [def]	0.61 ± 0.41 [abcde]
G_{R3}	11.46 ± 0.29 [ab]	11.46 ± 0.29 [h]	0.80 ± 0.05 [cdef]
G_{R4}	12.04 ± 1.56 [abc]	12.04 ± 1.55 [h]	0.75 ± 0.28 [bcdef]
G_{R5}	11.41 ± 0.60 [ab]	11.40 ± 0.60 [g]	0.95 ± 0.09 [ef]
G_{R6}	14.95 ± 0.05 [bc]	14.95 ± 0.05 [h]	0.85 ± 0.01 [def]
G_{R7}	16.91 ± 0.36 [c]	16.91 ± 0.36 [i]	1.13 ± 0.01 [ef]

Data (g/100 g DM) are expressed as mean values ± standard deviation. Means marked with the same letter are not statistically different from each other ($p < 0.05$). C_R: Control Red Sorghum, SC_R: Soak Control Red Sorghum, G_{R1}: Germination Sorghum Red 1 day, G_{R7}: Germination Sorghum Red 7 days; C_W: Control White Sorghum, SC_W: Soak Control White Sorghum, G_{W1}: Germination Sorghum White 1 days, G_{W7}: Germination Sorghum White 7 days.

In a study conducted by [29], on germinated lentils, it was reported that germination had a decreasing effect on the NDF amount, primarily due to the reduction in hemicellulose content, while cellulose and lignin amounts (ADF and ADL) increased. Similarly, in another study involving germinated cowpea samples, a decrease in hemicellulose and cellulose content, coupled with an increase in lignin content, was observed [30]. Furthermore, it was found that the crude fiber content increased by 212% in oats germinated for 72 h, as reported by [31]. Another study involving the germination of buckwheat and quinoa for 72 h revealed a 71% increase and a 52% increase, respectively, in total fiber content. These findings are attributed to the disruption of protein–carbohydrate interactions and structural alterations in polysaccharides within the seed cell wall. Additionally, the observed rise in fiber contents during the germination process can be explained by the loss of dry mass resulting from starch hydrolysis through enzymatic action triggered during germination, an increase in lignin content, and the degradation of cellular components such as lipids and proteins [21].

3.4. *Effect of Germination on Molecular Characterization of Sorghum Samples*

FT-IR results showing the change in molecular structures of red- and white-colored sorghum samples during germination are given in Figure 3.

The study data for the determination of the molecular structure of sorghum flour using the FT-IR method revealed the existence of important active regions in the wavenumber range of 1800 to 800 cm^{-1}. While FT-IR peaks reveal starch and proteins as the most basic components in sorghum, lipids and phenolic acids are characterized by peaks showing much lesser amounts (Figure 3) [5].

Due to the FT-IR spectra, the peaks were seen at a wavelength of approximately 3400 cm^{-1} for both sorghum samples (between 3420 and 3448 cm^{-1} for white sorghum samples and between 3400 and 3426 cm^{-1} for red sorghum samples). It has been defined as starch and protein molecules that are mainly responsible for the stretching actions of O-H and N-H bonds. The peaks seen at 1652 and 1538 cm^{-1} wavelengths characterize the amide I (C=O) and amide II (N-H, C-N) regions that form the basic structure of the protein molecule [5]. These peaks were detected in the wavelength ranges of 1647–1653 cm^{-1} (amide I) and 1540–1558 cm^{-1} (amide II) for both sorghum cultivars, respectively. The peaks in the wavelength range of 1400–900 cm^{-1} are characterized by another important basic component of sorghum: the starch molecule [5].

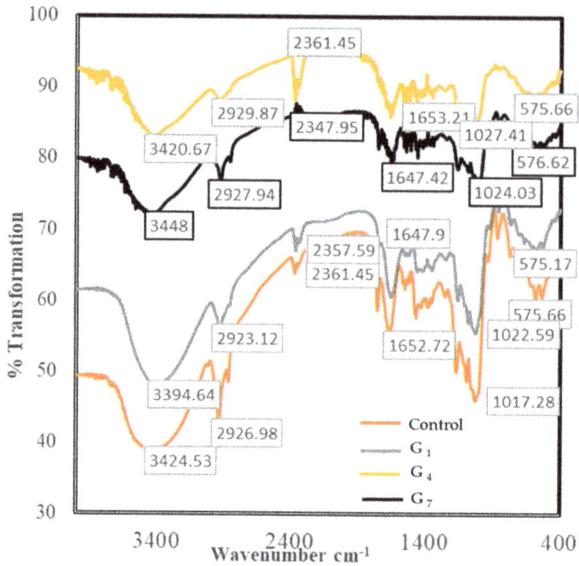

(**a**)

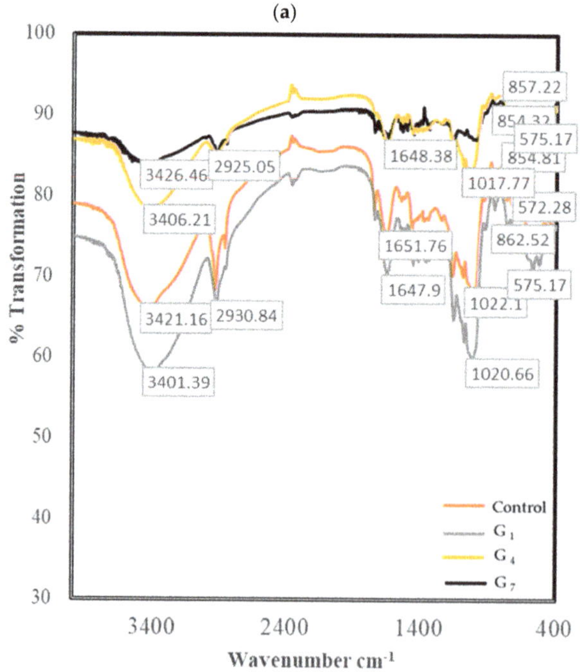

(**b**)

Figure 3. Effect of germination process on FT-IR of white (**a**) and red (**b**) sorghum samples.

FT-IR analysis shows that the germination process causes significant changes in the structures of carbohydrate and protein molecules of white and red sorghum samples. It is seen that this change has gained a more prominent feature depending on the increasing germination time, especially in white sorghum samples. These changes are very important in terms of improving the functional properties of germinated white sorghum samples.

Different intensity peaks seen in the wavelength range of 1400 to 900 cm^{-1} in the FT-IR spectrum were associated with the crystallinity level in sorghum starch due to the change in the amylose/amylopectin ratios forming the starch [32]. With germination, the band ratio in the starch molecule increased at the end of the seventh day compared to the raw material in white sorghum. This is due to the hydrolysis of starch by α- and β-amylase with germination [33]. However, this change was not observed in the red sorghum. This is thought to be due to botanical and genetic differences. Our study results are not similar to the 72 h germination of mung bean and the change in starch [34] but are similar to [33].

Infrared spectrum data show that the germination process causes a significant change in the molecular structure of white and red sorghum grains. Different germination times caused different rates of change in both sorghums. The germination process in white sorghum samples for 24 h caused the most change in the molecular structure of the samples compared to the control sample, while this period was determined as 4 days for red sorghum samples [35]. Groups of observers of germinated HomChaiya rice reported that it peaked at day 7 in the sequence of enzymatic hydrolysis of the outbreak, which occurred on day 5, which is in line with our study.

3.5. SEM Images of Sorghum Samples during Germination

SEM images of white and red sorghum samples are given in Figure 4 to better reflect the change in their molecular structures during the 7-day germination period.

The data show that there is a very strong destruction of starch granules and protein structures compared to control samples due to increased germination time. While it is possible to see individual starch granules surrounded by proteins, especially in the control samples, at the end of the 7-day germination period, the protein structures separate from the starch granules and lose their visibility. Starch granules, on the other hand, are eroded in the regions where the proteins are restricted. The study data are in agreement with the FT-IR results showing a decrease in the amount of protein and a change in the amount of starch. The effects of the germination process on red and white sorghum grains are in agreement with similar studies [36]. Enzymatic hydrolysis occurs during grain germination, and starch is released from the interconnection and entanglement of proteins and fibers, changing the structure of flour. Protein and fiber are hydrolyzed into amino acids and glucose [36]. The enzyme activity occurs during germination. Initially, starch hydrolysis starts after the enzyme adheres to the starch surface. The channels that allow the enzyme to spread into the granule center widen in the second stage as hydrolysis progresses and the number of tiny holes on the granule surface rises. The catalytic enzyme action degrades and modifies the granule surface in the final stage. Moreover, as a result of the change in molecular structure brought about by the rise in granule porosity, the crystallinity of the granules decreases [22].

In a previous study, it was reported that the polyhedral starch that adhered to the protein matrix in barley grain with germination was separated from the first day. This is due to the increase in enzymatic activity with germination. It was also reported in the study that the maximum activity of α-amylase occurs between the third and fifth days, while the protease can start from the fifth day of germination. Depending on the germination conditions and seed viability, the varying enzymatic activity has an impact on the various changes in the morphology of the grain throughout the germination period. According to several studies, the starch granules in cereals dissolve during germination, leaving them with rough surfaces and pinholes. The eroding of the granule surface or the digestion of the channels from the chosen locations on the surface towards the center are thought to be the causes of partial hydrolysis of starch. The uneven modification of the surface might be the result of variation in the susceptibility to hydrolytic enzyme activity during germination. The enzyme spreads to the solid surface and is adsorbed through the sites; finally, the catalytic reaction is produced. Adsorption can be affected by granule size, surface characteristics, and minor constituents, such as proteins and lipids on the surface [22,33]. The effect of germination on morphological traits is compatible with [22].

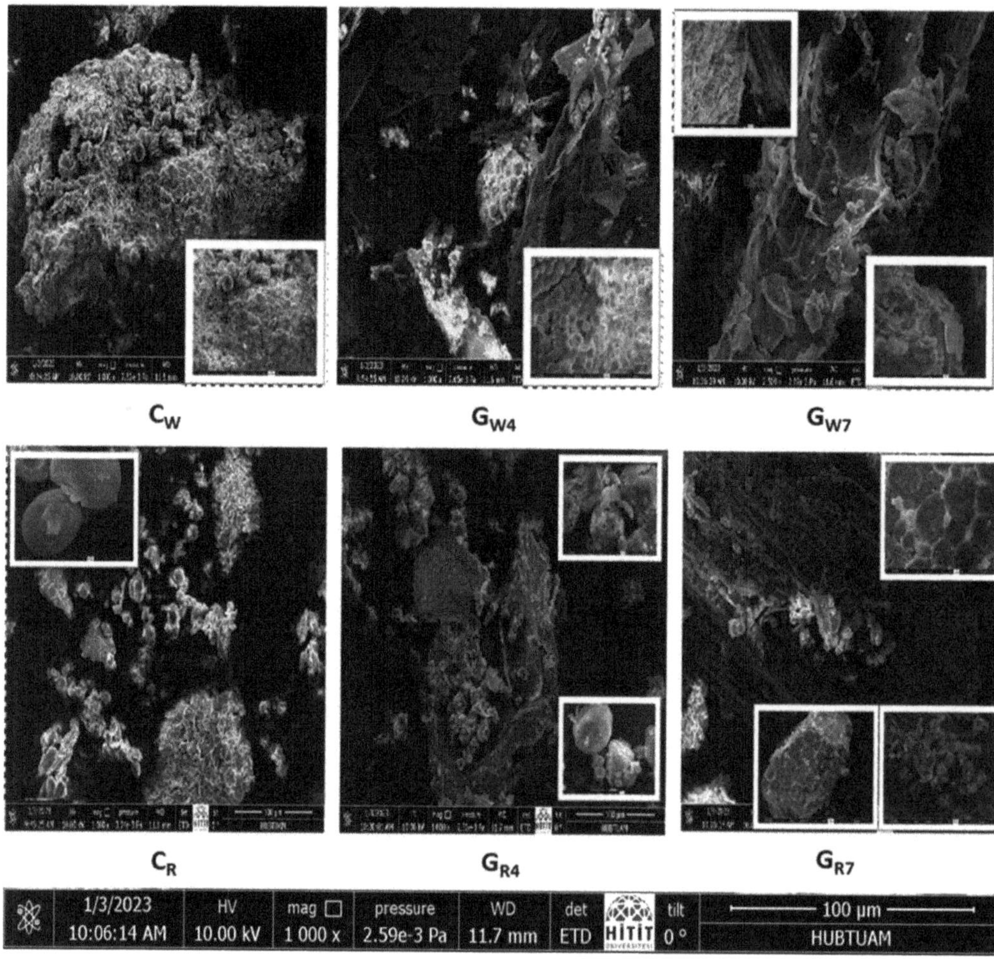

Figure 4. SEM images of white and red sorghum samples during germination process. C_W: Control White Sorghum, G_{W4}: Germination Sorghum White 4th day, G_{W7}: Germination Sorghum White 7th day, C_R: Control Red Sorghum, G_{R4}: Germination Sorghum Red 4th day, G_{R7}: Germination Sorghum Red 7th day.

3.6. Color Values of Sorghum Samples

The grain color is related to pigments in the testa and pericarp [37]. The phenolic compounds (tannins and anthocyanins) or starch granules are the color pigments in the sorghum pericarp [38]. The color changes during the germination in red and white sorghum are presented in Table 4. The L^* value indicates the lightness of the flour and ranges between 0 (black) and 100 (white). L^* values were in the range between 73.50 and 1.41 (red) and 80.27 and 2.25 (white) in ungerminated sorghum flour. The lightness values of red and white sorghum were decreased significantly with the germination process ($p < 0.05$) and were measured similar in both grains after 7 days of germination (Table 4). [39] stated that the decreasing lightness value associated with the increasing protein content during germination. L^* values of raw wheat (85.29) and sorghum (86.58) flour were comparable with the lightness of ungerminated white sorghum [40]. Flour color generally affects the color of finished products, and high L^* values are more desirable for many baking goods [41]. The L^* value (72.04 ± 1.38) of a cookie made from wheat flour was found higher compared to a cookie made from germinated (50.01 ± 1.99) and ungerminated (59.13 ± 2.45) minor millet flour [39]. Positive a^* and b^* values indicate redness and yellowness in the sample. a* values were 2.30 and 5.10 in raw white and red sorghum, respectively. The a^* values in white and red sorghum grains increased during the germination process, while the b^* values did not change significantly ($p > 0.05$).

Table 4. Effects of germination on CIE color parameters of red and white sorghum samples.

	L^*	a^*	b^*	C^*	h^o
C_W	80.27 ± 2.25 ^{ab}	2.30 ± 0.46 ^{c}	13.74 ± 0.74 ^{a}	13.93 ± 0.81 ^{a}	80.57 ± 1.36 ^{a}
SC_W	80.58 ± 1.58 ^{ab}	2.03 ± 0.30 ^{c}	14.09 ± 0.28 ^{a}	14.24 ± 0.30 ^{a}	81.81 ± 1.12 ^{a}
G_{W1}	82.28 ± 1.22 ^{a}	1.78 ± 0.17 ^{c}	14.15 ± 0.60 ^{a}	14.26 ± 0.62 ^{a}	82.83 ± 0.37 ^{a}
G_{W2}	82.25 ± 1.09 ^{a}	1.94 ± 0.14 ^{c}	13.63 ± 0.41 ^{a}	13.76 ± 0.42 ^{a}	81.93 ± 0.43 ^{a}
G_{W3}	80.61 ± 0.70 ^{ab}	2.28 ± 0.14 ^{c}	13.82 ± 0.19 ^{a}	14.01 ± 0.20 ^{a}	80.65 ± 0.50 ^{a}
G_{W4}	78.00 ± 1.34 ^{bc}	2.40 ± 0.22 ^{c}	13.85 ± 0.84 ^{a}	14.06 ± 0.81 ^{a}	80.11 ± 1.31 ^{a}
G_{W5}	75.44 ± 0.94 ^{cd}	3.39 ± 0.45 ^{b}	13.03 ± 0.22 ^{a}	13.47 ± 0.28 ^{a}	75.44 ± 1.76 ^{b}
G_{W6}	73.68 ± 0.33 ^{d}	3.76 ± 0.18 ^{ab}	12.97 ± 0.72 ^{a}	13.50 ± 0.66 ^{a}	73.80 ± 1.43 ^{bc}
G_{W7}	69.55 ± 1.36 ^{e}	4.47 ± 0.71 ^{a}	13.40 ± 0.92 ^{a}	14.15 ± 0.77 ^{a}	71.47 ± 3.50 ^{c}
C_R	73.50 ± 1.41 ^{abc}	5.10 ± 0.54 ^{b}	11.47 ± 0.68 ^{b}	12.55 ± 0.85 ^{b}	66.07 ± 1.02 ^{ab}
SC_R	71.35 ± 1.07 ^{bcd}	5.24 ± 0.20 ^{b}	12.36 ± 0.70 ^{ab}	13.42 ± 0.71 ^{ab}	67.02 ± 0.53 ^{a}
G_{R1}	70.34 ± 0.93 ^{cd}	5.74 ± 0.21 ^{b}	13.36 ± 0.30 ^{a}	14.54 ± 0.32 ^{a}	66.73 ± 0.63 ^{a}
G_{R2}	72.97 ± 1.14 ^{abc}	5.55 ± 0.19 ^{b}	12.23 ± 0.59 ^{ab}	13.43 ± 0.58 ^{ab}	65.57 ± 0.92 ^{ab}
G_{R3}	73.59 ± 1.05 ^{ab}	5.58 ± 0.45 ^{b}	12.52 ± 0.40 ^{ab}	13.71 ± 0.48 ^{ab}	66.01 ± 1.57 ^{ab}
G_{R4}	74.11 ± 0.68 ^{a}	5.47 ± 0.14 ^{b}	11.91 ± 0.54 ^{b}	13.10 ± 0.53 ^{b}	65.33 ± 0.81 ^{ab}
G_{R5}	73.73 ± 1.52 ^{ab}	5.97 ± 0.58 ^{b}	11.69 ± 0.43 ^{b}	13.13 ± 0.65 ^{b}	63.02 ± 1.46 ^{bc}
G_{R6}	70.66 ± 1.39 ^{cd}	6.95 ± 0.20 ^{ab}	11.95 ± 0.28 ^{b}	13.82 ± 0.21 ^{ab}	59.82 ± 1.14 ^{cd}
G_{R7}	69.54 ± 0.63 ^{d}	7.22 ± 0.70 ^{a}	12.03 ± 0.52 ^{b}	14.04 ± 0.39 ^{ab}	59.03 ± 3.20 ^{d}

Data are presented as the "mean ± standard deviation" of three independent experiments (n = 3). a–d: Lowercase letters indicate the effect of germination time on the CIE color value. Means followed by the same letter do not differ significantly at $p = 0.05$ according to Tukey multiple ranges.

3.7. Total Phenolic Content

The effects of germination on the red- and white-pigmented pericarp sorghum grains exhibited significant variation in TPC at $p < 0.05$ (Figure 5).

The phenolic content in red sorghum varied from 89.76 ± 8.31 mg GAE/100 g to 379.5 ± 6.92 mg GAE/100 g during the germination. After 72 h of germination, TPC was increased three times in red sorghum when compared to the ungerminated control sample (136.64 ± 3.76 mg GAE/100 g). Similarly, the phenolic content of white sorghum increased significantly ($p < 0.05$) from 52.84 ± 3.31 mg GAE/100 g (C_W) to 151.76 mg GAE/100g (G_{W7}). The red- and white-pigmented pericarp sorghum grains exhibited a significant difference in TPC, while similar increase ratios were observed for phenolic content after germination. [42] demonstrated that the black pericarp sorghum variety had the highest TPC at 11.50 ± 1.81 mg GAE /g, followed by the brown pericarp sorghum at 3.58 ± 1.63 mg GAE/g. However, the red- and white-pigmented pericarp varieties did not exhibit a significant difference in TPC. The results are comparable with the findings of [43], who reported a 39.74% increase in the TPC after 48 h of germination in sorghum. Many studies showed that germination processes clearly increase TPC in quinoa, amaranth, buckwheat, and millet [21,44,45]. Phenolic compounds can be found in free or bound forms in cereals. Free phenolic compounds are located in the pericarp and can usually be extracted with an organic solvent. It is thought that the total amount of phenolics produced by the germination process may be due to the oxidation of endogenous phenolics by the activity of enzymes such as peroxidase and polyphenol oxidase [46].

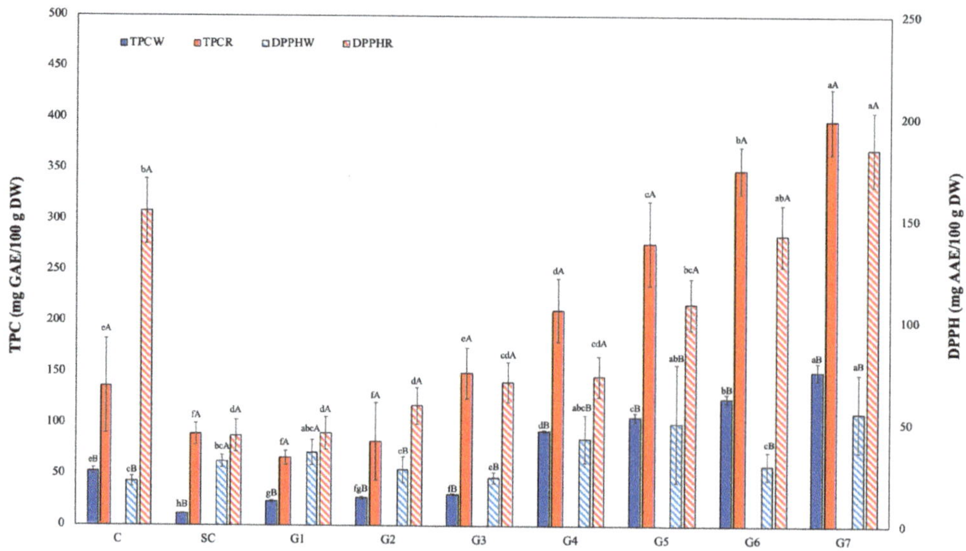

Figure 5. Effects of germination on total phenolic and antioxidant activity of sorghum samples. a–f: Lowercase letters indicate the effect of germination time on the TPC and DPPH value. Means followed by the same letter do not differ significantly at $p = 0.05$ according to Tukey multiple ranges; A–B: capital letters indicate the effect of sorghum variety at the same germination time. Means followed by the same letter do not differ significantly at $p = 0.05$, according to the t-test.

3.8. Radical Scavenging Activity

The DPPH free-radical-scavenging activity of red- and white-pigmented pericarp sorghum is presented in Figure 5.

The red sorghum grain had significantly higher antioxidant activity in comparison to the white sorghum grains at $p < 0.05$. The germination process was significantly effective in both red and white grains and reached its highest level (184.94 mg AAE/100g and 55.46 mg AAE/100g, respectively) after 7 days of germination. Previous studies have also indicated that germination has an enhancing impact on the DPPH scavenging activity of sorghum grains [43,47]. The DPPH scavenging activity of germinated wheat, barley, and millet grain was increased significantly from 1.37–1.64 g AAE/g to 3.19–3.76 g AAE/g after 72 h of germination [48]. The antioxidant activity (DPPH) of germinated amaranth, quinoa, and buckwheat grains also was significantly higher (35–178%) than that of ungerminated grains [21]. The antioxidant potential of cereal grains is mostly derived from phenolic compounds, which are also crucial in the prevention and management of degenerative illnesses. This study demonstrates that germination offers a novel strategy for advancing the development of sorghum seed as a useful food for human consumption.

4. Conclusions

Sorghum, due to its rich composition of bioactive compounds, low digestibility, high antioxidant capacity, and its resilience to drought conditions among cereal crops, is considered a highly important food resource for future human nutrition. This study investigates the impact of germination, specifically over a period of 7 days, on red and white sorghum grains, focusing on pivotal parameters associated with the germination process such as nutritional fiber content, protein and starch characteristics, and bioactive compounds.

During germination, there was a notable increase in protein and lipid content, accompanied by a decrease in starch content. White sorghum grains exhibited heightened levels of calcium and magnesium but experienced reductions in iron, potassium, and zinc. Conversely, red sorghum grains consistently displayed a decline in mineral content throughout germination. Prolonged germination durations correlated with elevated values of NDF, ADF, and ADL in both red and white sorghum samples when compared to control groups. This escalation in dietary fiber content holds significance for the functional attributes of sorghum grains with regards to health considerations.

Germination led to the separation of starch granules from their protein coatings and increased the damage to starch granules. Starch granules with altered molecular structures due to germination are transformed into healthier forms for nutrition. With a 7-day germination period, total phenolic content and antioxidant activity reached their maximum levels in both red and white sorghum, and the difference compared to the control group was statistically significant. After 72 h of germination, the TPC in germinated seeds exhibited a threefold increase compared to ungerminated seeds.

This study is highly significant for increasing the potential use of red and white sorghum varieties, which are currently only used as animal feed in most countries such as Turkey, in human nutrition by subjecting them to the germination process and determining the effects of germination at different durations on the functional composition of sorghum. During the 7-day germination period, an increase was observed in all parameters analyzed within the scope of the study, and the germination process positively affected the functional properties that determine the positive effects of both white and red sorghum samples on health. From a future perspective, digestibility experiments are needed to determine the optimum germination time more precisely. In particular, investigating the digestibility of sorghum proteins during germination is highly important.

Author Contributions: The article was devised, and literature search, data collection, and analysis were carried out by C.K., E.A., N.G. and S.U., who also wrote the manuscript. All authors have read and agreed to the published version of the manuscript.

Funding: This research is funded by Hitit University under Grant Number MUH19001.21.013.

Institutional Review Board Statement: Not applicable.

Informed Consent Statement: Not applicable.

Data Availability Statement: The original contributions presented in the study are included in the article, further inquiries can be directed to the corresponding author.

Acknowledgments: The authors would like to thank BOYYEM (Türkiye) for their support of dietary fiber analysis.

Conflicts of Interest: The authors declare that they have no conflicts of interest.

References

1. Anaemene, D.; Fadupin, G. Anti-Nutrient Reduction and Nutrient Retention Capacity of Fermentation, Germination and Combined Germination-Fermentation in Legume Processing. *Appl. Food Res.* **2022**, *2*, 100059. [CrossRef]
2. Hossain, M.S.; Islam, M.N.; Rahman, M.M.; Mostofa, M.G.; Khan, M.A.R. Sorghum: A Prospective Crop for Climatic Vulnerability, Food and Nutritional Security. *J. Agric. Food Res.* **2022**, *8*, 100300. [CrossRef]
3. Paiva, C.L.; Netto, D.A.M.; Queiroz, V.A.V.; Gloria, M.B.A. Germinated Sorghum (*Sorghum bicolor* L.) and Seedlings Show Expressive Contents of Putrescine. *LWT* **2022**, *161*, 113367. [CrossRef]
4. Rodríguez-España, M.; Figueroa-Hernández, C.Y.; de Dios Figueroa-Cárdenas, J.; Rayas-Duarte, P.; Hernández-Estrada, Z.J. Effects of Germination and Lactic Acid Fermentation on Nutritional and Rheological Properties of Sorghum: A Graphical Review. *Curr. Res. Food Sci.* **2022**, *5*, 807–812. [CrossRef] [PubMed]
5. Lin, H.; Bean, S.R.; Tilley, M.; Peiris, K.H.S.; Brabec, D. Qualitative and Quantitative Analysis of Sorghum Grain Composition Including Protein and Tannins Using ATR-FTIR Spectroscopy. *Food Anal. Methods* **2021**, *14*, 268–279. [CrossRef]
6. Toledo Oseguera, M.E.; Contreras-Jiménez, B.; Hernández-Becerra, E.; Rodriguez-Garcia, M.E. Physicochemical Changes of Starch during Malting Process of Sorghum Grain. *J. Cereal Sci.* **2020**, *95*, 103069. [CrossRef]

7. Li, C.; Oh, S.-G.; Lee, D.-H.; Baik, H.-W.; Chung, H.-J. Effect of Germination on the Structures and Physicochemical Properties of Starches from Brown Rice, Oat, Sorghum, and Millet. *Int. J. Biol. Macromol.* **2017**, *105*, 931–939. [CrossRef]
8. Meena, K.; Visarada, K.; Meena, D.K. *Sorghum bicolor* (L.) Moench a Multifarious Crop -Fodder to Therapeutic Potential and Biotechnological Applications: A Future Food for the Millennium. *Future Foods* **2022**, *6*, 100188. [CrossRef]
9. Rodriguez, D.A.; Lee, S.A.; Jones, C.K.; Htoo, J.K.; Stein, H.H. Digestibility of Amino Acids, Fiber, and Energy by Growing Pigs, and Concentrations of Digestible and Metabolizable Energy in Yellow Dent Corn, Hard Red Winter Wheat, and Sorghum May Be Influenced by Extrusion. *Anim. Feed Sci. Technol.* **2020**, *268*, 114602. [CrossRef]
10. Afify, A.E.-M.M.R.; El-Beltagi, H.S.; Abd El-Salam, S.M.; Omran, A.A. Protein Solubility, Digestibility and Fractionation after Germination of Sorghum Varieties. *PLoS ONE* **2012**, *7*, e31154. [CrossRef]
11. Balogun, R.O.; Bird, S.H.; Rowe, J.B. Germination Temperature and Time Affect in Vitro Fermentability of Sorghum Grain. *Anim. Feed Sci. Technol.* **2006**, *127*, 125–132. [CrossRef]
12. Liu, S.; Wang, W.; Lu, H.; Shu, Q.; Zhang, Y.; Chen, Q. New Perspectives on Physiological, Biochemical and Bioactive Components during Germination of Edible Seeds: A Review. *Trends Food Sci. Technol.* **2022**, *123*, 187–197. [CrossRef]
13. Gerrano, A.S.; Labuschagne, M.T.; van Biljon, A.; Shargie, N.G. Quantification of Mineral Composition and Total Protein Content in *Sorghum* [*Sorghum bicolor* (L.) Moench] Genotypes. *Cereal Res. Commun.* **2016**, *44*, 272–285. [CrossRef]
14. Van Soest, P.J.; Robertson, J.B.; Lewis, B.A. Methods for Dietary Fiber, Neutral Detergent Fiber, and Nonstarch Polysaccharides in Relation to Animal Nutrition. *J. Dairy Sci.* **1991**, *74*, 3583–3597. [CrossRef]
15. Turksoy, S.; Erturk, M.Y.; Kokini, J. Behavior of Semolina, Hard, Soft Wheat Flour Dough at Different Aging Times and Temperatures through LAOS Properties and Molecular Interactions of Proteins. *J. Food Eng.* **2021**, *301*, 110549. [CrossRef]
16. Güzel, N. Morphometric and Physico-Chemical Properties of Cornelian Cherry (*Cornus mas* L.) Grown in Çorum, Turkey. *Akad. Gıda* **2021**, *19*, 373–380. [CrossRef]
17. Singleton, V.L.; Rossi, J.A. Colorimetry of Total Phenolics with Phosphomolybdic-Phosphotungstic Acid Reagents. *Am J Enol Vitic.* **1965**, *16*, 144–158. [CrossRef]
18. Dicko, M.H.; Hilhorst, R.; Gruppen, H.; Traore, A.S.; Laane, C.; Van Berkel, W.J.H.; Voragen, A.G.J. Comparison of Content in Phenolic Compounds, Polyphenol Oxidase, and Peroxidase in Grains of Fifty Sorghum Varieties from Burkina Faso. *J. Agric. Food Chem.* **2002**, *50*, 3780–3788. [CrossRef]
19. Brand-Williams, W.; Cuvelier, M.E.; Berset, C. Use of a Free Radical Method to Evaluate Antioxidant Activity. *LWT-Food Sci. Technol.* **1995**, *28*, 25–30. [CrossRef]
20. Xu, M.; Jin, Z.; Simsek, S.; Hall, C.; Rao, J.; Chen, B. Effect of Germination on the Chemical Composition, Thermal, Pasting, and Moisture Sorption Properties of Flours from Chickpea, Lentil, and Yellow Pea. *Food Chem.* **2019**, *295*, 579–587. [CrossRef]
21. Thakur, P.; Kumar, K.; Ahmed, N.; Chauhan, D.; Eain Hyder Rizvi, Q.U.; Jan, S.; Singh, T.P.; Dhaliwal, H.S. Effect of Soaking and Germination Treatments on Nutritional, Anti-Nutritional, and Bioactive Properties of Amaranth (*Amaranthus hypochondriacus* L.), Quinoa (*Chenopodium quinoa* L.), and Buckwheat (*Fagopyrum esculentum* L.). *Curr. Res. Food Sci.* **2021**, *4*, 917–925. [CrossRef]
22. Oskaybaş-Emlek, B.; Özbey, A.; Kahraman, K. Effects of Germination on the Physicochemical and Nutritional Characteristics of Lentil and Its Utilization Potential in Cookie-Making. *Food Meas.* **2021**, *15*, 4245–4255. [CrossRef]
23. Salman Al-Kharkhi, M.H.; Ali Mousa, M. The effect of wheat germination processes on the nutritional parameters of wheat flour. *Plant Arch.* **2021**, *21*, 789–797. [CrossRef]
24. Kaur, N.; Singh, B.; Kaur, A.; Yadav, M.P.; Singh, N.; Ahlawat, A.K.; Singh, A.M. Effect of Growing Conditions on Proximate, Mineral, Amino Acid, Phenolic Composition and Antioxidant Properties of Wheatgrass from Different Wheat (*Triticum aestivum* L.) Varieties. *Food Chem.* **2021**, *341*, 128201. [CrossRef]
25. Chung, S.I.; Kang, M.Y. Comparative Studies on Physicochemical Properties of Rice with Germinated Dark Purple Giant Embryo Rice and Normal Embryo Rice. *Cereal Chem.* **2022**, *99*, 295–302. [CrossRef]
26. El-Beltagi, H.S.; Abd El-Salam, S.M.; Omran, A.A. Effect of Soaking, Cooking, Germination and Fermentation Processing on Proximate Analysis and Mineral Content of Three White Sorghum Varieties (*Sorghum bicolor* L. Moench). *Not. Bot. Horti Agrobot. Cluj-Napoca* **2012**, *40*, 92. [CrossRef]
27. Tizazu, S.; Urga, K.; Belay, A.; Abuye, C.; Retta, N. Effect of Germination on Mineral Bioavailability of Sorghum-Based Complementary Foods. *Afr. J. Food Agric. Nutr. Dev.* **2011**, *11*, 5083–5095. [CrossRef]
28. Nkhata, S.G.; Ayua, E.; Kamau, E.H.; Shingiro, J. Fermentation and Germination Improve Nutritional Value of Cereals and Legumes through Activation of Endogenous Enzymes. *Food Sci. Nutr.* **2018**, *6*, 2446–2458. [CrossRef] [PubMed]
29. Vidal-Valverde, C.; Frias, J. Changes in Carbohydrates during Germination of Lentils. *Z. Leb. Unters Forch.* **1992**, *194*, 461–464. [CrossRef]
30. Ologhobo, A.D.; Fetuga, B.L. Changes in Carbohydrate Contents of Germinating Cowpea Seeds. *Food Chem.* **1986**, *20*, 117–125. [CrossRef]
31. Rani, M.; Bandral, J.D.; Sood, M.; Sharma, S.; Gupta, S.; Chand, G. Effect of Germination on Physico-Chemical and Antinutritional Factors of Oats Flour. *Pharma Innov. J.* **2022**, *11*, 1424–1428.
32. Olamiti, G.; Takalani, T.K.; Beswa, D.; Jideani, A.I.O. Effect of Malting and Fermentation on Colour, Thermal Properties, Functional Groups and Crystallinity Level of Flours from Pearl Millet (*Pennisetum glaucum*) and Sorghum (*Sorghum bicolor*). *Heliyon* **2020**, *6*, e05467. [CrossRef] [PubMed]

33. Gutiérrez-Osnaya, L.J.; Hernández-Uribe, J.P.; Castro-Rosas, J.; Román-Gutiérrez, A.D.; Camacho-Díaz, B.H.; Palma-Rodríguez, H.M.; Rodríguez-Marín, M.L.; Hernández-Ávila, J.; Guzmán-Ortiz, F.A. Influence of Germination Time on the Morphological, Morphometric, Structural, and Physicochemical Characteristics of Esmeralda and Perla Barley Starch. *Int. J. Biol. Macromol.* **2020**, *149*, 262–270. [CrossRef] [PubMed]
34. Liu, Y.; Su, C.; Saleh, A.S.M.; Wu, H.; Zhao, K.; Zhang, G.; Jiang, H.; Yan, W.; Li, W. Effect of Germination Duration on Structural and Physicochemical Properties of Mung Bean Starch. *Int. J. Biol. Macromol.* **2020**, *154*, 706–713. [CrossRef] [PubMed]
35. Lekjing, S.; Venkatachalam, K. Effects of Germination Time and Kilning Temperature on the Malting Characteristics, Biochemical and Structural Properties of HomChaiya Rice. *RSC Adv.* **2020**, *10*, 16254–16265. [CrossRef] [PubMed]
36. Su, C.; Saleh, A.S.M.; Zhang, B.; Feng, D.; Zhao, J.; Guo, Y.; Zhao, J.; Li, W.; Yan, W. Effects of Germination Followed by Hot Air and Infrared Drying on Properties of Naked Barley Flour and Starch. *Int. J. Biol. Macromol.* **2020**, *165*, 2060–2070. [CrossRef] [PubMed]
37. Thilakarathna, R.C.N.; Madhusankha, G.D.M.P.; Navaratne, S.B. Potential Food Applications of Sorghum (*Sorghum Bicolor*) and Rapid Screening Methods of Nutritional Traits by Spectroscopic Platforms. *J. Food Sci.* **2022**, *87*, 36–51. [CrossRef]
38. Wulandari, E.; Sukarminah, E.; Rahayu, G.G. Effect of Time and Temperature Sorghum Germination on the Fineness and Color of Sorghum Sprout Flour. *IOP Conf. Ser. Earth Environ. Sci.* **2020**, *443*, 012083. [CrossRef]
39. Sharma, P.; Gujral, H.S. Antioxidant and Polyphenol Oxidase Activity of Germinated Barley and Its Milling Fractions. *Food Chem.* **2010**, *120*, 673–678. [CrossRef]
40. Patil, S.P.; Arya, S.S. Nutritional, Functional, Phytochemical and Structural Characterization of Gluten-Free Flours. *Food Meas.* **2017**, *11*, 1284–1294. [CrossRef]
41. Olojede, A.O.; Sanni, A.I.; Banwo, K. Effect of Legume Addition on the Physiochemical and Sensorial Attributes of Sorghum-Based Sourdough Bread. *LWT* **2020**, *118*, 108769. [CrossRef]
42. Rao, S.; Santhakumar, A.B.; Chinkwo, K.A.; Wu, G.; Johnson, S.K.; Blanchard, C.L. Characterization of Phenolic Compounds and Antioxidant Activity in Sorghum Grains. *J. Cereal Sci.* **2018**, *84*, 103–111. [CrossRef]
43. Li, R.; Wang, Q.; Zhao, G.; Peng, H.; Zhang, D.; Li, Z. Effects of Germination Time on Phenolics, Antioxidant Capacity, in Vitro Phenolic Bioaccessibility and Starch Digestibility in Sorghum. *Int. J. Food Sci. Tech.* **2022**, *57*, 5175–5185. [CrossRef]
44. Kumar, S.; Bamboriya, S.D.; Rani, K.; Meena, R.S.; Sheoran, S.; Loyal, A.; Kumawat, A.; Jhariya, M.K. Grain Legumes: A Diversified Diet for Sustainable Livelihood, Food, and Nutritional Security. In *Advances in Legumes for Sustainable Intensification*; Elsevier: Amsterdam, The Netherlands, 2022; pp. 157–178. ISBN 978-0-323-85797-0.
45. Altıkardes, E.; Guzel, N. Impact of Germination Pre-Treatments on Buckwheat and Quinoa: Mitigation of Anti-Nutrient Content and Enhancement of Antioxidant Properties. *Food Chem.* **2024**, *21*, 101182. [CrossRef]
46. Cáceres, P.J.; Martínez-Villaluenga, C.; Amigo, L.; Frias, J. Maximising the Phytochemical Content and Antioxidant Activity of Ecuadorian Brown Rice Sprouts through Optimal Germination Conditions. *Food Chem.* **2014**, *152*, 407–414. [CrossRef]
47. Singh, A.; Sharma, S.; Singh, B.; Kaur, G. In Vitro Nutrient Digestibility and Antioxidative Properties of Flour Prepared from Sorghum Germinated at Different Conditions. *J. Food Sci. Technol.* **2019**, *56*, 3077–3089. [CrossRef]
48. Kaur, H.; Gill, B.S. Changes in Physicochemical, Nutritional Characteristics and ATR–FTIR Molecular Interactions of Cereal Grains during Germination. *J. Food Sci. Technol.* **2021**, *58*, 2313–2324. [CrossRef]

foods

MDPI

Article

Resistant Starch Type 5 Formation by High Amylopectin Starch–Lipid Interaction

Fernanda G. Castro-Campos [1], Edgar A. Esquivel-Fajardo [1], Eduardo Morales-Sánchez [2], Mario E. Rodríguez-García [3], Oscar Yael Barron-Garcia [3,4], Cristian Felipe Ramirez-Gutierrez [5], Guadalupe Loarca-Piña [1] and Marcela Gaytán-Martínez [1,*]

1 Posgrado en Ciencia y Tecnología de los Alimentos, Facultad de Química, Universidad Autónoma de Querétaro, Centro Universitario, Santiago de Querétaro 76010, Mexico; fer.castro207@gmail.com (F.G.C.-C.); ed.esquivel.f@gmail.com (E.A.E.-F.); loarca@uaq.mx (G.L.-P.)
2 Instituto Politécnico Nacional, CICATA-IPN Unidad Querétaro, Cerro Blanco No. 141, Col. Colinas del Cimatario, Santiago de Querétaro 76090, Mexico; emoraless@ipn.mx
3 Centro de Física Aplicada y Tecnología Avanzada, Departamento de Nanotecnología, Universidad Nacional Autónoma de México, Campus Juriquilla, Querétaro 76230, Mexico; marioga@fata.unam.mx (M.E.R.-G.); yaelbarron@fata.unam.mx (O.Y.B.-G.)
4 División Industrial, Universidad Tecnológica de Querétaro, Av. Pie de la Cuesta 2501, Nacional, Santiago de Querétaro 76148, Mexico
5 Cuerpo Académico de Tecnologías de la Información y Comunicación Aplicada, Universidad Politécnica de Querétaro, El Marqués 76240, Mexico; cristian.ramirez@upq.edu.mx
* Correspondence: marcela.gaytan@uaq.mx

Abstract: The formation of resistant starch type 5 (RS5), primarily associated with amylose–lipid complexes, is typically attributed to starches with high-amylose content due to their affinity for lipid interactions. Recently, studies have also investigated the potential of amylopectin-rich starches to form amylopectin–lipid complexes (ALCs), expanding RS5 sources. This study explores the capacity of waxy corn starch (WS), which is rich in amylopectin, to develop ALCs with oleic acid (10% w/w) under different thermal and mechanical conditions. Specifically, WS was treated at temperatures of 80 °C, 85 °C, and boiling, with stirring times of 0 and 45 min. Results demonstrated significant ALC formation, reaching a peak complexation index (CI) of 59% under boiling conditions with 45 min of stirring. Differential scanning calorimetry (DSC) identified a notable endothermic transition at 110 °C, indicating strong ALC interactions. FTIR spectra further evidenced starch–lipid interactions through bands at 2970 cm^{-1} and 2888 cm^{-1}. X-ray diffraction (XRD) analysis confirmed the presence of orthorhombic nanocrystals in native WS, with ALC samples exhibiting a V-type diffraction pattern, supporting effective complexation. This study advances knowledge on starch–lipid interactions, suggesting ALCs as a promising RS5 form with potential food industry applications due to its structural resilience and associated health benefits.

Keywords: amylopectin–lipid complex; orthorhombic; oleic acid; waxy starch; gelatinization

1. Introduction

Resistant starch type 5 (RS5) is highly valued in the food industry for its versatility and functional properties [1]. Its popularity stems from its accessibility, cost-effectiveness, non-toxicity, biodegradability, and ability to selectively release bioactive compounds within the gastrointestinal tract, making it an excellent candidate for targeted nutritional applications [2]. Consequently, exploring cost-effective and efficient methods to produce this type of resistant starch is essential, particularly those that can also deliver enhanced health benefits. Amylose–lipid complexes, also known as RS5, are formed through interactions between fatty acids and the amylose helix. However, several factors significantly influence the formation of these complexes [3–5]. These factors include the amylose content of the starch, the type of lipid involved in the interaction, and the processing method used

[6–11]. Understanding these factors is crucial for comprehending how RS5 is formed and characterized during the various processes used to obtain it.

Starches with a high amylose content have been shown to produce a greater proportion of RS5 because amylose molecules exhibit a strong affinity for interacting with lipid compounds, leading to the formation of amylose–lipid complexes [12–15]. This is primarily due to the hydrophobic properties of the amylose helix, which enable it to interact and self-assemble with other amylose helices and non-polar lipids [16–18]. However, obtaining such starches can often involve genetic modification, which may make them a costly source of raw material [19]. Moreover, when analyzing interactions with other macromolecules and the effects of various processing methods, it is essential to recognize that starch should be considered a complex particle. In this context, the formation of amylose–lipid complexes particularly depend on the ability of gelatinized amylose to align into a single helix around the complex-forming element, a process facilitated by interactions between methylene groups and glycosidic bonds [20]. While the physical properties of amylose are widely acknowledged as crucial for the formation of amylose–lipid complexes, the temperatures typically used are often insufficient to achieve gelatinization of amylose-rich starches [21]. In this context, the parameters governing gelatinization are critical for the successful formation of RS5.

Regarding temperature, it has been reported that the formation of starch–lipid complexes depends on the size of the interacting macromolecule. The formation of RS5 is possible through thermal treatments at temperatures above 100 °C. This suggests that complex formation occurs during the gelatinization of starch; as such, temperatures allow the opening of the double helix structure of amylose chains or the fragmentation of amylopectin molecules. Therefore, starches rich in amylopectin may be a more accessible source for RS5 formation. In this context, the formation of these complexes using high-amylopectin rice starch through enzymatic debranching has been reported [22–24]. This process results in fractions with shorter chain lengths, and the studies highlight that these shorter chains can facilitate interactions with lipids. In another study, Luo et al. [25] reported that the size of the chains obtained by amylopectin debranching significantly affects their ability to interact with lipids. Specifically, wider helical fragments consisting of eight glucose units per turn were significantly more prone to form complexes with free fatty acids than shorter, narrower chains with fewer than eight glucose residues per turn. However, the enzymatic pathway for RS5 production is often costly and yields relatively low amounts, making it less attractive for industrial applications. An alternative approach could involve modifying amylose and amylopectin chains through thermal treatments, providing a more practical option [26,27]. The above suggests that an interaction between fatty acid and amylopectin-rich starch under specific conditions could result in an increased RS5 content. This opens new possibilities for creating RS5 from various starch sources, not limited to those high in amylose.

On the other hand, the type of fatty acid influences the formation of complexes with starch molecules and affects the content of RS5. Saturated fatty acids have been found to promote the formation of these complexes more effectively than unsaturated fatty acids, resulting in higher levels of RS5. Ai et al. [28] studied the effects of various types of lipids, including triglycerides (e.g., corn oil), phospholipids (e.g., soy lecithin), and free fatty acids with different chain lengths and degrees of unsaturation, on the enzymatic hydrolysis and thermal properties of starches from different sources such as normal corn, tapioca, waxy corn, and high-amylose corn. The authors reported that the type of lipid used significantly influenced the enzymatic hydrolysis rate and the thermal properties of the starch. Specifically, complexes formed with oleic acid exhibited lower degrees of hydrolysis. Oleic acid, a monounsaturated fatty acid, is known for its health benefits, including potential positive effects on cardiovascular health, could therefore enhance the nutritional value of resistant starch. In this context, the formation of resistant starch with essential fatty acids like oleic acid could confer additional health benefits [28–31].

These findings suggest a possible interaction between oleic acid and amylopectin-rich (waxy) starch under certain conditions, leading to a higher content of RS5 [26,32]. Understanding the amylopectin–lipid interaction could potentially lead to the development of new functional starches and ingredients for the food industry, offering potential applications in addition to health benefits.

Other studies have looked at the gelatinization process and the involvement of starch components in a physical way. According to Vega-Rojas et al. [33] and Esquivel-Fajardo et al. [34], the gelatinization of starch involves the solvation of the orthorhombic and/or hexagonal crystal structures. This solvation process requires a suitable temperature, water concentration, and exposure time. Therefore, the formation of RS5 can be influenced by the processing conditions, including time and temperature during the gelatinization. However, further studies are needed to fully understand the interactions between the main components of starch, and the mechanisms behind the formation of amylose–lipid and amylopectin–lipid complexes, as well as the possible presence of nanocrystals.

Therefore, the aim of this study was to investigate the formation of amylopectin–lipid complexes using high-amylopectin maize starch (waxy) to produce RS5. Oleic acid was employed as the complexing lipid under different conditions, including temperatures of 80, 85 °C, and boiling, and stirring times of 0 and 45 min. The interaction between amylopectin and oleic acid in forming RS5 was thoroughly evaluated using structural characterization techniques, including X-ray diffraction and transmission electron microscopy (TEM), scanning electron microscopy (SEM) for morphological analysis, FTIR vibrational spectroscopy, and resistant starch quantification.

2. Materials and Methods

2.1. Biological Material Results

In this study, a commercially available high amylopectin (waxy) corn starch (Amioca, CODE: 04401106) from Ingredion ® (Ingredion Mexico, S.A. de C.V., Queretaro, Qro City, Mexico) was used. The waxy corn starch contained 10.0 ± 0.3% moisture, 1.1 ± 0.2% ash, 0.2 ± 0.2% lipids, 0.3 ± 0.1% protein, 98.4 ± 1.2% total starch, 11.1± 0.3% amylose, 87.3 ± 0.3% amylopectin on a dry matter basis. Oleic acid (O1008 Sigma Aldrich®, St. Louis, MO, USA) was used as the test lipid.

2.2. Amylopectin–Lipid Complex Formation

To obtain the amylopectin–lipid complexes, 20 g of waxy corn starch was stirred in 30 mL of distilled water at room temperature. Separately, 220 mL of water was heated to 80, 85 °C, and boiling temperature. The previously hydrated starch was then added to the heated water. Once the mixture reached the desired temperature, oleic acid was added at a concentration of 10% (*w*/*w*), having been previously dissolved in 10 mL of ethanol.

The mixture of starch, water, and fatty acids was stirred continuously for 0, 15, 30, and 45 min for each treatment. After stirring, each mixture was cooled in an ice bath for 10 min. The samples were then centrifuged at 15,785× *g* for 10 min to remove excess water. The resulting precipitate was washed with ethanol to remove the free fatty acids. Finally, the samples were dried at 40 °C for 24 h, ground, sieved through a 60-mesh screen, and stored in plastic bags for further analysis (Figure 1).

2.3. Complexation Index

The complexation index (CI) was measured according to the method described by Chao et al. [12]. The amylopectin–lipid complex (0.4 g) was transferred into 50 mL tubes and diluted with water to a total weight of 5 g. The CI was determined as follows: The suspension was boiled in a water bath for 10 min with constant stirring. After heating, each sample was cooled at room temperature (25 ± 1 °C), then 25 mL of distilled water was added, and the sample was shaken for 2 min. The sample was then centrifuged at 3000× *g* for 15 min. A 500 μL portion of the supernatant was transferred to a test tube, followed by the addition of 15 mL of distilled water and 2 mL of iodine solution

(2.0% KI and 1.3% I_2 in distilled water). The absorbance at 620 nm was measured using a UV-VIS spectrophotometer (Thermo Scientific, Waltham, MA, USA. Multiskan Ascent, model 51118307). Starch without added lipids was used as a reference. Finally, the CI was calculated using the following equation (Equation (1)):

$$\%CI = \left(\frac{Abs_{reference} - Abs_{starch-lipid}}{Abs_{reference}}\right) \times 100, \quad (1)$$

where $Abs_{reference}$ is the absorbance of the reference starch solution without treatment, and $Abs_{starch\text{-}lipid}$ is the absorbance of the starch–lipid mixture.

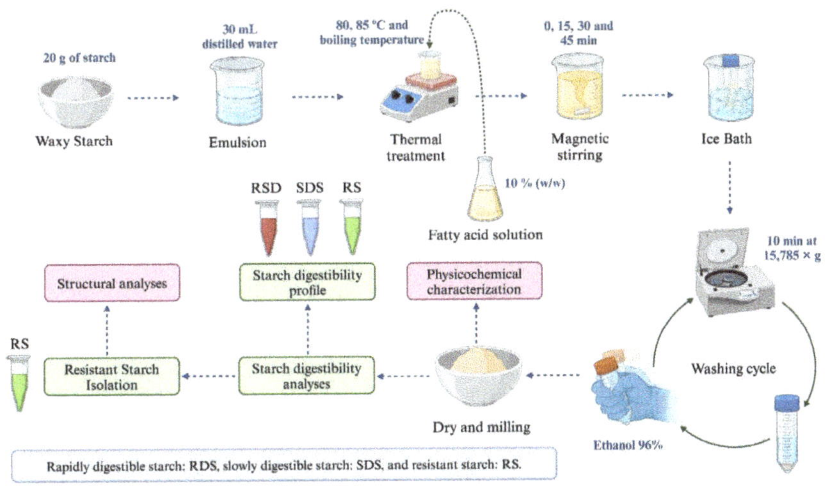

Figure 1. Flow diagram describing the production of amylopectin–lipid complexes.

2.4. Scanning Electron Microscopy Analysis

The morphology of native waxy corn starch, amylopectin–oleic acid complexes, and isolated RS5 was observed using a scanning electron microscope (JEOL, JSM-6060LV, Kenji Kazato and Kazuo Ito in Mitaka, Tokyo) with an accelerating voltage of 5 kV and a secondary electron detector. Samples were mounted on a sample holder with carbon tape and gold-coated prior to analysis [7].

2.5. Nanocrystals Isolation

For the isolation of nanocrystals from waxy corn starch, the method proposed by Rojas-Molina et al. [35] was employed. Specifically, 37 g of starch was suspended in 260 mL of sulfuric acid (H_2SO_4, 3.16 M). The solution was maintained at 40 °C for 7 days with constant stirring. The crystalline fraction of starch was recovered from the solution by centrifugation at 15,785× *g*. Subsequently, the resulting pellet was washed with distilled water until a pH of 5 was reached. Finally, to prevent solvation of the nanocrystals, an ammonia treatment was performed to adjust the pH to 8–8.5, and the sample was stored at 4 °C.

2.6. Transmission Electronic Microscopy (TEM)

The aim of acquiring transmission electron microscopy (TEM) images of nanocrystals was to determine the crystal directions and correlate them with the powder diffraction data reported by Rodriguez-Garcia et al. [36] to ascertain the crystal structure.

TEM images of the isolated nanocrystals (INs) isolated from waxy corn starch were obtained using a JEOL JEM-2200 FS transmission electron microscope (TEM) according to the method proposed by Rojas-Molina et al. [35]. The sample was ultrasonically dispersed in isopropyl alcohol and deposited onto a 3 mm copper grid using a capillary tube. The gird was then placed in a plasma cleaner (Glow Discharge Plasma Cleaner, SKU: M-PDC-32G9, NY, USA) to remove volatile organics before imagining. TEM measurements were conducted at an accelerating voltage of 80 kV. The crystal size of the nanocrystal was determined using the Digital Micrograph V 4.0 software (Gatan Inc. Pleasanton, CA, USA). An array mask was applied to clean the FFT signal and calculate the interplanar distances.

2.7. X-Ray Diffraction Patterns

X-ray diffraction patterns of the waxy corn starch amylopectin–lipid complex and the isolated RS were obtained using a Rigaku diffractometer (Ultima IV model, Cedar Park, TX, USA) equipped with a D/tex Ultra detector at 35 Kv and 15 mA. CuKα radiation (λ = 1.5405 Å) was used, scanning the sample from 5 to 35 ° (2θ scale) with a step of 0.02° [36].

2.8. Differential Scanning Calorimeter Analysis

The waxy corn starch sample and the amylopectin–lipid complex were analyzed using a differential scanning calorimeter (DSC 1 Star® System, Mettler Toledo, Greifensee, Switzerland). The thermograms were obtained according to the method published by Cervantes-Ramírez et al. [6], with slight modifications. Briefly, 0.5 g of each sample was mixed with the distilled water to achieve 60% moisture content on a dry basis. The mixtures were then stabilized at room temperature in sealed bags for 30 min. Subsequently, 10 $\pm$ 0.1 mg of each stabilized sample was weighed into an aluminum pan (40 µL), and the pan was sealed with the T-press instrument. The sample was heated from 30 to 150 °C at 10 °C/min. These procedures were performed in duplicate.

2.9. Vibrational Characterization

The functional groups for the waxy corn starch components and amylopectin–lipid complexes were determined using the method proposed by Cervantes-Ramírez et al. [6] and Cabrera-Ramirez, et al. [7], using a Perkin Elmer IR spectrometer (Spectrum Two model, Waltham, MA, USA) equipped with a diamond-coated attenuated total reflectance (ATR) accessory.

2.10. Rapidly Digestible Starch, Slowly Digestible Starch, and Resistant Starch

The in vitro digestibility of starch was determined according to the method described in the literature [37], with some modifications. Starch and the amylopectin–lipid complex (100 mg each) were added to a centrifuge tube containing 12 mL of sodium acetate buffer (pH: 5.2). The mixture was shaken for 20 s. A 0.5 mL aliquot, labeled GF (glucose-free) was then withdrawn. Subsequently, 300 µL of amyloglucosidase and 100 µL of pancreatic amylase were added. After enzyme addition, the mixture was incubated at 37 °C with constant stirring. Aliquots of 0.5 mL were withdrawn after 20, 120, and 180 min. Each aliquot was transferred to another tube, where 4.5 mL of ethanol (99%) was added, and the mixture was centrifuged at 3000$\times$ g for 10 min. Once the reaction was complete, a portion of the obtained pellet was collected from the samples treated at 80, 85 °C, and boiling temperature with stirring times of 0 and 45 min, to characterize the resistant starch fraction. Free glucose was then measured in all treatments using the glucose-peroxidase kit (GAGO20, Sigma Aldrich). Each sample was analyzed in triplicate, and the proportions of rapidly digestible starch (RDS), slowly digestible starch (SDS), and resistant starch (RS) were calculated using the following equations:

$$\%RSD = \left(\frac{(G_{20} - FG)}{TS} \right) \times 0.9 \times 100, \quad (2)$$

$$\%SDS = \left(\frac{(G_{120} - G_{20})}{TS} \right) \times 0.9 \times 100, \quad (3)$$

$$\%RS = \left(\frac{TS - RSD - SDS)}{TS}\right) \times 0.9 \times 100, \quad (4)$$

where *FG* represents the free glucose content, *G20* for the released glucose content (mg) after 20 min of incubation, *G120* for the released glucose content (mg) after 120 min, and *TS* (total starch) for the total released glucose (mg) after 180 min. *RS* refers to the resistant starch content.

In order to determine the structural changes suffered by the RS fraction after digestion process, a second test was performed according to the method described above. However, after the enzymatic treatment and before RS quantification, the resulting pellet was recovered and analyzed with RDX for all samples.

2.11. Design and Statistical Analysis

Data are presented as mean ± standard deviation of independent experiments performed in duplicate. Statistical analysis was performed using Minitab 16 (Minitab Inc., 2010, State College, PA, USA). Analysis of variance (ANOVA) was performed using Tukey–Kramer's test to determine significant differences at $p < 0.05$ for multiple variables. Pearson correlation was performed to find out the relationship between the studied variables and to determine the treatment that provided the highest formation of RS5.

3. Results and Discussion

3.1. Complexation Index (CI)

Figure 2 presents the complexation index percentage for each treatment, which indirectly indicates the amylopectin–oleic acid complex formation. Overall, the results show that there is an almost linear relationship between the percentage of complexation index and the increase in temperature and stirring time. Thus, higher temperatures (boiling temperature) and longer stirring times (45 min) result in increased complexation index percentages, indicating greater formation of the waxy starch–oleic acid complex. Significantly higher complex formation was observed at 85 °C (Figure 2B) and boiling temperature (Figure 2C) with a stirring time of 45 min. The observed increase in the complex index can be attributed to the gelatinization of the waxy starch. Previous studies have reported that gelatinization of amylopectin-rich starch occurs between 62 and 73 °C [38]. In this sense, the treatment at 80 °C is within the gelatinization temperature range, suggesting that the waxy starch used in the experiment is partially gelatinized at this temperature. When the temperature increases up to the boiling temperature, the degree of gelatinization of the starch increases further. Consequently, the amylopectin chains are more efficiently released, leading to the opening of the double helix structure of the branching chains and helical opening [39], which increases the complex index (Figure 2C). Furthermore, Figure 2 shows that stirring time significantly influences the complexation index, in addition to the effects of heating temperature. This observation suggests that the duration of stirring, likely impacting the structural arrangement of amylopectin, plays a role as critical as starch gelatinization in enhancing the complexation index.

The observed effects of temperature and stirring time suggest that these parameters play a crucial role in the interaction between amylopectin and oleic acid, facilitating the incorporation of the fatty acid and the formation of the amylopectin–oleic acid complex [40,41]. This interaction may result in the physical entrapment of oleic acid in the amylopectin chains, leading to the formation of a self-assembles complex [16]. The results also indicate that hydrolysis of amylopectin (major component of starch) significantly impacts the proportion of starch that can interact with a lipid chain to form a complex. This hydrolysis can be induced by mechanical agitation such as constant stirring.

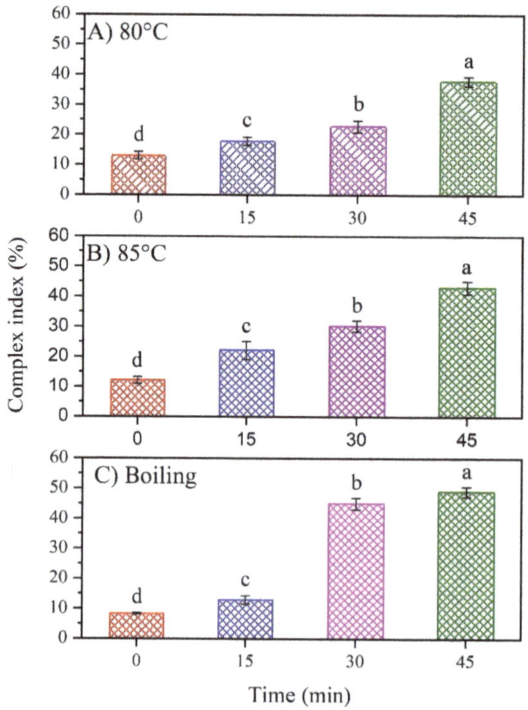

Figure 2. Complexation index of amylopectin and oleic acid at (**A**) 80 °C, (**B**) 85 °C, and (**C**) boiling temperature at different stirring times. Different letters in the same column are statistically different ($p < 0.05$). Table A1 in Appendix A shows the statistical analysis of temperature parameters.

Constant stirring leads to a more substantial mechanical breakage of the amylopectin and amylose molecules, resulting in smaller fractions whose helical structures are easier to dissociate. Previous studies have reported that the helical structure of these components can be opened or swelled at certain temperatures, allowing the incorporation of lipids [39]. Moreover, the helical opening of amylopectin plays a crucial role in complexes formation [25]. Accordingly, the present study shows that increased stirring time and temperature enhance the interaction between amylopectin and oleic acid due to the partial hydrolysis of amylopectin.

This study demonstrates the importance of controlling temperature and interaction time between fatty acids and the amylopectin or amylose components of waxy starch to promote the formation of amylopectin–lipid complexes. The type of fatty acid must also be considered, as its molecular structure can influence its incorporation into amylopectin chains. For instance, oleic acid is an 18-carbon monounsaturated fatty acid whose cis-double bond introduces a link in the hydrocarbon chain, causing chain twisting and steric hindrance that may impede the formation of amylopectin–lipid complexes [17].

3.2. Scanning Electron Microscopy Analysis (SEM)

Figure 3 shows the SEM images used to study the morphological changes of the waxy starch granules at different temperatures and stirring times. Additionally, Figure 3 shows the SEM images of the ALC samples after enzymatic digestion, allowing for the examination of the morphology of the digestion-resistant part.

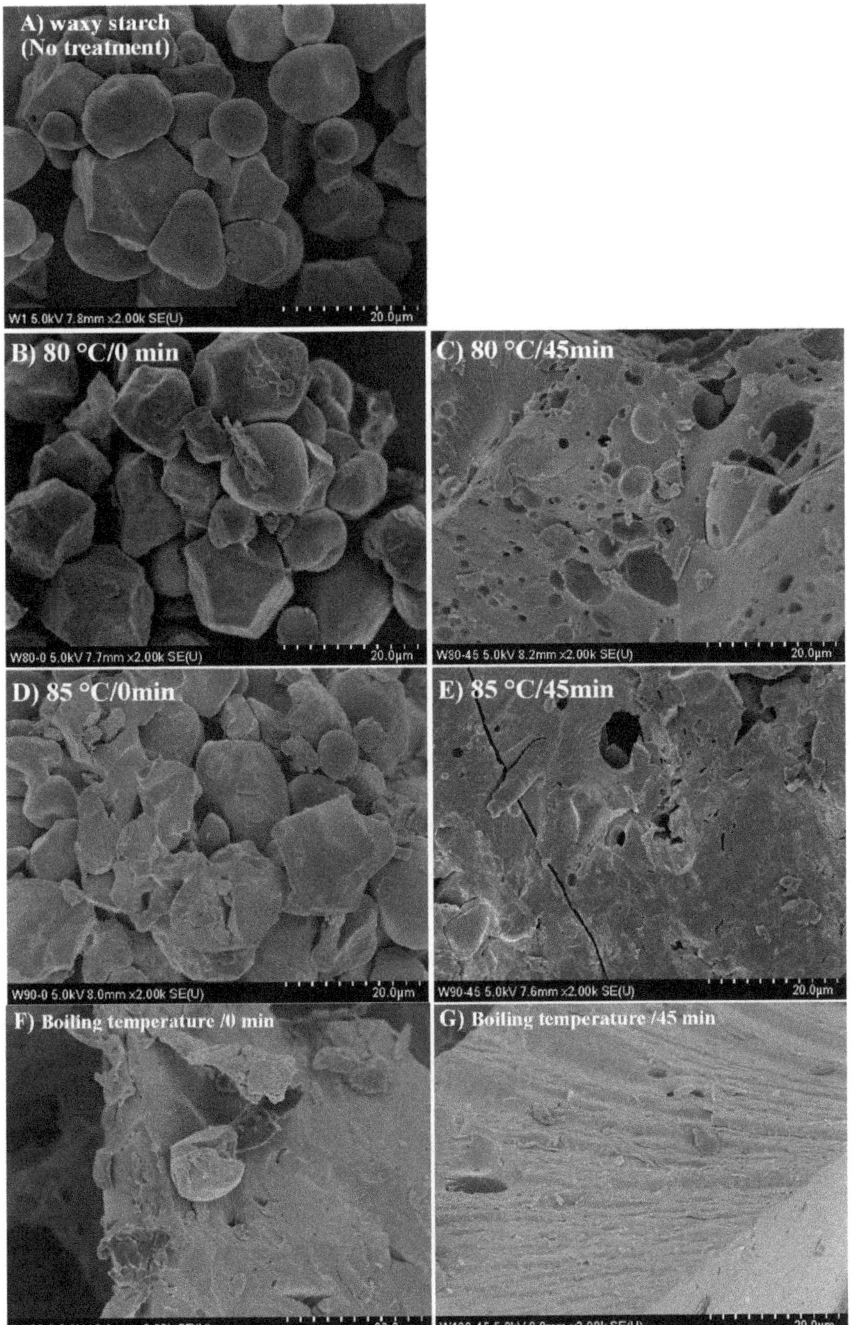

Figure 3. *Cont.*

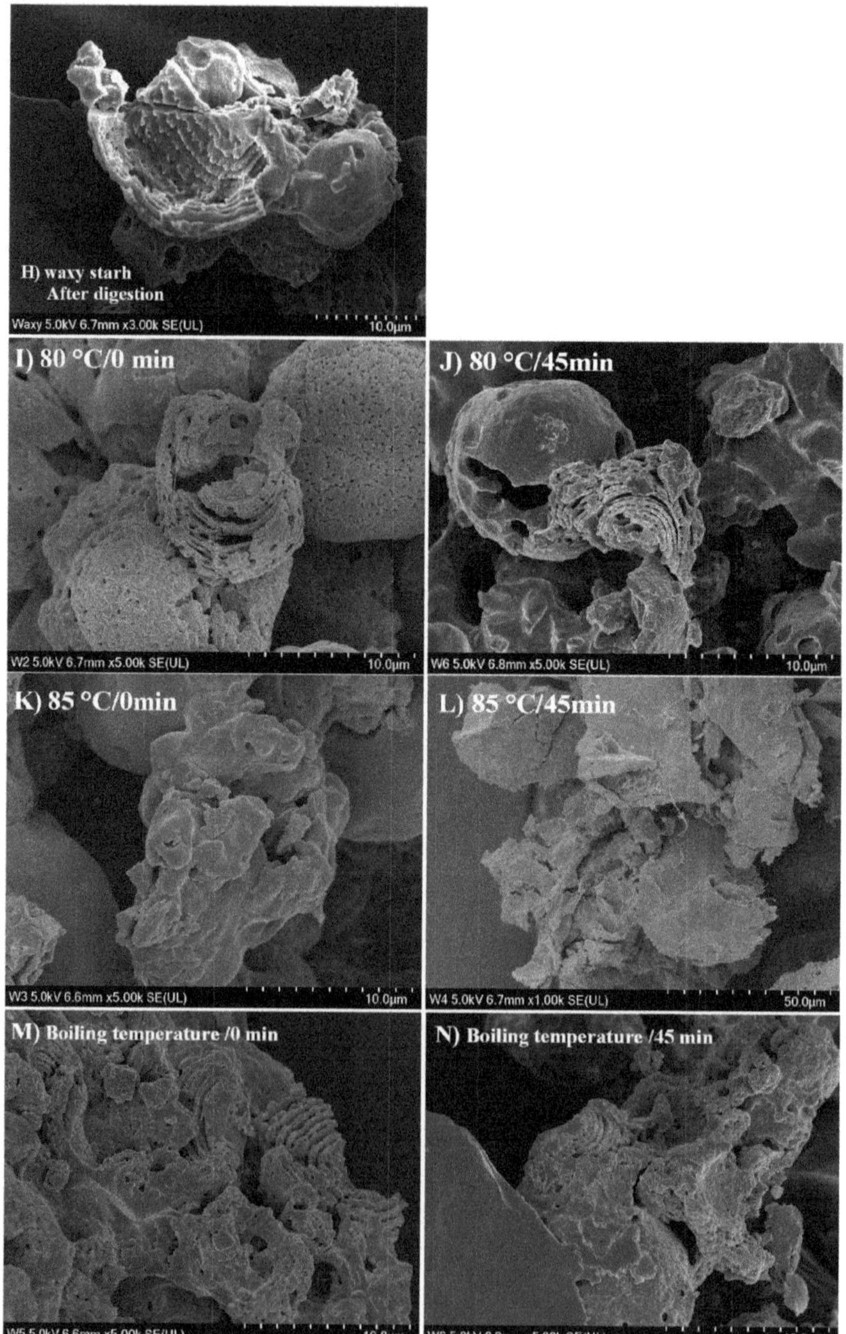

Figure 3. SEM images of the (**A**) native waxy starch and the amylopectin–oleic acid complex at different temperatures and stirring times (**B**) 80 °C at 0 min and (**C**) 45 min; (**D**) 85 °C at 0 min and (**E**) 45 min; and (**F**) boiling temperature at 0 min and (**G**) 45 min. Figure 3 also shows the (**H**) native

waxy starch after enzymatic digestion and the amylopectin–oleic acid complex at different temperatures after enzymatic digestion (**I**) 80 °C/0 min, (**J**) 80 °C/45 min, (**K**) 85 °C/0 min, (**L**) 85 °C/45 min, (**M**) boiling temperature/0 min; and (**N**) boiling temperature/45 min.

In its native form (Figure 3A), waxy starch exhibits irregular round and polygonal granules with sizes around 20 μm, which is consistent with the morphology of waxy maize starch granules reported previously [42]. Pores are also visible in some granules, which may be more susceptible to enzymatic attack. This susceptibility is related to the resistant content in their natural state. After the thermal treatment at 80 °C and 0 min (Figure 3B), the starch granules are deformed, and this deformation was also seen in the other treatments at 85 °C and boiling temperature, the gelatinization temperature is around 70 °C.

However, it should be mentioned that a longer time is required for the complete gelatinization of the starch. Therefore, gel and some coated granules can be seen in the treatments with 45 min of constant stirring (Figure 3C). This coating may be related to the surface layer or oleic acid coating of the starch granule, which could indicate the interaction of the main components of starch, in this case, amylopectin interacting with oleic acid on the surface of the granule (Figure 3D,E).

No whole grains were observed in the treatment at boiling temperature, due to the gelatinization of the starch by the thermal treatment, which leads to gel formation in which oleic acid molecules can immerse (Figure 3F). In this sense, the obtained images show a physical interaction between the starch component and the oleic acid. These results are consistent with the CI results (Figure 2), which showed a time-dependent increase, suggesting a homogeneous complex formation after a longer stirring time, in this case up to 45 min.

In addition, Figure 3 shows the waxy starch and ALC complex after the enzymatic attack and the possible fraction that can reach the colon. The untreated waxy starch was used as a control to study the effects of the amylopectin–oleic acid complexes (Figure 3H) on the morphology of the starch during enzymatic digestion. The micrographs reveal visible damage caused by the α-amylase and glucosidase enzymes in the waxy starch. The micrographs show clear pore formation in the granules indicating the enzymatic attack. However, despite this some lamellae structures that make up the starch granules and some starch granules remain.

When treated at 80 °C with stirring for 0 min (Figure 3I), the enzymatic damage is evident from the presence of pores on the surface of the granules and the lamellae as evidence of enzymatic attack, but when the stirring time is extended to 45 min (Figure 3J), the enzymatic damage becomes visible by the presence of large pores on the granule and the smooth surfaces, but the starch retains its structure, indicating resistance to enzymatic attack and associated with the coat seen in the same treatment before digestion. Nevertheless, it can be said that the damage is not severe and may be associated with partial gelatinization or protection from interaction with oleic acid.

In the treatment at 85 °C with 0 min of stirring (Figure 3K), after digestion, there is no presence of starch granules, and the remaining part has no pores, which can be associated with the gelatinization of the starch granules and consequently the formation of type 3 resistant starch; moreover, amylopectin–oleic acid complexes are also formed during the treatment. However, as can be seen above, when the stirring time was increased (85 °C for 45 min stirring), only some surfaces of the starch granules showed pores, and these micrographs demonstrate the resistance to enzymatic attack.

Finally, the micrographs of the treatments at boiling temperature and 0 min of stirring after the enzymatic attack show smooth surfaces and no whole granules. These micrographs demonstrate the resistance to the enzymatic attack, and the resistance can be related to the resistant starch type 5 through the amylopectin–oleic acid interaction. Above all, the treatment at 45 min of stirring shows a smooth surface without enzymatic damage, indicating the formation of RS, but a minimal part of the surface of the starch granules is damaged with pores, indicating that it is a starch resistant to enzymatic attack, but this treatment shows a greater resistance to enzymatic attack and this resistance may be associated with the presence of an amylopectin–oleic acid complex.

It is important to note that the amylopectin–oleic acid interactions were verified by DSC results, which showed that endotherms are associated with the presence of amylopectin–oleic acid complexes. The treatments conducted at 85 °C and boiling temperature with 45 min stirring had higher enthalpy than the other treatments and showed type I and II endotherms, indicating the most stable starch–lipid interactions. The results suggest that the formation of the amylopectin–oleic acid interaction induces enzymatic resistance.

These results show that the interaction between lipids and the major components of starch can be achieved by starch with high amylopectin content, representing a fast, simple, and inexpensive alternative to obtain these complexes. Consequently, amylopectin–lipid complexes obtained from waxy starch can be used to formulate a new ingredient for the food industry while maintaining the health benefits of resistant starch.

3.3. TEM Analysis

Interpreting X-ray diffraction patterns for both modified and unmodified starches remains challenging, as it is not fully understood which components—such as fats, proteins, or crystalline structures—are affected by complexation. Starch functions as a composite material containing a range of components, with crystalline structures varying by source: cereals typically exhibit an orthorhombic crystal structure (type A), while tubers generally present a hexagonal one (type B). Each component, from fats to proteins to structural arrangements, contributes uniquely to the diffraction pattern [34]. Given these complexities, TEM analysis, which allows localized measurements within the starch granule, could offer critical insights by identifying which components and structures are specifically present and in what form. This capability could be key to clarifying how each constituent contributes to the overall structural and functional properties of the starch. Figure 4 shows the TEM images of isolated nanocrystals of native maize starch with high amylopectin content (waxy), showing isolated nanocrystals (100 nm scale). These structures consist of highly crystalline, elongated regions with an average crystallite size of 30 ± 6 nm in length, 10 ± 1 nm in width, and a notably thin profile (2–4 nm). Figure 4B shows a selected area with isolated, highly ordered nanocrystals around 10 nm in size. This specific area was used to determine the interplanar distance.

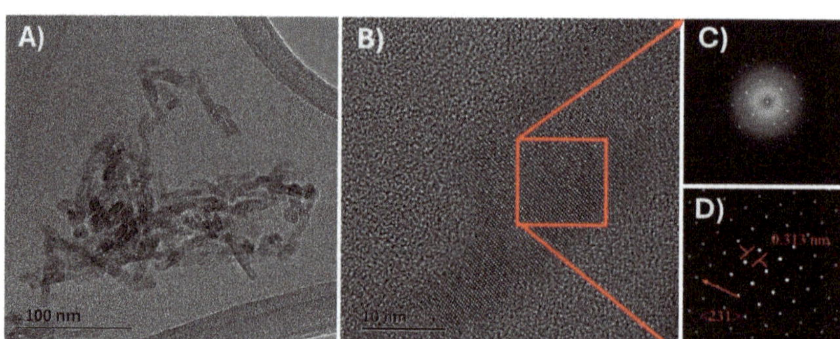

Figure 4. (**A**) Isolated nanocrystals, (**B**) red squares represent characteristic sections of the isolated nanocrystals, and (**C**) the representation of these sections in FT space and (**D**) the representation in inverse FT space.

Figure 4C shows the investigated region in the nanocrystals where the FFT (Fast Fourier Transform) was applied using Digital Micrograph Software 3.5 to convert a signal in the spatial domain into its frequency domain representation. This can reveal information about the crystal structure of the sample, such as the orientation and the interplanar distance of the crystal. Figure 4D shows the results of applying a mask to the FFT. This makes it possible to focus on specific areas of an image, reduce noise, and obtain more meaningful and precise results that can be used to study crystal structures and features in detail according to the method proposed by Rojas-Molina et al. [35].

Figure 4D displays an interplanar distance of 0.313 nm along the <231> direction, confirming that maize starch with high amylopectin content (waxy) consists of nanocrystals with an orthorhombic structure. Comparing the distances obtained through TEM analysis with those reported by Rodriguez-Garcia et al. [36] for the orthorhombic crystal structure further supports this conclusion. These results confirm that the nanocrystals in this starch possess an orthorhombic structure, highlighting the importance of determining whether the starch modification proposed in this work influences this crystal arrangement.

3.4. Structural Analysis

Figure 5 shows the X-ray diffraction patterns of the analyzed samples. In Figure 5A–C, the black line corresponds to the pattern obtained for untreated waxy starch, whereas the red and green lines correspond to the thermal treatments at different stirring times. In Figure 5D–F, the black lines correspond to the waxy starch after digestion without prior treatment, and the red and green lines correspond to the treated samples after digestion. The dashed lines indicate the positions of the indexed diffraction peaks for the starch. The diffraction pattern of untreated waxy starch matches the type A starch diffraction pattern, which corresponds to an orthorhombic crystal structure, as reported by Rodriguez-Garcia et al. [36]. The waxy starch diffraction pattern exhibits broad peaks and continuous halo, suggesting an amorphous phase; however, peak broadening may also result from crystallite size effects [43]. In this context, TEM analysis (Figure 4A) reveals that waxy starch contains highly ordered nanometer-scale structures (nanocrystals). These nanocrystals have not been previously reported or considered, and thus their role in forming complexes with amylose/amylopectin and lipids has not been accounted for.

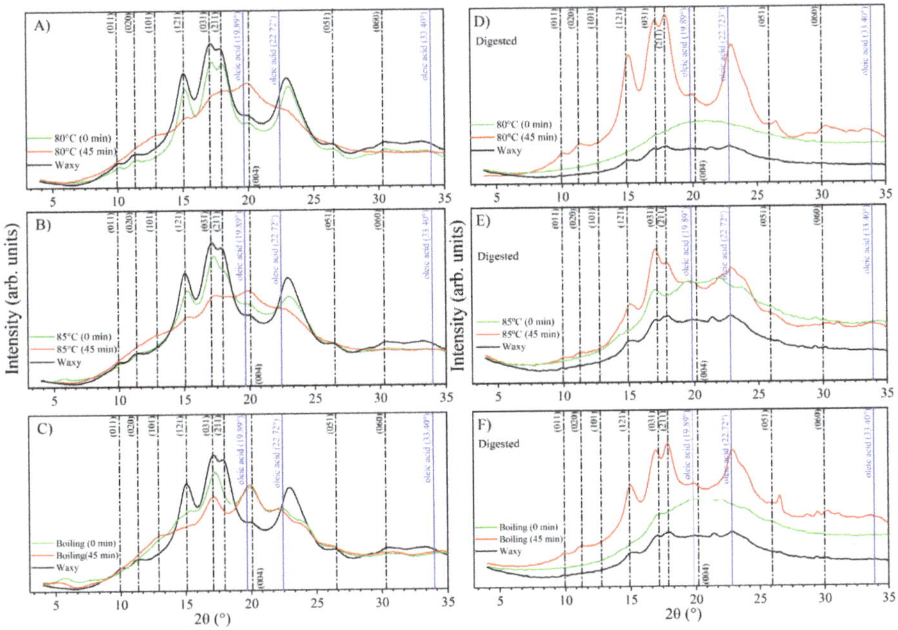

Figure 5. X-ray diffraction patterns for complexes formed with waxy starch and oleic acid at (**A**) 80 °C, (**B**) 85 °C, and (**C**) boiling temperature at different stirring times. Figure (**D–F**) show ALC treatments after digestion (RS fraction) at 80 °C, 85 °C, and boiling temperature, respectively, at different stirring times.

At 0 min of stirring, the diffraction pattern retains features typical of type A starch, though with noticeably reduced intensity. As treatment temperature rises, prominent peaks associated with the (031) and (211) planes begin to broaden and overlap, suggesting partial disruption of the crystalline arrangement. This broadening indicates the onset of gelatinization (shown by the green lines in Figure 5A–C), where structural integrity starts to break down. However, the presence of distinct peaks reveals that a portion of the crystalline structure persists, likely due to insufficient hydrolysis to fully dissolve the ordered regions [44]. This retained crystallinity may influence subsequent interactions in complex formation with amylose and amylopectin. At the prolonged stirring times of 45 min, the diffraction pattern reveals a predominantly amorphous structure, characterized by an increased amorphous halo and a lack of distinct peaks. This effect is evident for all three treatment temperatures—80 °C, 85 °C, and boiling—as illustrated by the red lines in Figure 5A–C. The marked reduction in relative intensities indicates a significant loss of crystallinity. The combination of heat treatment and mechanical agitation disrupts starch crystallinity, leading to an increased degree of gelatinization. The diffraction patterns demonstrate that under these conditions, much of the original ordered structure breaks down, resulting in an amorphous matrix that lacks the defined peaks associated with a crystalline phase.

As previously mentioned, amylose and amylopectin can interact with lipids to form complexes that exhibit a characteristic V-type diffraction pattern. In this context, Sun et al. [41] investigated the effects of chain length and degree of unsaturation of the fatty acid on the structure, physicochemical properties, and in vitro digestibility of maize starch complexes. They found that these complexes produced a diffraction pattern with peaks at 7.5°, 13.5°, and 20.5° on the 2θ scale. Additionally, Marinopoulou et al. [30] identified crystalline complexes that exhibited characteristic peaks at 13.2° and 19.8°, indicating a V-type crystalline structure associated with the complexed amylose. Their findings highlight the presence of these peaks in the X-ray diffraction patterns of Hylon VII–fatty acid complexes, further reinforcing the notion of a distinct crystalline arrangement in these starch–lipid interactions [30]. However, the diffraction patterns reported by Marinopoulou et al. [30] correspond to partially gelatinized starch. This makes it challenging to distinguish between the peaks generated by the partially gelatinized starch and those produced by the formed complexes, particularly in samples treated at lower temperatures. Moreover, none of the authors indexed the V-type diffraction patterns, indicating that there are no associated crystallographic planes or Miller indices to explain the characteristic peaks observed in the diffraction pattern.

The diffraction patterns for treatments conducted at 80 °C and 85 °C, with stirring for 45 min, as well as for boiling temperatures across both stirring times, revealed a peak at 19.9° (Figure 5A–C). Although the peaks at 7.5° and 13.5° are not distinctly visible, this may be attributed to their overlap with the amorphous phase of the partially gelatinized starch. Given that the peak at 20.5° is associated with the V-type structure [2,30,45], the diffraction patterns for these treatments may suggest the formation of amylopectin–lipid complexes.

However, it is important to note that the previously mentioned crystalline structure has not been reported, nor has the V-type diffraction pattern been indexed. This suggests that the crystalline structure of the amylose–lipid complex does not form. In fact, the V-type structure was first proposed by Katz [46] to describe gelatinized starch, specifically amylose in a simple helical arrangement. Although the V-type pattern does not correspond to a new crystalline structure, the presence of peaks at 12° and 20° on the 2θ scale indicates that a short-range rearrangement occurs during the starch–lipid complexation. Therefore, the true structure of starch–lipid complexes remains an open question. Nevertheless, the results from the diffraction patterns in this study indicate that waxy starch requires a higher temperature (above 85 °C) and a longer stirring time to induce interaction between amylopectin and fatty acids, thereby facilitating the formation of amylopectin–lipid complexes.

Additionally, this study demonstrates that gelatinization causes amylose and amylopectin molecules to leach out, resulting in partially solvated nanocrystals with an orthorhombic structure. This process leads to an irreversible transition from ordered to disordered structures [33,34]. However, despite this transition, some crystal structures remain intact. It is important to note that even after boiling and stirring for 45 min, the starch is only partially gelatinized, as indicated in Figures 3N and 5C. The partial gelatinization of starch during thermal treatment has also been reported by Rojas-Molina et al. [35]. This suggests that complex formation occurs only with gelatinized starch, particularly with those chains that have a wider molecular weight [37].

On the other hand, the diffraction patterns of the digested starch fractions indicate a conversion to a continuous amorphous halo for native waxy starch and samples treated with 0 min of stirring. According to Nieves-Hernández et al. [47], enzymatic activity does not cause significant alterations in starch crystallinity. Therefore, the changes observed in the diffraction patterns of waxy starch may be attributed to the method used to quantify resistant starch (RS), which involves thermal and chemical reactions. Additionally, gelatinization leads to the leaching of amylose and amylopectin and the partial solvation of nanocrystals [34]. As a result, samples that underwent no stirring treatment at low temperatures (80° and 85 °C) are more likely to exhibit this amorphous halo, reflecting a loss of atomic ordering in the system due to prior partial gelatinization.

However, the crystalline structure persists in all samples treated with 45 min of stirring, even after digestion. The diffraction patterns for these samples (Figure 5D–F) reveal strong, well-defined peaks at 15.08°, 17.12°, 17.86°, and 22.94° on the 2θ scale. These peaks correspond to an orthorhombic crystalline structure as reported by Rodriguez-Garcia et al. [36]. This phenomenon may be attributed to the formation of amylopectin–lipid complexes. After gelatinization, the formation of these complexes competes with the reorganization of starch molecules. Consequently, an inclusion complex can arise from the interaction of amylose or amylopectin with surrounding elements such as other starch molecules or fatty acids [48]. The presence of well-defined diffraction patterns in these treatments suggests that a network composed of gelatinized starch and amylopectin–lipid complexes forms around the partially gelatinized starch, effectively protecting the nanocrystals from enzymatic activity and thermal treatment. According to this theory, it is the amylopectin–oleic acid complex and the highly compacted crystal structures (nanocrystals of pyroglucans) that reach the large intestine as the resistant fraction. Furthermore, these nanocrystals are part of RS2, as suggested by Rojas-Molina et al. [35].

However, additional studies are necessary to validate this theory. Then, based on SEM measurements and diffraction pattern data, the proposed kinetic model for amylopectin–lipid complex formation is shown in Figure 6. Initially, partial gelatinization of the starch occurs. Following this, the fatty acid is added, and with mechanical stirring, the fatty acid chain interacts with the gelatinized starch fraction. The fatty acid can then position itself either on the outer or inner regions of the amylopectin chain, forming the complex. Thus, the resistant starch fraction formed by partially gelatinized starch consists of a combination of pyroglucan nanocrystals and amylopectin–oleic acid complexes.

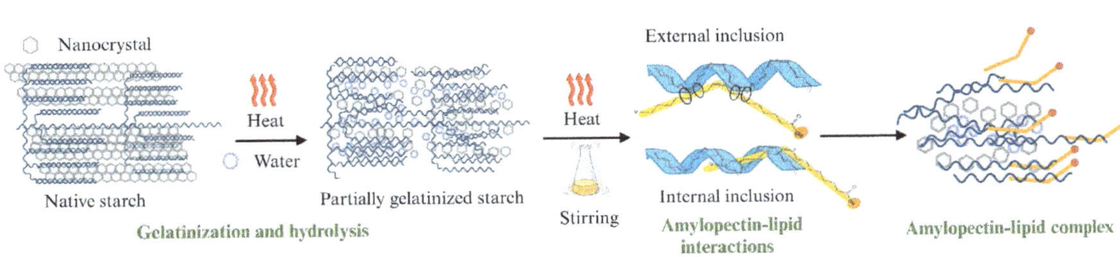

Figure 6. Proposed kinetic model for the formation of the amylopectin–lipid complex.

3.5. Thermal Properties

Depending on their structural arrangement, two types of starch–lipid interactions are considered: type I and II. Since each of these interactions have been described as a function of their endothermic transition with respect to the corresponding amylose–lipid interaction, these complexes can be studied by DSC, and therefore classified as type I at 90–110 °C (endotherm 1) and type II at 110–130 °C (endotherm 2) [9]. Figure 7 shows the thermal profiles of waxy starch, and the complex formed after thermal treatment at different stirring times. DSC analysis was used to determine the interactions between the amylopectin molecule and the oleic acid, and the following endotherms were found: E1: Starch gelatinization temperature due in part to solvation of the orthorhombic nanocrystals [9]; E2: Amylopectin–oleic acid complex type I; and E3: Amylopectin–oleic acid complex type II and thermal oxidation of oleic acid [49,50].

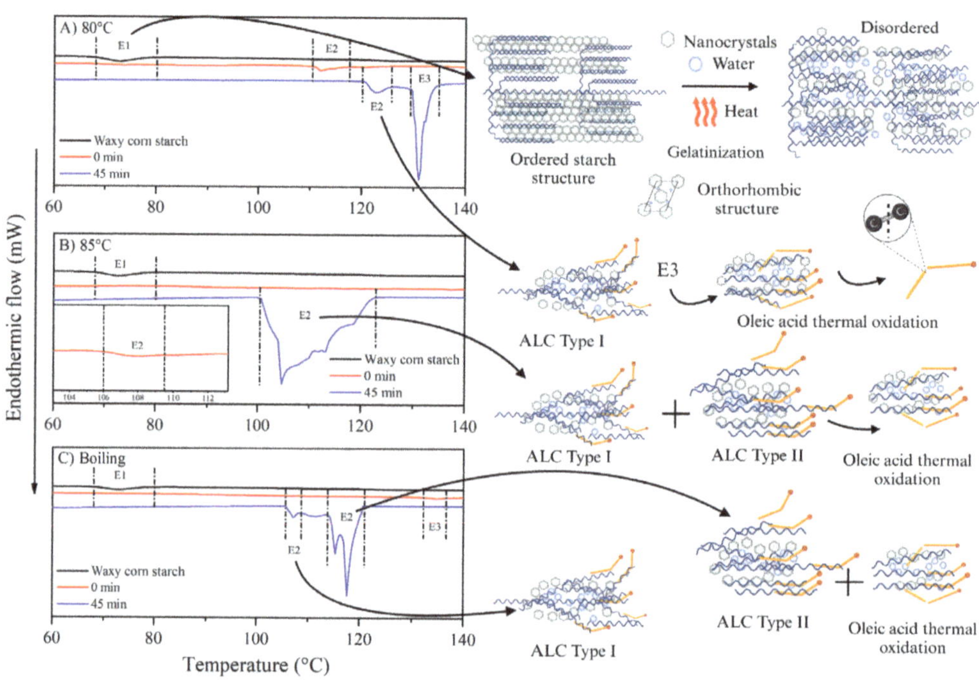

Figure 7. Thermograms of amylopectin–lipid complexes at different temperatures (**A**) 80 °C, (**B**) 85 °C, and (**C**) boiling temperature, with 0 and 45 min of stirring. * ALC—Amylopectin–lipid complex.

The formation of amylopectin–oleic acid complexes depends on the method used and the set temperature. However, the necessary conditions for this complex formation still need to be clarified [51]. Waxy starch without treatment has a gelatinization temperature from 58.6 to 76.3 °C with an enthalpy of 7.13 J/g. This transition corresponds to the gelatinization of waxy starch (E1). The shift in gelatinization transition (69.53 to 81.1 °C) was observed when the treatment was carried out at 80 °C and stirred for 0 min. However, the enthalpy decreased (7.14 to 1.13 J/g), indicating that partial gelatinization of the starch occurred, resulting in the release of amylopectin. The treatments at 85 °C and 0 min stirring time showed the presence of a second endotherm (E2), ranging from 106.26 to 108.72 °C, and the treatment at boiling temperature with 0 min stirring time showed three endotherms: E1 (68 to 76 °C, ΔH: 0.54 J/g), E2, and E3.

The presence of E1 in the treatments indicates the presence of residual non-gelatinized starch. While the presence of E2 corresponds to the formation of type I amylopectin–oleic acid complexes. However, the left shift of this endotherm indicates the presence of type II amylopectin–oleic acid complexes. In addition, endotherm 3 (E3) could indicate the thermal oxidation of oleic acid. The formation of type I complexes could be related to temperature, as heat treatments above the gelatinization temperature and constant stirring can promote solvation of the nanocrystals and leaching of the amylopectin chains [33,34], allowing the helices in the amylopectin branches to interact with the oleic acid.

On the other hand, the stirring time can promote the debranching/hydrolysis of amylopectin, releasing smaller molecules (dextrin) that can interact more efficiently with oleic acid [9,52] and consequently, form complexes with greater thermal stability, as shown by the treatments with longer stirring time and higher temperature.

The treatment at boiling temperature and 45 min of stirring showed the formation of type II amylopectin–oleic acid complexes with the highest enthalpy (13.02 J/g). These results demonstrate the existence of type II amylopectin–oleic acid complexes and show the thermal stability of the interaction. The endothermic E2 transition was observed at 80 and 85 °C and 0 min of stirring, while the 100–122 °C transition at the 85 °C treatment with 45 min of stirring was undefined and can be explained by a mixture of type I and II complexes. A third transition (E3) can be seen in the thermograms. This transition is undefined and is observed at 110 and 130 °C. On the other hand, the thermal oxidation reactions of free oleic acid involve the cleavage of the C=C double bond on carbon 9 of the aliphatic chain at 120–140 °C. Therefore, this transition can be associated with the presence of an amylopectin–lipid complex, the possible thermal oxidation of oleic acid (E3) and, at the same time, a dissociation of oleic acid-amylopectin complexes in the range of 100–122 °C [49,50].

The results obtained for the thermal properties show that the formation of amylopectin–oleic acid complexes increases at temperatures above 80 °C, and that the interaction time influences the stability of the complex formed. This behavior is related to the CI results, which show that the strongest interaction occurs at 85 °C and at the boiling temperature, both at 45 min of stirring. Furthermore, the thermal properties confirmed the interaction between amylopectin and oleic acid and emphasized the higher stability of the complexes at boiling temperature and 45 min of stirring than in the other treatments used in this study.

3.6. Fourier Transform Infrared Spectroscopy (FTIR)

Figure 8A–C shows the infrared spectra of oleic acid, untreated waxy starch, and the amylopectin–lipid complexes formed from waxy corn starch at 80, 85 °C, and boiling temperature at 0 and 45 min of stirring. Figure 8A displays the infrared spectrum of oleic acid and untreated waxy starch. These spectra correspond to the characteristic fingerprints of these materials. The spectra of the treated samples exhibit the characteristic bands associated with untreated starch. However, several new bands appear at 2970, 2888, 1464, 1379, and 1251 cm^{-1}. These additional bands may indicate modifications in the molecular structure of the starch, suggesting potential interactions or chemical changes resulting from the treatment process. The bands at 2970 cm^{-1} (a) and 2888 cm^{-1} (b) correspond to the stretching of C–H bonds in the methyl (–CH_3) and methylene (–CH_2) groups of oleic acid [2,26,53]. It is important to note that these bands appear shifted to higher wavenumbers compared to the typical position of oleic acid, suggesting that there is an interaction between the polymeric matrix and the fatty acid chains via these functional groups. This implies that it is not a simple mixture but rather an interaction-driven complex formation. Similarly, the 1464 cm^{-1} (c) band, related to C–H bending vibrations in methylene groups, and the 1379 cm^{1} (d) band, associated with CH_3 bending, appear shifted to lower wavenumbers and increase in intensity with higher thermal treatment and stirring time. This further supports the hypothesis of a chemical interaction and confirms the formation of a complex between the starch and fatty acid. The 1240 cm^{-1} (e) band, tied to C–O stretching, indicates potential hydrogen bonding or esterification between hydroxyl groups in starch and oleic acid, enhancing structural stability [6,52]. Finally, the intense band at 1710 cm^{-1}, characteristic of oleic acid, was almost absent in the spectrum of the treated samples. This absence is likely due to the washing step applied to treated samples to remove any free oleic acid residue from the mixture. Therefore, studies reporting this band as an indicator of complexation may, in fact, be detecting residual free oleic acid in the system. In this context, the lack of intensity in this band serves as further evidence of complex formation [54], a phenomenon similarly observed in other complex systems, such as oleic acid with nanoparticles and chitosan, where the normal mode also disappears [55].

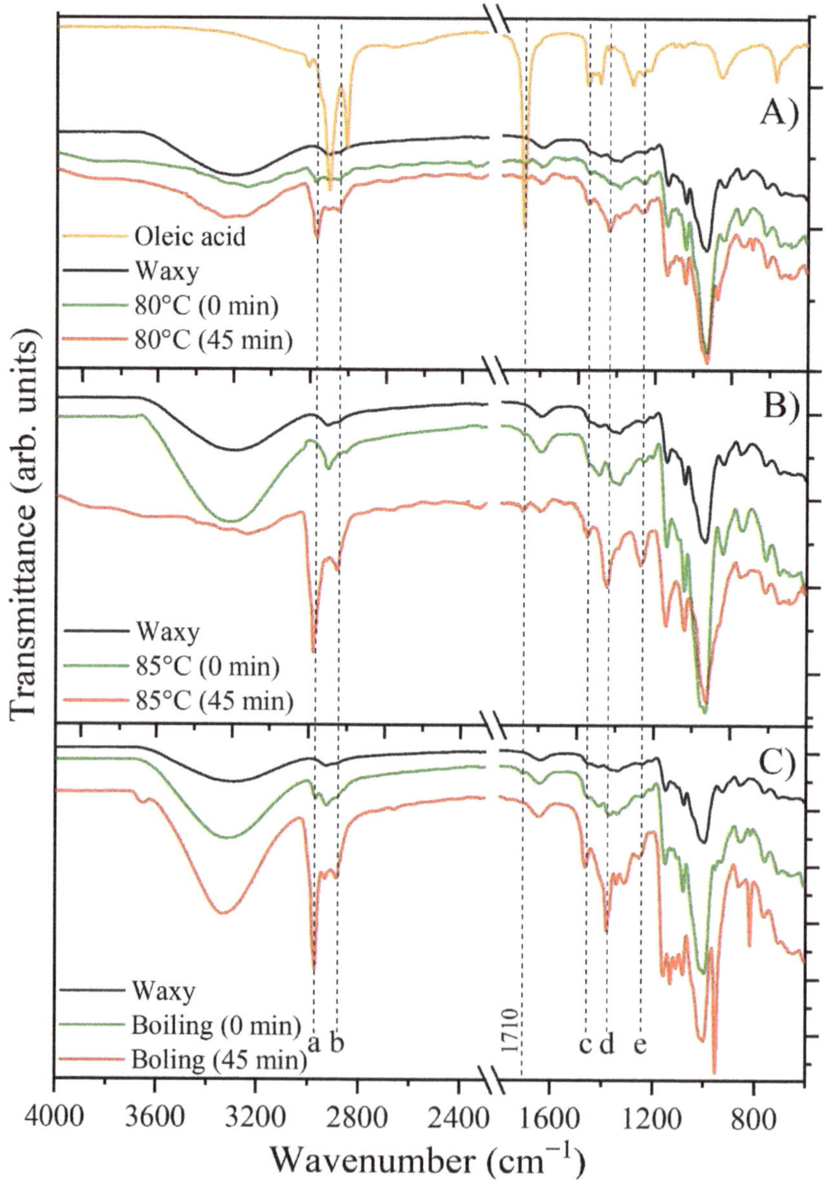

Figure 8. FTIR-ATR spectra for complexes formed from oleic acid and waxy starch at different temperatures (**A**) 80 °C, (**B**) 85 °C, and (**C**) boiling temperature. Different letters (a, b, c, d, and e) correspond to the main bands associated with amylopectin–lipid complexes formation.

Furthermore, as the temperature and stirring time increase, the intensity of these bands also rises, suggesting a greater degree of complexation. Thus, it can be said that the formation of complexes is favored by longer stirring times at 85 °C and at boiling temperature.

3.7. Rapidly Digested Starch, Slowly Digested Starch, and Resistant Starch Quantification

Table 1 shows the percentage of rapidly digestible starch (RDS), slowly digestible starch (SDS), and resistant starch (RS) formed from the amylopectin–lipid complexes and the waxy starch. The quantification of the different digestible starch fractions shows significant differences in the RDS content between the treatments at 80 and 85 °C and the treatments at boiling temperature and at different stirring times (0 and 45 min). The highest content of rapidly digestible starch (90.07 ± 1.1%) was observed in the treatment with 0 min of stirring. However, no differences ($p > 0.05$) in SDS content were observed between the treatments.

Table 1. Content of rapidly digested starch (RDS), slowly digested starch (SDS), and resistant starch (RS) in the different treatments for the formation of amylopectin–lipid complexes with waxy corn starch.

Treatment		RDS (%)	SDS (%)	RS (%)
Temperature (°C)	Time (min)			
Native		61.50 ± 0.51	3.51 ± 0.22	20.83 ± 0.3
80	0	90.07 ± 1.10 [a]	2.31 ± 1.63 [d]	6.4 ± 0.74 [d]
	45	35.81 ± 0.10 [d]	31.87 ± 1.22 [a]	27.74 ± 0.23 [b]
85	0	79.98 ± 2.70 [b]	6.89 ± 0.18 [d]	12.63 ± 0.39 [c]
	45	44.26 ± 3.20 [c]	21.24 ± 0.51 [c]	26.61 ± 0.45 [b]
Boiling	0	50.39 ± 2.1 [c]	32.90 ± 0.31 [a]	11.05 ± 1.90 [c]
	45	24.56 ± 1.2 [e]	25.61 ± 1.30 [b]	49.33 ± 1.61 [a]

The mean ± standard deviation of two replicates is shown. Different letters in the same column are statistically different ($p < 0.05$).

The increase in RS content was related to the increase in treatment temperature and stirring time. However, no significant differences were found between the treatments at 80 and 85 °C and 45 min stirring time (27.74 ± 0.23 and 26.61 ± 0.45%, respectively). On the other hand, the treatment at boiling temperature and 45 min of stirring time was statistically different from the other treatments as it had the highest amount of RS (49.33 ± 1.61%).

The RS values at 80 °C and 0 min of stirring were lower than the other treatments because gelatinization of waxy starch occurs in the range of 68 to 75 °C. Therefore, the treatment at 80 °C resulted in the gelatinization of starch, while the stirring time of 0 min does not show an effect on the formation of RS. Although an amylopectin–oleic acid interaction was observed in the CI and DSC results, this interaction between amylopectin and oleic acid may help to resist the digestion process of starch because the amylopectin–oleic acid complex restricts the access of enzymes so that the complexes are resistant to digestion [56,57].

Meanwhile, the effect of temperature on RS content was not significant in the 80 and 85 °C treatments, but a significant difference was observed in the boiling temperature treatments. Interestingly, the samples that were stirred for 45 min were significantly different from the samples that were stirred for 0 min, suggesting that stirring time has a greater effect on RS formation than temperature.

It has previously been shown that waxy starch has a low ability to interact with fatty acids due to the length of its branches. These branches represent a steric hindrance to interaction with fatty acid chains [58]. However, the results of this study suggest that the effect of temperature on RS content is due to thermally mediated gelatinization of the starch, which leads to leaching of the amylopectin chains (which are fragmented with constant stirring to form further complexes with the oleic acid). In this sense, the formation of amylopectin–oleic acid complexes is reflected in the DSC results (Figure 7), especially in the endotherms observed for the type I and type II complexes. Remarkably, the treatment at boiling temperature and a stirring time of 45 min showed an enthalpy of 13.02 J/g in the DSC results, indicating that it is a stable complex. In agreement with the RS results, the treatment at boiling temperature and 45 min of stirring time showed the highest resistance to digestion among all treatments. The observed effect may be related to the fact that type I amylopectin–oleic acid–lipid complexes can dissociate after boiling, suggesting that the resistance to digestion in boiling temperature treatments may originate from type II amylopectin–oleic acid complexes [56,57]. Finally, waxy starch showed a 1.3-fold increase in RS at boiling temperature and 45 min stirring time compared to untreated starch, confirming an increasing formation of amylopectin–oleic acid complexes.

Therefore, CI is associated with interaction time and increase in RS ($r = 0.86$, $p < 0.05$), in contrast, RDS is negatively associated with stirring time ($r = -0.82$, $p < 0.05$). The increase in RS is associated with CI ($r = 0.85$, $p < 0.05$), indicating the formation of an interaction between amylopectin and oleic acid. From the above results, it can be concluded that at a controlled temperature and constant stirring time, the formation of amylopectin–oleic acid complexes is possible, and these complexes lead to an increase in RS5.

4. Conclusions

Waxy starch, which is high in amylopectin, can form complexes with oleic acid either through self-assembly or by positioning itself along the outer regions of the amylopectin chain. This occurs under controlled stirring times and at temperatures above 85 °C. The structural characteristics of these complexes were identified by the formation of V-type diffraction patterns. For native waxy starch, both X-ray XRD and TEM revealed an orthorhombic crystalline structure, which experiences partial degradation during the gelatinization process.

The formation of the amylopectin–oleic acid complex occurs as soon as the starch gelatinizes, because during gelatinization, the starch granules swell and break up, releasing amylopectin molecules. Simultaneously, the constant stirring plays a decisive role in the modification of the amylopectin chains. This hydrolysis can be induced by constant mechanical agitation.

X-ray diffraction patterns and TEM images revealed the presence of nanocrystals with orthorhombic crystal structure in waxy starch, which form part of RS5 for partially gelatinized starch. According to the XRD analysis of the resistant fraction of ALC, the presence of well-defined diffraction patterns after digestion suggests that for partially gelatinized starch, a network composed of gelatinized starch and amylopectin–lipid complexes protects the nanocrystals from enzymatic activity and thermal treatment. Thus, the total resistant fraction is the sum of the pyroglucan nanocrystals (RS_2) and the amylopectin–oleic acid complex, as well as some whole grains that resist all treatments. Finally, these results have shown that the formation of starch–lipid complexes can be achieved with starch with high amylopectin content and represents a fast, simple alternative.

Author Contributions: Conceptualization, F.G.C.-C., M.E.R.-G. and M.G.-M.; data curation, F.G.C.-C., E.M.-S., E.A.E.-F. and M.G.-M.; formal analysis, F.G.C.-C., E.M.-S., M.E.R.-G., E.A.E.-F., O.Y.B.-G., G.L.-P. and M.G.-M.; funding acquisition, M.E.R.-G. and M.G.-M.; investigation, F.G.C.-C., E.M.-S., M.E.R.-G. and M.G.-M.; methodology, F.G.C.-C., M.E.R.-G. and M.G.-M.; project administration, M.E.R.-G. and M.G.-M.; software, C.F.R.-G.; supervision, M.E.R.-G. and M.G.-M.; validation, E.A.E.-F., O.Y.B.-G., C.F.R.-G. and M.G.-M.; visualization, F.G.C.-C. and E.A.E.-F.; writing—original draft, F.G.C.-C., M.E.R.-G., E.A.E.-F., O.Y.B.-G. and M.G.-M.; writing—review and editing, F.G.C.-C., M.E.R.-G., C.F.R.-G. and M.G.-M. All authors have read and agreed to the published version of the manuscript.

Funding: This research was funded by Química Somos Todos-UAQ 2023.

Institutional Review Board Statement: Not applicable.

Informed Consent Statement: Not applicable.

Data Availability Statement: The original contributions presented in the study are included in the article, further inquiries can be directed to the corresponding author.

Acknowledgments: Author Fernanda G. Castro-Campos was supported by a scholarship from CONAHCYT (712858). The Química Somos Todos-UAQ-2023 supported this research. Laboratorio Nacional de Caracterización de Materiales at CFATA-UNAM-Conahcyt, Mexico supported this project. The authors thank the SEM support of Manuel Flores.

Conflicts of Interest: The authors declare no conflicts of interest.

Appendix A

Table A1. Results of Tukey's multiple comparisons test, showing the mean differences, 95% confidence intervals, and adjusted p-values for the comparisons between temperature treatments (80 °C, 85 °C, and boiling). Statistically significant differences are indicated by asterisks, with **** representing $p < 0.0001$ and ** representing $p < 0.01$.

Tukey's Multiple Comparisons Test	Mean Diff.	95.00% CI of Diff.	Adjusted p-Value	
85 °C vs. 80 °C	3.978	2.475 to 5.481	****	<0.0001
Boiling vs. 80 °C	5.963	4.460 to 7.466	****	<0.0001
Boiling vs. 85 °C	1.985	0.4820 to 3.488	**	0.0073

References

1. Farooq, M.A.; Yu, J. Recent Advances in Physical Processing Techniques to Enhance the Resistant Starch Content in Foods: A Review. *Foods* **2024**, *13*, 2770. [CrossRef]
2. Marinopoulou, A.; Papastergiadis, E.; Raphaelides, S.N.; Kontominas, M.G. Structural Characterization and Thermal Properties of Amylose-Fatty Acid Complexes Prepared at Different Temperatures. *Food Hydrocoll.* **2016**, *58*, 224–234. [CrossRef]
3. Bello-Perez, L.A.; Flores-Silva, P.C.; Agama-Acevedo, E.; Tovar, J. Starch Digestibility: Past, Present, and Future. *J. Sci. Food Agric.* **2020**, *100*, 5009–5016. [CrossRef]
4. Öztürk, S.; Mutlu, S. Chapter 8—Physicochemical Properties, Modifications, and Applications of Resistant Starches. In *Starches for Food Application*; Silva Clerici, M.T.P., Schmiele, M., Eds.; Academic Press: Cambridge, MA, USA, 2019; pp. 297–332. ISBN 978-0-12-809440-2.
5. Zhao, X.; Mei, T.; Cui, B. Lipids-Modified Starch: Advances in Structural Characteristic, Physicochemical Property, and Application. *Food Res. Int.* **2024**, *197*, 115146. [CrossRef]
6. Cervantes-Ramírez, J.E.; Cabrera-Ramirez, A.H.; Morales-Sánchez, E.; Rodriguez-García, M.E.; Reyes-Vega, M.d.l.L.; Ramírez-Jiménez, A.K.; Contreras-Jiménez, B.L.; Gaytán-Martínez, M. Amylose-Lipid Complex Formation from Extruded Maize Starch Mixed with Fatty Acids. *Carbohydr. Polym.* **2020**, *246*, 116555. [CrossRef]
7. Cabrera-Ramírez, A.H.; Castro-Campos, F.G.; Gaytán-Martínez, M.; Morales-Sánchez, E. Relationship between the Corneous and Floury Endosperm Content and the Popped Sorghum Quality. *J. Cereal Sci.* **2020**, *95*, 102999. [CrossRef]
8. Oliveira, L.C.; Barros, J.H.T.; Rosell, C.M.; Steel, C.J. Physical and Thermal Properties and X-Ray Diffraction of Corn Flour Systems as Affected by Whole Grain Wheat Flour and Extrusion Conditions. *Starch Stärke* **2017**, *69*, 1600299. [CrossRef]
9. Tufvesson, F.; Wahlgren, M.; Eliasson, A.-C. Formation of Amylose-Lipid Complexes and Effects of Temperature Treatment. Part 1. Monoglycerides. *Starch Stärke* **2003**, *55*, 61–71. [CrossRef]
10. Sun, S.; Hong, Y.; Gu, Z.; Cheng, L.; Ban, X.; Li, Z.; Li, C. Different Starch Varieties Influence the Complexing State and Digestibility of the Resulting Starch-Lipid Complexes. *Food Hydrocoll.* **2023**, *141*, 108679. [CrossRef]
11. Liu, Q.; Luo, H.; Liang, D.; Zheng, Y.; Shen, H.; Li, W. Effect of Electron Beam Irradiation Pretreatment and Different Fatty Acid Types on the Formation, Structural Characteristics and Functional Properties of Starch-Lipid Complexes. *Carbohydr. Polym.* **2024**, *337*, 122187. [CrossRef]
12. Chao, C.; Yu, J.; Wang, S.; Copeland, L.; Wang, S. Mechanisms Underlying the Formation of Complexes between Maize Starch and Lipids. *J. Agric. Food Chem.* **2018**, *66*, 272–278. [CrossRef]
13. Guraya, H.S.; Kadan, R.S.; Champagne, E.T. Effect of Rice Starch-Lipid Complexes on In Vitro Digestibility, Complexing Index, and Viscosity. *Cereal Chem.* **1997**, *74*, 561–565. [CrossRef]

14. Hu, H.; Jiang, H.; Sang, S.; McClements, D.J.; Jiang, L.; Wen, J.; Jin, Z.; Qiu, C. Research Advances in Origin, Applications, and Interactions of Resistant Starch: Utilization for Creation of Healthier Functional Food Products. *Trends Food Sci. Technol.* **2024**, *148*, 104519. [CrossRef]
15. Zhou, T.; Wang, Q.; Hu, Z.; Huang, J.; Zheng, X.; Tang, Y.; Xiang, D.; Peng, L.; Sun, Y.; Zou, L.; et al. The Composition, Internal Interactions, Auxiliary Preparation Methods, and Applications of Type Five Resistant Starch: A Review. *Food Hydrocoll.* **2024**, *160*, 110835. [CrossRef]
16. Gutiérrez, T.J.; Tovar, J. Update of the Concept of Type 5 Resistant Starch (RS5): Self-Assembled Starch V-Type Complexes. *Trends Food Sci. Technol.* **2021**, *109*, 711–724. [CrossRef]
17. Putseys, J.A.; Lamberts, L.; Delcour, J.A. Amylose-Inclusion Complexes: Formation, Identity and Physico-Chemical Properties. *J. Cereal Sci.* **2010**, *51*, 238–247. [CrossRef]
18. Chen, Z.; Hu, A.; Ihsan, A.; Zheng, J. The Formation, Structure, and Physicochemical Characteristics of Starch-Lipid Complexes and the Impact of Ultrasound on Their Properties: A Review. *Trends Food Sci. Technol.* **2024**, *148*, 104515. [CrossRef]
19. Li, H.; Gidley, M.J.; Dhital, S. High-Amylose Starches to Bridge the "Fiber Gap": Development, Structure, and Nutritional Functionality. *Compr. Rev. Food Sci. Food Saf.* **2019**, *18*, 362–379. [CrossRef]
20. Ai, Y.; Jane, J. Understanding Starch Structure and Functionality. In *Starch in Food: Structure, Function and Applications*; Woodhead Publishing: Sawston, UK, 2018; pp. 151–178. [CrossRef]
21. Castro-Campos, F.G.; Morales-Sánchez, E.; Cabrera-Ramírez, Á.H.; Martinez, M.M.; Rodríguez-García, M.E.; Gaytán-Martínez, M. High Amylose Starch Thermally Processed by Ohmic Heating: Electrical, Thermal, and Microstructural Characterization. *Innov. Food Sci. Emerg. Technol.* **2023**, *87*, 103417. [CrossRef]
22. Hu, X.; Huang, Z.; Zeng, Z.; Deng, C.; Luo, S.; Liu, C. Improving Resistance of Crystallized Starch by Narrowing Molecular Weight Distribution. *Food Hydrocoll.* **2020**, *103*, 105641. [CrossRef]
23. Roman, L.; Gomez, M.; Hamaker, B.R.; Martinez, M.M. Shear Scission through Extrusion Diminishes Inter-Molecular Interactions of Starch Molecules during Storage. *J. Food Eng.* **2018**, *238*, 134–140. [CrossRef]
24. Gerschenson, L.N.; Rojas, A.M.; Fissore, E.N. Chapter 3—Carbohydrates. In *Nutraceutical and Functional Food Components*, 2nd ed.; Galanakis, C.M., Ed.; Academic Press: Cambridge, MA, USA, 2022; pp. 49–126. ISBN 978-0-323-85052-0.
25. Luo, S.; Zeng, Z.; Mei, Y.; Huang, K.; Wu, J.; Liu, C.; Hu, X. Improving Ordered Arrangement of the Short-Chain Amylose-Lipid Complex by Narrowing Molecular Weight Distribution of Short-Chain Amylose. *Carbohydr. Polym.* **2020**, *240*, 116359. [CrossRef] [PubMed]
26. Sun, S.; Hong, Y.; Gu, Z.; Cheng, L.; Li, Z.; Li, C. An Investigation into the Structure and Digestibility of Starch-Oleic Acid Complexes Prepared under Various Complexing Temperatures. *Int. J. Biol. Macromol.* **2019**, *138*, 966–974. [CrossRef] [PubMed]
27. Compart, J.; Singh, A.; Fettke, J.; Apriyanto, A. Customizing Starch Properties: A Review of Starch Modifications and Their Applications. *Polymers* **2023**, *15*, 3491. [CrossRef]
28. Ai, Y.; Hasjim, J.; Jane, J.L. Effects of Lipids on Enzymatic Hydrolysis and Physical Properties of Starch. *Carbohydr. Polym.* **2013**, *92*, 120–127. [CrossRef]
29. Kawai, K.; Takato, S.; Sasaki, T.; Kajiwara, K. Complex Formation, Thermal Properties, and in-Vitro Digestibility of Gelatinized Potato Starch–Fatty Acid Mixtures. *Food Hydrocoll.* **2012**, *27*, 228–234. [CrossRef]
30. Marinopoulou, A.; Papastergiadis, E.; Raphaelides, S.N. An Investigation into the Structure, Morphology and Thermal Properties of Amylomaize Starch-Fatty Acid Complexes Prepared at Different Temperatures. *Food Res. Int.* **2016**, *90*, 111–120. [CrossRef]
31. Zarski, A.; Kapusniak, K.; Ptak, S.; Rudlicka, M.; Coseri, S.; Kapusniak, J. Functionalization Methods of Starch and Its Derivatives: From Old Limitations to New Possibilities. *Polymers* **2024**, *16*, 597. [CrossRef]
32. Lim, J.H.; Kim, H.R.; Choi, S.J.; Park, C.-S.; Moon, T.W. Complexation of Amylosucrase-Modified Waxy Corn Starch with Fatty Acids: Determination of Their Physicochemical Properties and Digestibilities. *J. Food Sci.* **2019**, *84*, 1362–1370. [CrossRef]
33. Vega-Rojas, L.J.; Londoño-Restrepo, S.M.; Rodriguez-García, M.E. Study of Morphological, Structural, Thermal, and Pasting Properties of Flour and Isolated Starch from Unripe Plantain (Musa Paradisiaca). *Int. J. Biol. Macromol.* **2021**, *183*, 1723–1731. [CrossRef]
34. Esquivel-Fajardo, E.A.; Martinez-Ascencio, E.U.; Oseguera-Toledo, M.E.; Londoño-Restrepo, S.M.; Rodriguez-García, M.E. Influence of Physicochemical Changes of the Avocado Starch throughout Its Pasting Profile: Combined Extraction. *Carbohydr. Polym.* **2022**, *281*, 119048. [CrossRef] [PubMed]
35. Rojas-Molina, I.; Nieves-Hernandez, M.G.; Gutierrez-Cortez, E.; Barrón-García, O.Y.; Gaytán-Martínez, M.; Rodriguez-Garcia, M.E. Physicochemical Changes in Starch during the Conversion of Corn to Tortilla in the Traditional Nixtamalization Process Associated with RS_2. *Food Chem.* **2024**, *439*, 138088. [CrossRef] [PubMed]
36. Rodriguez-Garcia, M.E.; Hernandez-Landaverde, M.A.; Delgado, J.M.; Ramirez-Gutierrez, C.F.; Ramirez-Cardona, M.; Millan-Malo, B.M.; Londoño-Restrepo, S.M. Crystalline Structures of the Main Components of Starch. *Curr. Opin. Food Sci.* **2021**, *37*, 107–111. [CrossRef]
37. Ding, Y.; Luo, F.; Lin, Q. Insights into the Relations between the Molecular Structures and Digestion Properties of Retrograded Starch after Ultrasonic Treatment. *Food Chem.* **2019**, *294*, 248–259. [CrossRef]
38. Li, Z.; Wei, C. Morphology, Structure, Properties and Applications of Starch Ghost: A Review. *Int. J. Biol. Macromol.* **2020**, *163*, 2084–2096. [CrossRef]

39. Nishiyama, Y.; Mazeau, K.; Morin, M.; Cardoso, M.B.; Chanzy, H.; Putaux, J.-L. Molecular and Crystal Structure of 7-Fold V-Amylose Complexed with 2-Propanol. *Macromolecules* **2010**, *43*, 8628–8636. [CrossRef]
40. Lee, M.H.; Kim, H.R.; Lim, W.S.; Kang, M.C.; Choi, H.D.; Hong, J.S. Formation of Debranched Wheat Starch-Fatty Acid Inclusion Complexes Using Saturated Fatty Acids with Different Chain Length. *LWT* **2021**, *141*, 110867. [CrossRef]
41. Sun, S.; Jin, Y.; Hong, Y.; Gu, Z.; Cheng, L.; Li, Z.; Li, C. Effects of Fatty Acids with Various Chain Lengths and Degrees of Unsaturation on the Structure, Physicochemical Properties and Digestibility of Maize Starch-Fatty Acid Complexes. *Food Hydrocoll.* **2021**, *110*, 106224. [CrossRef]
42. Liu, W.; Li, Y.; Goff, H.D.; Nsor-Atindana, J.; Zhong, F. Distribution of Octenylsuccinic Groups in Modified Waxy Maize Starch: An Analysis at Granular Level. *Food Hydrocoll.* **2018**, *84*, 210–218. [CrossRef]
43. Ramirez-Gutierrez, C.F.; Contreras-Jiménez, B.L.; Londoño-Restrepo, S.M. Characterization of Starches Isolated from Mexican Pulse Crops: Structural, Physicochemical, and Rheological Properties. *Int. J. Biol. Macromol.* **2024**, *268*, 131576. [CrossRef]
44. Morales-Sánchez, E.; Gaytán-Martínez, M.; Rodriguez-García, M.E.; Millan-Malo, B.M.; Cabrera-Ramírez, A.H. Behavior of Pasting Properties of Ohmic-Heated Corn Starch Versus Moisture and Temperature Applied. *Starch Stärke* **2024**, *76*, 2200245. [CrossRef]
45. Marinopoulou, A.; Papastergiadis, E.; Raphaelides, S.N.; Kontominas, M.G. Morphological Characteristics, Oxidative Stability and Enzymic Hydrolysis of Amylose-Fatty Acid Complexes. *Carbohydr. Polym.* **2016**, *141*, 106–115. [CrossRef] [PubMed]
46. Katz, J.R. On the Change in X-Ray Pattern When Starch Preparations Are Dried (as a Way of Characterizing These Substances with x-Rays). *Recl. Trav. Chim. Des Pays-Bas* **1937**, *56*, 766–772. [CrossRef]
47. Nieves-Hernández, M.G.; Correa-Piña, B.L.; Esquivel-Fajardo, E.A.; Barrón-García, O.Y.; Gaytán-Martínez, M.; Rodriguez-Garcia, M.E. Study of Morphological, Structural, Pasting, Thermal, and Vibrational Changes in Maize and Isolated Maize Starch during Germination. *J. Cereal Sci.* **2023**, *111*, 103685. [CrossRef]
48. Tan, L.; Kong, L. Starch-Guest Inclusion Complexes: Formation, Structure, and Enzymatic Digestion. *Crit. Rev. Food Sci. Nutr.* **2020**, *60*, 780–790. [CrossRef]
49. Chen, H.; Wang, Y.; Cao, P.; Liu, Y. Thermal Oxidation Rate of Oleic Acid Increased Dramatically at 140 °C Studied Using Electron Spin Resonance and GC–MS/MS. *J. Am. Oil Chem. Soc.* **2019**, *96*, 937–944. [CrossRef]
50. Chen, H.; Cao, P.; Li, B.; Sun, D.; Wang, Y.; Li, J.; Liu, Y. Effect of Water Content on Thermal Oxidation of Oleic Acid Investigated by Combination of EPR Spectroscopy and SPME-GC-MS/MS. *Food Chem.* **2017**, *221*, 1434–1441. [CrossRef]
51. Wang, S.; Chao, C.; Cai, J.; Niu, B.; Copeland, L.; Wang, S. Starch–Lipid and Starch–Lipid–Protein Complexes: A Comprehensive Review. *Compr. Rev. Food Sci. Food Saf.* **2020**, *19*, 1056–1079. [CrossRef]
52. Cabrera-Ramírez, A.H.; Cervantes-Ramírez, E.; Morales-Sánchez, E.; Rodriguez-García, M.E.; Reyes-Vega, M.d.l.L.; Gaytán-Martínez, M. Effect of Extrusion on the Crystalline Structure of Starch during RS5 Formation. *Polysaccharides* **2021**, *2*, 187–201. [CrossRef]
53. Assi, A.; Michael-Jubeli, R.; Baillet-Guffroy, A.; Tfayli, A. Characterization of Free Fatty Acids Photo-Oxidation under UV Radiations: A Stepwise Raman Study. *J. Raman Spectrosc.* **2023**, *54*, 24–36. [CrossRef]
54. Wang, J.; Yu, J.; Copeland, L.; Wang, S. Revisiting the Formation of Starch–Monoglyceride–Protein Complexes: Effects of Octenyl Succinic Anhydride Modification. *J. Agric. Food Chem.* **2023**, *71*, 19033–19044. [CrossRef] [PubMed]
55. Ibarra, J.; Melendres, J.; Almada, M.; Burboa, M.G.; Taboada, P.; Juárez, J.; Valdez, M.A. Synthesis and Characterization of Magnetite/PLGA/Chitosan Nanoparticles. *Mater. Res. Express* **2015**, *2*, 095010. [CrossRef]
56. Amoako, D.B.; Awika, J.M. Resistant Starch Formation through Intrahelical V-Complexes between Polymeric Proanthocyanidins and Amylose. *Food Chem.* **2019**, *285*, 326–333. [CrossRef] [PubMed]
57. Li, C.; Hu, Y.; Li, E. Effects of Amylose and Amylopectin Chain-Length Distribution on the Kinetics of Long-Term Rice Starch Retrogradation. *Food Hydrocoll.* **2021**, *111*, 106239. [CrossRef]
58. Wang, H.; Zhang, B.; Chen, L.; Li, X. Understanding the Structure and Digestibility of Heat-Moisture Treated Starch. *Int. J. Biol. Macromol.* **2016**, *88*, 1–8. [CrossRef]

foods

MDPI

Article

Screening of Nutritionally Important Components in Standard and Ancient Cereals

Vesna Dragičević [1,*], Milena Simić [1], Vesna Kandić Raftery [1], Jelena Vukadinović [1], Margarita Dodevska [2], Sanja Đurović [3] and Milan Brankov [1]

[1] Maize Research Institute "Zemun Polje", Slobodana Bajića 1, 11185 Zemun Polje, Serbia; smilena@mrizp.rs (M.S.); vkandic@mrizp.rs (V.K.R.); jmesarovic@mrizp.rs (J.V.); mbrankov@mrizp.rs (M.B.)
[2] Institute of Public Health of Serbia "Dr. Milan Jovanović Batut", Dr Subotića Starijeg 5, 11000 Belgrade, Serbia; margarita_dodevska@batut.org.rs
[3] Institute for Plant Protection and Environment, Teodora Drajzera 9, 11040 Belgrade, Serbia; stojakovicsm@yahoo.com
* Correspondence: vdragicevic@mrizp.rs; Tel.: +381-113756704

Abstract: Sustainable nutrition and food production involve dietary habits and farming systems which are eco-friendly, created to provide highly nutritious staple crops which could serve as a functional food at the same time. This research sought to provide a comprehensive analysis of whole-grain cereals, and some ancient grains toward important macro- (protein), micro-nutrients (mineral elements), and bioactive compounds, such as dietary fiber (arabinoxylan and β-glucan) and antioxidants (phytic acid, total glutathione, yellow pigment, and phenolic compounds) to provide functionality in a sustainable diet. Genotypes, such as durum wheat, triticale, spelt, emmer wheat, and barley, could be considered important and sustainable sources of protein (ranging 11.10–15.00%), as well as prebiotic fiber (β-glucan and arabinoxylan, ranging 0.11–4.59% and 0.51–6.47%, respectively), essential elements, and various antioxidants. Ancient grains can be considered as a source of highly available essential elements. Special attention should be given to the Cimmyt spelt 1, which is high in yellow pigment (5.01 $\mu g \cdot g^{-1}$) and has a capacity to reduce DPPH radicals (186.2 $\mu mol\ TE \cdot g^{-1}$), particularly Zn (70.25 $mg \cdot kg^{-1}$). The presence of phenolics, dihydro-*p*-coumaric acid, naringin, quercetin, epicatechin in grains of oats (Sopot), as well as catechin in barley grains (Apolon and Osvit) underline their unique chemical profile, making them a desirable genetic pool for breeding genotypes. This research provides a comprehensive assessment of different nutritional aspects of various cereals (some of which are commonly used, while the others are rarely used in diet), indicating their importance as nutraceuticals. It also provides a genetic background that could be translated the genotypes with even more profound effects on human health.

Keywords: cereals; mineral elements; antioxidants; phytic acid; phenolics; dietary fiber

1. Introduction

Sustainable nutrition and sustainable agriculture imply eco-friendly dietary habits and farming systems, created to provide highly nutritious staple crops that will fulfill requirements for essential macro- and micro-nutrients and could be used as functional food at the same time. Modern genotypes, designed to have high yields, largely failed to meet the nutritional demand of humans and animals [1], due to the poor soil quality and climate changes. Thus, sustainable solutions are necessary. Irrespective of the increasing market demand and research on functional products, which are not only able to meet basic requirements, but also have a whole range of health benefits [2–4], crop diversity and gene pool diversity are able to provide reliable solutions for producing nutrient–dense food [5,6].

According to the FAO Statistical Yearbook [7], cereals are the leading crops produced worldwide (with 3.1 billion tons produced in 2022), whereas the production level of wheat increased to38% in 2022 (compared to 808 million tons in 2000) and barley production reached 155 million tons. Whole-grain cereals provide many important nutrients which may reduce the risk of developing various chronic diseases and age-related conditions, like obesity, cardiovascular diseases, metabolic syndrome, type II diabetes, and cancer [8–10]. From this standpoint, much more attention is being drawn to ancient wheat genotypes, such as spelt, emmer wheat, and khorasan wheat, which are relatively high in protein compared to modern wheat genotypes [9–12]. Regardless of the lower yields (compared to the modern wheat genotypes), their sustainability is one of the most important qualities, as they are highly adaptable to various environments, particularly to low-input and organic agriculture, growing on marginal soils and in dry conditions. Accordingly, old genotypes and cereals with colored grains are also able to better exploit environmental conditions and accumulate greater amounts of phytonutrients in grains [13–15].

To target the nutraceutical properties of cereals, it is important to identify health promoting traits, on the one hand, and antinutrients, such as phytic acid, on the other hand. Phytic acid can bind mineral elements, protein, and starch, compromising their bio-availability, so the ratio between phytic acid and mineral elements is the simplest and best measure of potential bioavailability [16,17]. However, phytic acid also exerts dietary benefits through antioxidant activity, thereby preventing lipid peroxidation [18], reducing the risk of certain cancers, cardio-vascular diseases, renal stones, etc. Its concentration in grains depends mostly on genotype, as well as on its interaction with the environment and growing conditions [14,15,19]. Lowering the concentration of phytic acid and simultaneously increasing the concentration of essential elements, such as Ca, Mg, Fe, Zn, and Mn, in grains of staple crops are goals of the bio-fortification strategy, aimed to combat malnutrition globally [20]. Various approaches, such as transgenic, conventional breeding, and agronomical approaches, are used to achieve this goal, increasing the concentration of targeted minerals and vitamins in edible parts of plants. The main focus of bio-fortification is on staple crops like rice, wheat, maize, sorghum, lupin, common bean, potato, sweet potato, and tomato. The exploitation of various fertilizers and substances that promote the absorption and accumulation of essential elements in crops, including the selection and development of desirable genotypes that can accumulate high levels of mineral nutrients are important parts of the strategy.

Sustainable plant-based protein sources are of particular importance today. While legumes are considered a main plant protein source, cereals as staple crops are targeted to elevate protein concentration in their grains too. Therefore, genetic tools should be applied to provide a more sustainable protein supply from cereal grains [21]. Moreover, whole-grain cereals are an important source of soluble and insoluble functional dietary fiber (e.g., cellulose, arabinoxylan, β-glucan, xyloglucan, and fructan), with sound health benefits, including cardio-protective association [22–24]. Despite this, the general consumption of dietary fiber from grains is considerably lower than the recommended levels worldwide. The most prominent dietary fibers with health benefits, as well as those that play a prebiotic role, are arabinoxylan and β-glucan [25,26]. For this reason, different breeding programs have been developed that have included research into genetic pools and translated desirable traits into improved cereal genotypes [2,27].

Bearing in mind the health-promoting bioactive compounds, whole-grain cereals are a rich source of various antioxidants in the human diet [8] and can inhibit Maillard reaction products [28]. Among cereals, ancient grains and modern genotypes with colored grains are the richest sources of phenolic compounds [11,12,29]. Their concentration and composition depend mainly on genotype and agroecological conditions. While ferulic and *p*-coumaric acids are the most common phenolics present in cereal grains [10–12,29], many others compounds, such as flavonoids, isoflavones, flavanones, and flavanols (catechin and epicatechin), could also be found in the cereal grains. Phenolic compounds from cereal grains also have a prebiotic role, supporting the development of beneficial microorganisms in the human gut [30–32].

Thus far, according to the literature review, most of the research on the nutritional quality of cereal grains is based on the examination of certain classes of compounds such as macronutrients, mineral elements, antioxidants, vitamins, etc. To our knowledge, this study is the first report on the concentration of thirty-three simultaneously evaluated parameters in whole-grain cereals, including some ancient grains, such as spelt and emmer wheat. Therefore, this study sought to evaluate the content of important macro- (protein) and micro-nutrients (mineral elements), as well as bioactive compounds, such as dietary fiber (arabinoxylan and β-glucan) and antioxidants (phytic acid, total glutathione, yellow pigment, and the specific profile of phenolic compounds) in various cereals. Furthermore, to determine the possible unique nutritive fingerprint of tested whole-grain cereals, principal component analysis was applied. This article provides extensive research on modern genotypes and ancient grains about their nutraceuticals properties, also giving essential information for breeders on the development of genotypes with enhanced grain composition.

2. Materials and Methods

2.1. Grain Material and Sample Preparation

Twenty genotypes of cereals (durum wheat, bread wheat, barley, emmer wheat, spelt wheat, triticale, rye, and oats) from the Gene Bank of the Maize Research Institute "Zemun Polje" (MRIZP), including some standard and ancient cereals, were chosen to evaluate their chemical composition from anutritional viewpoint. Grains from all genotypes which were harvested in 2022 are characterized in Supplementary Table S1.

The experiment was conducted at the experimental field of Maize Research Institute "Zemun Polje", Belgrade (44°52′ N, 20°19′ E, 81 m a.s.l.), Serbia, on a slightly calcareous Chernozem. Texture is silt clay loam, containing the following: 51.0% sand, 31.0% silt, 18.0% clay, 3.5% organic matter, 7.0 pH KCl, and 7.5 pH H_2O. Mineral composition revealed 68.25 mg·kg^{-1} of available N, 65.00 mg·kg^{-1} of available P, 105.85 mg·kg^{-1} of K, 264.00 mg·kg^{-1} of Mg, 2854.25 mg·kg^{-1} of Ca, 10.28 mg·kg^{-1} of Fe, 111.50 mg·kg^{-1} of Mn, 6.23 mg·kg^{-1} of Cu, and 3.46 mg·kg^{-1} of Zn in the 0-to-30 cm soil layer. The preceding crop was maize (*Zea mays* L.). Accordingly to the soil analysis, fertilization included incorporation of the 250 kg urea ha^{-1} in the spring of 2022.

Grain samples (approximately 0.5 kg) were dried in a ventilation oven at 60 °C (EU instruments, EUGE425, Novo Mesto, Slovenia, EU) to obtain uniform moisture content (12%) and then milled on Perten 120—Hägersten, Sweden (particle size < 500 μm).

2.2. Chemicals and Reagents

Gallic acid (GA), catechin (CAT), dihydrocaffeic acid (diCA), epicatechin (EPI), dihydro-p-coumaric acid (di-p-CoumA), naringin (NA), daidzein (DA), cinnamic acid (CIN), naringenin (NG), chlorogenic acid (ChlA), caffeic acid (CA), p-coumaric acid (p-CoumA), ferulic acid (FA), isoferulic acid (IFA), quercetin (QUE), as well as acetone, formic acid, methanol, all HPLC-grade, were obtained from Sigma-Aldrich (Munich, Germany). From the same supplier, Folin–Ciocalteu reagent, butanol, sodium acetate, sodium carbonate, and trichloroacetic acid (TCA) were also purchased. TRIS (2-amino-2-hydroxymethyl-propane-1,3-diol), DTNB (2,2′-dinitro-5,5′-dithiobenzoic acid), iron (III) chloride hexahydrate, DPPH• (2,2-diphenyl-1-picrylhydrazyl), 5′-sulfosalicylic acid, TROLOX (6-hydroxy-2,5,7,8-tetramethylchroman-2-carboxylic acid), and TPTZ (2,4,6-Tris(2-pyridyl)-s-triazine) were purchased from Merck (Darmstadt, Germany). Ammoniumheptamolybdate tetrahydrate, ammonium metavanadate hydrochloric acid, and nitric acid were purified. Ultrapure water, with a conductivity of 0.055 μS/cm (TKA, Niederelbert, Germany), was used to prepare the appropriate standard solutions and samples, while Target2® syringe filters (PTF membrane, 0.45 μm, diameterof17 mm) were used to filter the extracts. Enzymatic assay kits K-XYLOSE and K-BGLU were purchased from Megazyme (Bray, Ireland).

2.3. Determination of Protein Content

The protein content from whole grains was determined using an infrared analyzer (Infraneo, Chopin Technologies, Villeneuve-la-Garenne Cedex, France). For the analysis, a sample of approximately 0.5 kg was passed through the sliding cell and the near-infrared spectra range was 700 nm to 1100 nm using a high-resolution monochromator (0.1 nm). The protein content was reported as the mean value of two replications and expressed in g per 100 g (%) of the sample.

2.4. Determination of the Concentration of Phytic Phosphorus, Inorganic Phosphorus, and Total Glutathione

Phytic acid (Pphy), inorganic phosphorus (Pi), and total glutathione (GSH) from the analyzed samples (approximately 0.5 g) were extracted with the 5% TCA solution (10 mL) using avertical shaker (30 min, at room temperature). After centrifugation (12,000 rpm for 15 min (Dynamica—Model Velocity 18R Versatile Centrifuge, Rotor TA15-24-2), at 4 °C, the obtained clear supernatant was used for further analysis.

2.4.1. Determination of Phytic Acid and Inorganic Phosphorus

Pphy and Pi were determined using the method of Dragičević et al. [33]. For phytic acid determination, the distilled water (1.25 mL) and Wade reagent containing the 0.3 g of $FeCl_3 \times 6H_2O$ and 3 g of 5′-sulfosalicylic acid in one liter (0.5 mL) were added to previously obtained supernatant (0.25 mL). After developing the pink color, absorbance was measured using the spectrophotometer (Biochrom Libra S22 UV/Vis Spectrophotometer—Biochrom, UK) at λ = 500 nm against a blank probe (containing 5% TCA). For Pi determination, an aliquot of previously obtained supernatant (1 mL) was mixed with distilled water (1 mL) and Pi reagent (0.5 mL). The Pi reagent consisted of 6% ammoniumheptamolybdate (500 mL) and 0.25% ammonium metavanadate mixed with 4.74 M nitric acid (500 mL). Absorbance was measured using a spectrophotometer at λ = 400 nm against a blank probe (containing 5% TCA). The Pphy concentration and the Pi concentration were reported as the mean values of three replications and expressed in mg per g of the DW sample.

2.4.2. Determination of Total Glutathione

To determine total glutathione (GSH) [34], an aliquot of previously obtained supernatant (0.5 mL) was mixed with distilled water (0.5 mL) and reagent consisting of 0.1 M TRIS (pH = 7) and 1.5 mM of DTNB (1 mL). After centrifugation, the absorbance of the extract was measured at 412 nm against a blank probe (containing 5% TCA). The GSH concentration was calculated based on the measured absorbance values of the samples and the molar absorption coefficient of 1.36×10^4 dm^3 mol^{-1} cm^{-1} at 412 nm. The GSH concentration was reported as the mean value of three replications and expressed in nmol per g of the DW sample.

2.5. Determination of Total Phosphorus

Total phosphorus (P) determination was achieved using the method ISO 13730 [35] with some modifications. The analyzed samples were firstly subjected to dry incineration. Then, after homogenization with nitric acid (4.81 M), followed by the heating in a boiling water bath (30 min), the extracts were cooled and filtered. After adding the Pi reagent (used for determining inorganic phosphorus) samples were left to stand for 15 min, and then the absorbance was measured at 430 nm against a blank probe (containing nitric acid and Pi reagent) using a spectrophotometer. The P concentration was reported as the mean value of three replications and expressed in $g \cdot kg^{-1}$ of the DW samples.

2.6. Determination of Yellow Pigment

The concentration of yellow pigment (YP) was determined using the method of Vančetović et al. [36]. After extracting approximately 2 g of flour with 10 mL of saturated 1-butanol using a vertical shaker (30 min, at room temperature) and conducting centrifugation at 10,000 rpm for 5 min, the absorbance was measured at λ = 436 nm. The β-carotene (βCE) concentration was reported as the mean value of three replications and expressed in μg per g of the DW samples.

2.7. Identification and Determination of Total Phenolic Compounds and Antioxidant Activity

The total phenolic compounds (TPCs) were determined and the antioxidant activity was obtained according to the procedure described by Singleton et al. [37]. Approximately 2 g of flour was extracted with 10 mL of 70% methanol (MeOH) using an ultrasound bath (A-Sonic, frequency of 400 Hz) at 2 × 30 min and after centrifugation at 10,000 rpm (5 min), a clear supernatant was used for the determination of both total phenolic concentration (TPC) and antioxidant activity. An aliquot of 0.2 mL was mixed with 0.5 mL of distilled water. Then, 0.25 mL of Folin–Ciocalteu reagent (2 M relative to the acid) and 1.25 mL of Na_2CO_3 (20%) were added and the cuvettes were vigorously shaken and left in the dark for 40 min. After that, the absorbance was measured at 722 nm against a blank probe (containing MeOH). The concentration of the TPCs was reported as the mean value of four replications and expressed in mg of gallic acid equivalents (GAEs) per g of the DW sample.

The total phenolic compounds (TPCs) were determined and the antioxidant activity was obtained according to the procedure described by Singleton et al. [37]. Approximately 2 g of flour was extracted with 10 mL of 70% methanol (MeOH) using an ultrasound bath (A-Sonic, frequency of 400 Hz) at 2 × 30 min and after centrifugation at 10,000 rpm (5 min), a clear supernatant was used for the determination of both total phenolic concentration (TPC) and antioxidant activity. An aliquot of 0.2 mL was mixed with 0.5 mL of distilled water. Then, 0.25 mL of Folin–Ciocalteu reagent (2 M relative to the acid) and 1.25 mL of Na_2CO_3 (20%) were added and the cuvettes were vigorously shaken and left in the dark for 40 min. After that, the absorbance was measured at 722 nm against a blank probe (containing MeOH). The concentration of the TPCs was reported as the mean value of four replications and expressed in mg of gallic acid equivalents (GAEs) per g of the DW sample.

DPPH radical scavenging activity was performed using the method described by Brand-Williams et al. and Wong et al. [38,39]. The obtained clear supernatant was diluted 1:100 in MeOH prior to the analysis and 200 µL of aliquots was mixed with 3.8 mL of DPPH• reagent (0.1 mM). After incubation (30 min in dark, 25 °C), the absorbance was measured at 517 nm against a blank probe (containing MeOH). The DPPH radical scavenging activity was expressed in µmol of TROLOX equivalents (TEs) per g of the DW sample and was reported as the mean value of four replications.

Antioxidant activity was also determined using the ferric-reducing antioxidant power assay (FRAP) method [39,40]. The obtained clear supernatant (0.15 mL) was mixed with 4.5 mL of FRAP reagent (2.5 mL of 10 mM TPTZ was dissolved in 40 mM HCl and then mixed with a 2.5 mL solution of 20 mM $FeCl_3 \times 6\,H_2O$ in acetate buffer pH = 3.6). After the incubation (30 min, 37 °C), absorbance was measured at 593 nm against a blank probe (0.15 mL MeOH + 4.5 mL FRAP reagent). The antioxidant activity was expressed in µmol Fe^{2+} per g of DW sample and it was reported as the mean value of four replications.

2.8. Determination of Arabinoxylan and β-Glucan

Arabinoxylan was determined by a spectrophotometric method using the enzymatic assay kit K-xylose (Megazyme, Bray, Ireland). The assay was performed according to the instruction manual of the kit producer. Generally, the conversion of α-D-xylose to β-D-xylose is catalyzed by xylose mutarotase. The obtained β-D-xylose is then oxidized by NAD+ to D-xylonic acid in the presence of β-xylose dehydrogenase (pH 7.5). The obtained amount of NADH formed in this reaction is proportional to the D-xylose concentration and it is measured by the increase in absorbance at 340 nm. The concentration of arabinoxylan was calculated according to Formula (1):

$$\text{Arabinoxylan} = \text{Concentration of D} - \text{xy lose } (\text{g}/100\text{ g}) \times \frac{100}{62}\ (\text{g}/100\text{ g}) \quad (1)$$

β-glucan was quantified by a spectrophotometric method [41] using the enzymatic assay kit K-BGLU (Megazyme, Bray, Ireland). The samples were suspended and hydrated in a buffer solution (pH 6.5) and then incubated with purified lichenase enzyme and filtered. Subsequently, an aliquot of the filtrate was hydrolyzed to completion with purified β-glucosidase. The D-glucose produced was assayed using a glucose oxidase/peroxidase reagent and the absorbance was measured at 510 nm.

2.9. HPLC Analysis

The identification and quantification of phenolic compounds (gallic acid (GA), catechin (CAT), dihydrocaffeic acid—(DiCA), epicatechin (EPI), dihydro-p-coumaric acid (di-p-CoumA), naringin (NA), daidzein (DA), cinnamic acid (CIN), naringenin (NG), chlorogenic acid (ChlA), caffeic acid (CA), p-coumaric acid (p-CoumA), ferulic acid (FA), isoferulic acid (IFA), and quercetin (QUE)) were determined in flour using high-performance liquid chromatography (HPLC system Shimadzu Nexera XR) equipped with column Zorbax SB C18 (4.6 × 250 mm, pore diameter 5 µm). The mobile phase A was 0.1% HCOOH in H_2O, while the mobile phase B was 0.1% HCOOH in MeOH. The used linear gradient was adopted [42], starting from 95% A to 70% A after 25 min, 60% A after 35 min, 52% A after 40 min, 30% A after 50 min, 0% A after 55 min, 95% A after 65 min, and 95% A after 75 min. The flow rate was 1 mL min^{-1}, with 10 µL of the injected sample being used. The column temperature was set at 25 °C and the wavelengths were 254 nm, 280 nm, and 325 nm. The LabSolutions Shimadzu program was used for instrument control, as well as for data acquisition and analysis. Calibration curves were used for the quantitative calculation of each phenolic compound. The concentration was reported as the mean value of three replications and expressed in µg per g of the DW sample.

2.10. Mineral Element Analysis

The concentration of potassium (K), sodium (Na), magnesium (Mg), calcium (Ca), copper (Cu), iron (Fe), manganese (Mn), selenium (Se), zinc (Zn), and nickel (Ni) in flour of samples was determined by plasma inductively spectrometry coupled with optical emission (ICP-OES) using a ICP 7400 Duo Thermo Scientific, UV/VIS 214–766 nm. Samples were prepared after dry incineration, except for Se where the acid hydrolysis process was applied, according to the procedure described by the Association of Official Analytical Chemists [43]. Quantifying mineral composition was performed based on Multi–Element Plasma Standard Solution 4 (Thermo Scientific, Berlin, Germany) and the results were expressed in$g \cdot kg^{-1}$ of dry grain samples for K, as well as $mg \cdot kg^{-1}$ for Na, Mg, Ca, Cu, Fe, Mn, Se, Zn, and Ni. The Se concentration was under the limit of the detection, while trace amounts of Ni were only determined in three samples of oats, so they were not included in the results.

2.11. Statistical Analysis

To assess the data on tested biochemical traits, one-factorial analysis of variance (ANOVA) was applied using the SPSS for Windows Version 15.0 (SPSS, 2006). In order to determine the differences between the samples, the Tukey HSD test was performed at a 0.95 confidence level ($p \leq 0.05$). Moreover, MATLAB (R2011a) with PLS Toolbox software package (v.6.2.1) was used for principal component analysis (PCA) and agglomerative hierarchical cluster analysis (HCA). For these analyses, the obtained data were mean-centered and auto-scaled, and the singular value decomposition (SVD) algorithm was employed (at a 95% confidence level) for Hotelling T2 limits.

3. Results

3.1. The Concentration of Protein, Antioxidants, and Dietary Fiber in the Standard and Ancient Grains

Examined genotypes expressed the significant variability in the concentration of the following important nutrients: protein, Pi, Pphy, YP, TPCs, GSH, β-glucan, and arabinoxylan (Table 1). Greater differences in protein concentration were between durum wheat genotypes, where Dur1 had the greatest concentration (15%). The lowest concentration was in spelt wheat (9.85%). Bar2 had the highest concentration of Pi, while Bar1 was the highest in GSH and β-glucan, and Bar2 had the lowest Pi and highest β-glucan. Among all tested samples, the greatest concentration of Pphy and YP was noticed in spelt grain, while the lowest Pphy value was found in Oats3 and the lowest concentration of YP was in Trit1. Oat samples Oats2 and Oats1 showed a high TPC and arabinoxylan concentration, respectively. Contrary, the lowest TPC concentration was found in BW1 and the lowest arabinoxylan concentration was noticed in the sample of spelt.

3.2. Essential Elements and Their Ratios with Phytic Acid in the Standard and Ancient Grains

The significant differences among tested genotypes in the accumulation of essential elements (P, K, Ca, Mg, Na, Fe, Mn, Zn, and Cu) (Table 2) could also be ascertained. Among the tested essential elements, the greatest concentration of Cu was recorded in grains of Em2, while the lowest value was found in the grains of Bar2. The spelt grain had the highest concentration of P, K, Mg, Mn, and Zn. The lowest concentration of P was found in BW1, while the lowest concentration of Mn and Zn was recorded in the Bar3 grain. The lowest concentration of K, Mg, and Ca was noticed in the grain of Em2, whereas the highest Ca concentration was found in the Oats1 grain. The highest concentration of Na and the lowest Fe concentration were noticed intheTrit2 grain. Moreover, the lowest concentration of Na was found in Oats2, and the highest Fe concentration was observed in the Rye1 grain.

According to the cluster analysis encompassing all examined elements, the following clusters were present (Figure 1). The first consisted of only Oats2 and the second one was Em2. The third comprised all other genotypes, consisting of two sub-clusters (the first one consisted of Oats1 and Oats3, while the second one consisted of all the other genotypes).

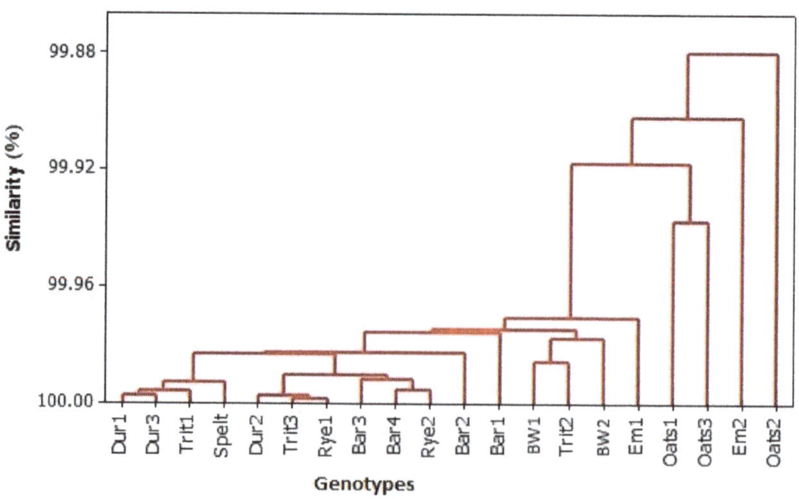

Figure 1. Dendrogram presenting the concentration of mineral elements in standard and ancient grains.

Among all tested genotypes, the lowest ratio between phytic acid and essential elements (Phy/Ca, Phy/Mg, Phy/Fe, and Phy/Mn) was in Oats3, while the lowest Phy/Zn was in spelt (Supplementary Table S2). The greatest Phy/Ca and Phy/Mg values were in Em2 grain; the highest Phy/Mn and Phy/Zn values were in Bar3, whereas the greatest Phy/Fe value was in the Trit2 grain. According to cluster analysis (Figure 2), two clusters were present, with spelt independent of other genotypes that formed three sub-clusters. Regarding the first sub-cluster, samples of Dur1, Dur2, and Em2 formed the first subgroup; the samples of Dur3, Bar3, Rye2, and Bar4 created the second subgroup; and the Em1, Oats2, and Oats3 were grouped in the third subgroup. The second sub-cluster consisted of BW1, BW2, Trit1, Trit2, Trit3, and Oats1, while the samples of Bar1 and Rye1 established the third sub-cluster.

Table 1. The concentration of protein, inorganic phosphorus (Pi), phytic phosphorus (Pphy), yellow pigment (YP), total phenolics (TPCs), total glutathione (GSH) and dietary fiber (β-glucan and arabinoxylan) in the examined standard and ancient grains.

	Protein	Pi	Pphy	YP	TPCs	GSH	β-glucan	Arabinoxylan
	%	mg·g^{-1}		μg·g^{-1}		nmol·g^{-1}	g 100·g^{-1}	
Dur1	15.03 a ± 0.32 *	0.37 k ± 0.019	4.80 bc ± 0.02	2.75 hi ± 0.06	1402. 7 h ± 3.51	77.86 k ± 2.80	0.11 h ± 0.01	5.55 i ± 0.21
Dur2	11.93 ef ± 0.24	0.50 i ± 0.005	4.54 defg ± 0.02	4.70 b ± 0.15	1240.9 i ± 9.23	85.05 jk ± 5.99	0.13 h ± 0.02	5.39 j ± 0.18
Dur3	12.60 bcd ± 0.48	0.48 i ± 0.003	4.50 fg ± 0.06	4.39 c ± 0.04	926.0 k ± 14.69	96.86 jk ± 5.36	0.14 h ± 0.01	4.81 j ± 0.31
BW1	10.50 ij ± 0.22	0.39 jk ± 0.013	4.48 fg ± 0.00	3.40 f ± 0.09	735.4 l ± 7.00	130.57 i ± 5.99	0.17 h ± 0.02	6.19 h ± 0.26
BW2	10.53 hij ± 0.26	0.37 k ± 0.005	4.73 cde ± 0.06	3.05 g ± 0.10	1550.1 f ± 1.48	107.41 ij ± 9.98	0.19 h ± 0.04	6.74 f ± 0.35
Trit1	12.53 cde ± 0.26	0.41 j ± 0.012	5.02 ab ± 0.06	2.54 j ± 0.06	789.0 l ± 8.90	226.82 f ± 2.38	0.18 h ± 0.01	6.50 fg ± 0.30
Trit2	12.03 de ± 0.36	0.55 h ± 0.003	4.67 cdefg ± 0.12	2.66 ij ± 0.06	1115.9 j ± 8.08	179.30 g ± 2.78	0.23 h ± 0.03	6.47 g ± 0.25
Trit3	13.00 bc ± 0.83	0.49 i ± 0.005	4.68 cdef ± 0.09	2.90 gh ± 0.02	925.8 k ± 5.45	188.87 g ± 5.99	0.13 h ± 0.01	6.32 gh ± 0.21
Bar1	11.13 g ± 0.54	0.87 b ± 0.010	4.46 g ± 0.00	3.77 e ± 0.11	1115.1 j ± 11.18	607.34 a ± 21.16	4.15 b ± 0.28	4.25 l ± 0.34
Bar2	14.50 a ± 0.24	0.91 a ± 0.015	4.50 fg ± 0.01	3.67 e ± 0.04	1784.5 d ± 6.19	399.74 d ± 4.36	4.59 a ± 0.38	3.86 m ± 0.37
Bar3	11.23 g ± 0.27	0.72 e ± 0.019	4.70 cdef ± 0.17	3.61 e ± 0.13	924.1 k ± 8.88	364.56 e ± 7.59	3.78 c ± 0.40	4.54 k ± 0.30
Bar4	13.17 b ± 0.53	0.60 g ± 0.010	4.19 h ± 0.05	4.16 d ± 0.02	2175.2 b ± 4.18	498.73 b ± 2.80	3.44 cd ± 0.18	6.26 gh ± 0.33
Em1	10.12 jk ± 0.18	0.54 h ± 0.011	4.53 efg ± 0.07	2.63 ij ± 0.15	1486.6 g ± 1.61	155.36 h ± 2.76	0.13 h ± 0.03	3.84 m ± 0.32
Em2	10.10 jk ± 0.30	0.69 ef ± 0.011	4.76 cd ± 0.04	3.65 e ± 0.02	1631.4 e ± 11.87	170.73 gh ± 3.37	0.17 h ± 0.04	4.81 j ± 0.30
Spelt	9.85 k ± 0.23	0.82 c ± 0.013	5.22 a ± 0.02	5.01 a ± 0.06	1735.5 d ± 4.66	217.64 f ± 4.37	0.39 gh ± 0.08	0.51 n ± 0.17
Rye1	11.03 ghi ± 0.61	0.72 e ± 0.005	4.13 h ± 0.01	4.19 d ± 0.15	1261.2 i ± 7.46	397.13 d ± 0.57	0.62 fg ± 0.04	7.25 e ± 0.33
Rye2	11.07 ghi ± 0.51	0.56 h ± 0.007	4.04 h ± 0.10	4.04 d ± 0.02	1082.8 j ± 11.87	370.17 e ± 8.37	0.85 f ± 0.13	7.62 d ± 0.28
Oats1	11.40 fg ± 0.32	0.63 g ± 0.008	4.06 h ± 0.09	4.73 b ± 0.10	2359.1 a ± 3.98	473.78 c ± 9.37	3.05 e ± 0.17	12.69 a ± 0.33
Oats2	11.10 gh ± 0.29	0.76 d ± 0.000	4.00 h ± 0.11	3.26 f ± 0.15	2404.6 a ± 10.25	481.18 bc ± 8.37	3.70 c ± 0.39	12.04 b ± 0.25
Oats3	11.13 g ± 0.84	0.67 f ± 0.022	3.49 i ± 0.07	2.81 hi ± 0.02	1981.7 c ± 1.18	461.39 c ± 6.94	3.31 de ± 0.48	10.65 c ± 0.73
Tukey HSD ($p < 0.05$)	0.11	0.006	0.041	0.036	9.934	4.246	0.066	0.048

* Average ± standard error; the values signed with the same letter are not significantly different at the 0.05 level of significance.

Table 2. Essential elements in the grain of the examined standard and ancient grains.

	P	K	Ca	Mg	Na	Fe	Mn	Zn	Cu
	g·kg^{-1}					mg·kg^{-1}			
Dur1	6.12 de ± 0.05 *	2.99 h ± 0.02	397.0 k ± 4.94	1274.0 b ± 8.82	73.50 c ± 3.45	51.65 d ± 1.76	30.40 f ± 0.98	38.20 b ± 1.63	7.10 b ± 0.24
Dur2	6.25 cd ± 0.04	3.22 e ± 0.03	486.5 fg ± 7.86	1183.5 g ± 9.37	56.95 gh ± 3.74	44.85 ef ± 1.27	28.80 g ± 1.63	32.30 efg ± 1.80	6.20 de ± 0.33
Dur3	5.41 gh ± 0.06	2.68 kl ± 0.03	377.0 l ± 8.08	1097.5 k ± 9.74	54.50 hi ± 1.86	40.60 g ± 1.71	22.65 i ± 1.10	33.30 de ± 1.39	6.60 c ± 0.33
BW1	4.94 k ± 0.00	3.02 fg ± 0.02	491.5 ef ± 8.31	1090.5 k ± 7.90	62.65 fg ± 0.50	38.65 g ± 1.59	32.60 cde ± 1.47	32.40 efg ± 1.31	5.30 f ± 0.33
BW2	5.25 hi ± 0.05	3.03 f ± 0.02	478.0 g ± 7.63	1256.5 c ± 9.37	66.90 def ± 1.54	39.90 g ± 1.47	34.00 bc ± 1.39	30.05 h ± 0.94	6.45 cd ± 0.29
Trit1	5.73 f ± 0.04	2.73 ij ± 0.02	336.5 m ± 7.41	1202.5 f ± 9.74	81.35 b ± 2.57	36.50 h ± 1.55	31.85 e ± 1.59	32.65 ef ± 1.43	7.05 b ± 0.37
Trit2	5.69 f ± 0.04	3.39 c ± 0.03	442.0 i ± 4.94	1238.5 d ± 12.31	92.95 a ± 1.88	23.50 k ± 0.82	26.30 h ± 1.14	23.85 k ± 1.02	3.60 ij ± 0.33
Trit3	5.76 f ± 0.05	2.99 gh ± 0.02	462.5 h ± 10.10	1148.0 i ± 9.55	62.00 fg ± 1.59	28.40 j ± 1.14	32.10 e ± 1.14	34.95 c ± 1.67	4.85 g ± 0.45
Bar1	5.71 f ± 0.02	2.52 m ± 0.03	401.5 k ± 5.61	1171.5 gh ± 8.27	88.6 a ± 1.72	55.70 c ± 1.14	13.40 l ± 0.57	19.30 l ± 1.14	5.55 f ± 0.45
Bar2	5.80 f ± 0.05	2.67 kl ± 0.02	497.5 e ± 6.96	1092.0 k ± 7.72	81.85 b ± 1.19	34.85 hi ± 1.67	13.60 l ± 0.65	31.15 gh ± 1.43	3.80 i ± 0.33
Bar3	5.74 f ± 0.06	3.23 e ± 0.02	455.0 h ± 9.43	1134.0 j ± 8.45	65.00 ef ± 2.12	24.40 k ± 0.49	10.70 m ± 0.73	18.70 l ± 0.98	3.35 j ± 0.29
Bar4	6.07 e ± 0.01	3.33 d ± 0.03	547.0 d ± 9.88	1174.5 g ± 9.37	57.00 gh ± 1.59	45.05 ef ± 0.94	15.75 k ± 0.86	27.25 i ± 1.18	7.35 ab ± 0.37
Em1	6.56 b ± 0.03	2.70 jk ± 0.02	368.5 l ± 7.86	1220.0 e ± 8.08	61.95 fg ± 0.82	59.65 b ± 2.16	34.20 b ± 1.63	35.00 c ± 1.55	7.55 a ± 0.45
Em2	6.41 bc ± 0.07	2.24 n ± 0.02	258.5 n ± 8.31	1039.0 l ± 7.35	68.20 cdef ± 2.65	49.85 f ± 1.92	32.50 de ± 1.39	38.00 b ± 1.22	7.70 a ± 0.41
Spelt	7.94 a ± 0.08	3.76 a ± 0.03	538.0 d ± 7.63	1568.5 a ± 11.57	63.10 fg ± 1.91	57.50 c ± 2.37	39.20 a ± 1.88	70.25 a ± 2.33	7.15 b ± 0.53
Rye1	5.10 ijk ± 0.01	2.67 l ± 0.02	414.0 j ± 6.29	1040.5 l ± 7.53	72.00 cd ± 1.59	62.70 a ± 1.47	23.60 i ± 1.22	31.35 fgh ± 0.94	6.20 de ± 0.24
Rye2	5.08 ijk ± 0.05	2.76 i ± 0.03	464.0 h ± 6.29	1040.5 l ± 7.90	73.85 c ± 1.41	45.55 e ± 1.02	21.05 j ± 1.18	32.55 ef ± 1.59	6.15 de ± 0.45
Oats1	5.47 g ± 0.08	3.02 fg ± 0.02	816.5 a ± 7.41	1161.5 h ± 11.94	50.00 i ± 2.12	34.00 i ± 1.14	33.60 bcd ± 0.73	34.40 cd ± 0.98	4.55 gh ± 0.37
Oats2	5.15 ij ± 0.06	3.48 b ± 0.01	761.5 b ± 7.86	1140.0 ij ± 7.35	42.00 j ± 1.59	43.00 f ± 1.63	26.85 h ± 1.10	25.45 j ± 1.02	4.35 h ± 0.12
Oats3	5.01 jk ± 0.04	2.51 m ± 0.02	729.0 c ± 8.98	1097.0 k ± 9.55	71.00 cde ± 1.06	61.00 ab ± 2.86	33.90 bcd ± 0.82	32.10 efg ± 1.63	6.00 e ± 0.41
Tukey HSD ($p < 0.05$)	0.032	0.006	1.87	2.32	1.17	0.39	0.26	0.25	0.07

* Average ± standard error; the values signed with the same letter are not significantly different at the 0.05 level of significance.

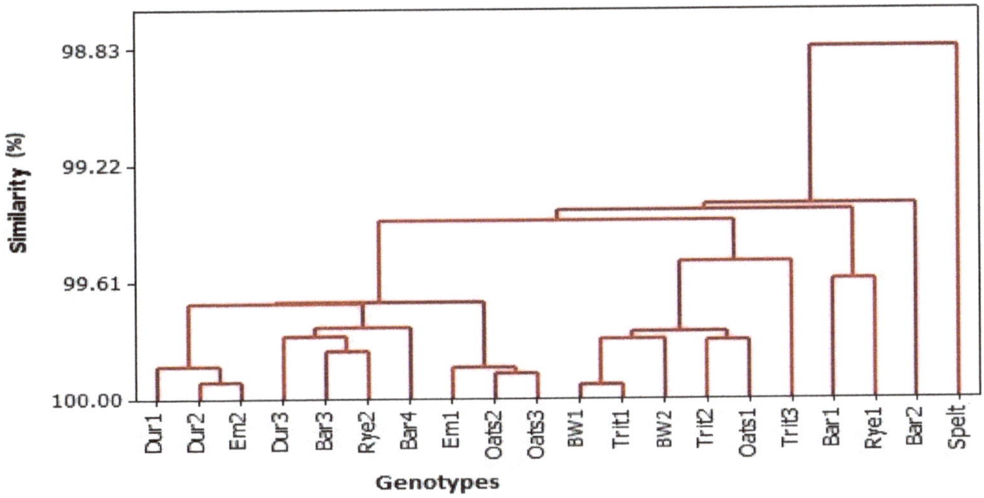

Figure 2. Dendrogram presenting ratio of Phy and essential elements in standardgrains and ancient grains.

3.3. Phenolic Acids and Antioxidant Status in the Standard and Ancient Grains

There was great variability in the concentration of certain phenolic compounds among the tested genotypes, while some of them had concentrations under the limit of detection. Oat grains were observed to have the highest concentration in the majority of examined phenolic compounds (Table 3). According to the obtained results, Oats1 had the highest concentration of EPI, NA, DA, and QE; Oats2 had the highest concentration of PA, CA, and FA; andOats3 had the greatest concentration of p-CoumA. The highest concentration of NG was found in Em2. The highest concentration of diCA, IFA, and CIN was observed in samples Trit2, Trit3, and BW1, respectively. Barley genotypes were also relatively high in phenolic compounds, where Bar2 had the greatest concentration of GA and CAT and Bar4 had the greatest concentration of ChlA.

When the antioxidant activity was considered, the greatest value of capacity to reduce DPPH radical was noticed for durum wheat genotypes, particularly for Dur1, although the greatest FRAP values were obtained in oat samples, especially in Oats2 (Table 4).

According to the cluster analysis, which encompassed all examined antioxidants, including the capacity to reduce DPPH radicals, as well as FRAP, two clusters were formed (Figure 3). The samples Trit1 and Oats3 constituted the first cluster among all tested samples. The second cluster encompassed all other genotypes, which were divided into two sub-clusters. The first one consisted of samples Dur1 and Dur2. Under the second sub-cluster, two sub-sub-clusters were formed, encompassing Dur3, Em2, all barley genotypes, both rye genotypes, as well as Oats1 and Oats2 in the first sub-sub-cluster. The second sub-sub-cluster included both bread wheat genotypes Trit2, Trit3, spelt, and Em1.

Table 3. Phenolic compounds in the grains of examined standard and ancient grains: gallic acid (GA), Cctechin (CAT), dihydrocaffeic acid (DiCA), epicatechin (EPI), dihydro-*p*-coumaric acid (di-p-CoumA), naringin (NA), daidzein (DA), cinnamic acid (CIN), naringenin (NG), chlorogenic acid (ChlA), caffeic acid (CA), p-coumaric acid (p-CoumA), ferulic acid (FA), isoferulic acid (IFA), and quercetin (QUE).

	GA	CAT	EPI	DiCA	di-p-CoumA	NA	NG	DA
Dur1	n.d.	n.d.	n.d.	74.60 ± 0.93	36.77 ± 0.54	38.53 ± 0.48	n.d.	n.d.
Dur2	n.d.	n.d.	n.d.	59.03 ± 0.85	n.d.	37.04 ± 0.53	n.d.	n.d.
Dur3	n.d.	n.d.	38.51 ± 0.06	54.19 ± 0.20	43.35 ± 0.61	40.92 ± 0.51	n.d.	n.d.
BW1	n.d.	n.d.	n.d.	65.68 ± 0.84	45.19 ± 0.25	38.05 ± 0.22	n.d.	25.48 ± 0.34
BW2	n.d.	31.07 ± 0.70	38.72 ± 0.32	83.28 ± 0.56	61.72 ± 0.78	40.33 ± 0.50	46.42 ± 0.58	n.d.
Trit1	n.d.	n.d.	n.d.	87.32 ± 1.10	32.08 ± 0.17	39.96 ± 0.51	n.d.	26.15 ± 0.38
Trit2	n.d.	n.d.	n.d.	113.7 ± 1.47	31.79 ± 0.17	39.31 ± 0.50	n.d.	25.57 ± 0.37
Trit3	n.d.	n.d.	39.30 ± 0.03	87.92 ± 0.72	36.82 ± 0.17	38.01 ± 0.48	n.d.	25.63 ± 0.37
Bar1	33.36 ± 5.88 *	443.29 ± 5.80	44.61 ± 0.12	38.89 ± 0.12	48.18 ± 0.25	38.31 ± 0.21	n.d.	28.54 ± 0.37
Bar2	35.13 ± 6.09	459.19 ± 6.04	41.86 ± 0.06	39.81 ± 0.04	42.65 ± 0.20	38.30 ± 0.49	46.66 ± 0.56	26.89 ± 0.59
Bar3	31.52 ± 4.71	360.52 ± 4.65	39.13 ± 0.10	37.45 ± 0.06	42.13 ± 0.18	34.02 ± 0.45	n.d.	27.68 ± 0.40
Bar4	31.16 ± 5.57	426.00 ± 5.46	45.31 ± 0.33	50.07 ± 0.22	62.54 ± 0.45	36.69 ± 0.48	n.d.	27.43 ± 0.40
Em1	n.d.	n.d.	n.d.	105.9 ± 1.36	30.03 ± 0.13	38.52 ± 0.50	49.51 ± 0.71	28.77 ± 0.62
Em2	n.d.	n.d.	n.d.	97.42 ± 1.25	28.45 ± 0.13	36.24 ± 0.47	52.33 ± 0.76	26.45 ± 0.65
Spelt	n.d.	n.d.	42.34 ± 0.25	64.28 ± 0.52	22.45 ± 0.45	35.22 ± 0.51	n.d.	n.d.
Rye1	n.d.	35.57 ± 0.99	n.d.	79.45 ± 1.15	n.d.	36.44 ± 0.53	n.d.	n.d.
Rye2	n.d.	33.84 ± 1.17	n.d.	92.59 ± 1.34	n.d.	37.78 ± 0.55	n.d.	n.d.
Oats1	31.48 ± 0.45	n.d.	121.69 ± 2.28	n.d.	231.6 ± 2.71	50.68 ± 0.66	47.10 ± 0.64	38.30 ± 0.63
Oats2	32.90 ± 0.47	n.d.	100.26 ± 3.05	n.d.	280.42 ± 3.47	48.22 ± 0.61	42.35 ± 0.64	32.48 ± 0.55
Oats3	34.23 ± 0.49	n.d.	114.25 ± 2.06	n.d.	210.2 ± 2.45	45.84 ± 0.60	45.22 ± 0.61	35.45 ± 0.59

Table 3. *Cont.*

	ChlA			CA			*p*-CoumA			FA			CIN			IFA			QUE		
Dur1		n.d.			n.d.		30.59	±	0.03	28.34	±	0.28	21.44	±	0.31	28.74	±	0.35	8.87	±	24.44
Dur2		n.d.			n.d.		30.81	±	0.03	29.23	±	0.27		n.d.		28.60	±	0.32	10.38	±	21.89
Dur3		n.d.			n.d.		27.62	±	0.02	29.34	±	0.33	21.55	±	0.31	29.18	±	0.40	6.63	±	15.97
BW1	27.95	±	0.02	28.58	±	0.05	27.26	±	0.05	30.90	±	0.28	21.48	±	0.37	28.02	±	0.31	10.23	±	12.75
BW2	27.77	±	0.42		n.d.		29.93	±	0.02	28.72	±	0.31	21.21	±	0.33	28.31	±	0.38	6.83	±	26.98
Trit1		n.d.			n.d.		29.20	±	0.03	30.18	±	0.33		n.d.		31.40	±	0.42	7.64	±	13.88
Trit2		n.d.			n.d.		27.60	±	0.03	29.57	±	0.30		n.d.		29.00	±	0.37	8.29	±	19.79
Trit3		n.d.			n.d.		28.72	±	0.06	30.46	±	0.33		n.d.		33.43	±	0.43	9.39	±	15.60
Bar1	96.68	±	0.98	29.85	±	0.08	27.44	±	0.45	33.80	±	0.42	21.21	±	1.27		n.d.		14.59	±	19.55
Bar2	73.30	±	0.63	29.07	±	0.04	29.95	±	0.45	32.26	±	0.40		n.d.			n.d.		15.40	±	31.26
Bar3	89.14	±	0.87	29.08	±	0.03	28.31	±	0.42	30.39	±	0.38		n.d.			n.d.		14.84	±	16.11
Bar4	100.6	±	1.13	31.13	±	0.03	32.04	±	0.47	33.13	±	0.42		n.d.			n.d.		13.94	±	37.96
Em1		n.d.			n.d.		28.45	±	0.42	30.11	±	0.39		n.d.			n.d.		8.47	±	26.17
Em2		n.d.			n.d.		31.45	±	0.48	34.22	±	0.45		n.d.			n.d.		7.25	±	28.54
Spelt		n.d.		n.d.			34.55	±	0.09	28.56	±	0.31		n.d.		27.89	±	0.38	6.45	±	30.69
Rye1		n.d.		n.d.			36.85	±	0.11	30.45	±	0.31		n.d.		28.85	±	0.36	8.45	±	21.86
Rye2		n.d.		n.d.			32.83	±	0.06	28.35	±	0.29		n.d.		30.53	±	0.37	9.51	±	18.75
Oats1		n.d.		41.55	±	0.56	34.26	±	0.66	76.28	±	0.68		n.d.		27.48	±	0.07	31.39	±	41.30
Oats2		n.d.		45.88	±	0.55	36.23	±	0.69	78.42	±	0.76		n.d.		26.56	±	0.02	25.45	±	42.18
Oats3		n.d.		38.22	±	0.29	42.56	±	0.41	60.22	±	0.47		n.d.		27.45	±	0.02	28.56	±	35.44

* Average ± standard error; n.d.—not detectable.

Table 4. DPPH radical scavenging activity and ferric-reducing antioxidant power (FRAP) of extracts in the examined standard grains and ancient grains.

Genotypes	DPPH			FRAP		
	μmol TE·g^{-1}			μmol Fe^{2+} eq·g^{-1}		
Dur1	201.1 a	±	1.88 *	20.81 m	±	0.29
Dur2	183.7 cd	±	4.95	29.87 g	±	0.47
Dur3	192.8 b	±	7.87	25.03 j	±	0.47
BW1	173.6 e	±	3.75	26.97 i	±	0.41
BW2	155.4 fgh	±	0.79	30.45 fg	±	0.38
Trit1	135.4 k	±	4.77	36.45 de	±	0.55
Trit2	149.3 ghi	±	4.33	23.40 k	±	0.38
Trit3	111.8 l	±	2.92	18.78 n	±	0.28
Bar1	117.8 l	±	5.99	36.12 e	±	0.57
Bar2	141.7 jk	±	3.32	37.39 cd	±	0.52
Bar3	150.2 gh	±	4.76	28.79 h	±	0.45
Bar4	183.5 cd	±	2.24	31.31 f	±	0.42
Em1	177.1 de	±	0.86	22.22 l	±	0.29
Em2	180.5 cde	±	6.36	25.61 j	±	0.45
Spelt	187.3 bc	±	2.50	28.32 h	±	0.40
Rye1	159.4 f	±	4.00	23.09 kl	±	0.36
Rye2	177.7 de	±	6.36	20.01 m	±	0.38
Oats1	142.4 ijk	±	2.13	42.15 b	±	0.56
Oats2	148.2 hij	±	5.49	45.32 a	±	0.67
Oats3	156.1 fg	±	0.63	37.99 c	±	0.48
Tukey HSD ($p < 0.05$)	1.33			0.198		

* Average ± standard error; the values signed with the same letter are not significantly different at the 0.05 level of significance.

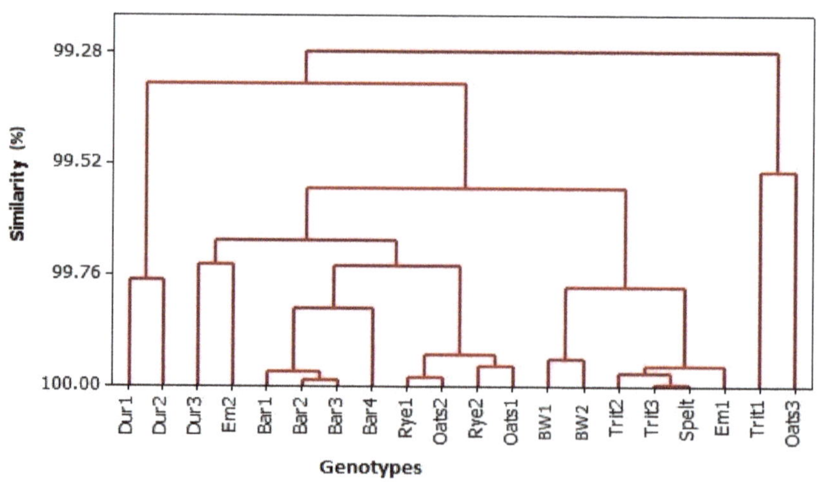

Figure 3. Dendrogram presenting concentrations of antioxidants in the grain of examined standard and ancient grains.

3.4. PC Analysis

To access and evaluate the possible linkage between tested genotypes and phytochemicals present in all examined genotypes, PCA was performed. PC analysis showed a six-component model which explained 87.51% of the total variability. The first axis had a total variability of 34.1%, while the second axis had 18.9%, the third axis had 12.6%, and the fourth axis had 10.7% of total variability. PCA showed that the concentration of arabinoxylan, Ca, FA, and QUE correlated with the first axis in a significantly positive way, while Pphy correlated with the first axis a in significantly negative way. Zn and p-CoumA correlated with the second axis in a significantly positive way. Pi, GSH, βGlu, and P correlated with the third axis in a significantly positive way, while Mn and NA correlated with the same axis in a significantly negative way; Fe and Cu correlated with the fourth axis in a significantly positive way, while K and Mg correlated with the same axis in a significantly negative way (Supplementary Table S3). The best separation after PCA resulted in the model consisting of the first two axes (Figure 4). Results revealed that all three oat genotypes were mostly determined by the concentration of arabinoxylan, NA, and QUE. The best separation after PCA resulted in the model consisting of the first two axes (Figure 4). Results revealed that all three oat genotypes were mostly determined by the concentration of arabinoxylan, NA, and QUE, while the Bar1, Bar2, and Bar3 genotypes, together with Trit2, were found to be associated with protein content. The Bar4 genotype was associated with the Pi concentration. To a lesser extent, barley was found to be associated with B-gluc and GSH. According to the PCA, the oat and barley genotypes, with the exception of Bar4, were separated from all other species.

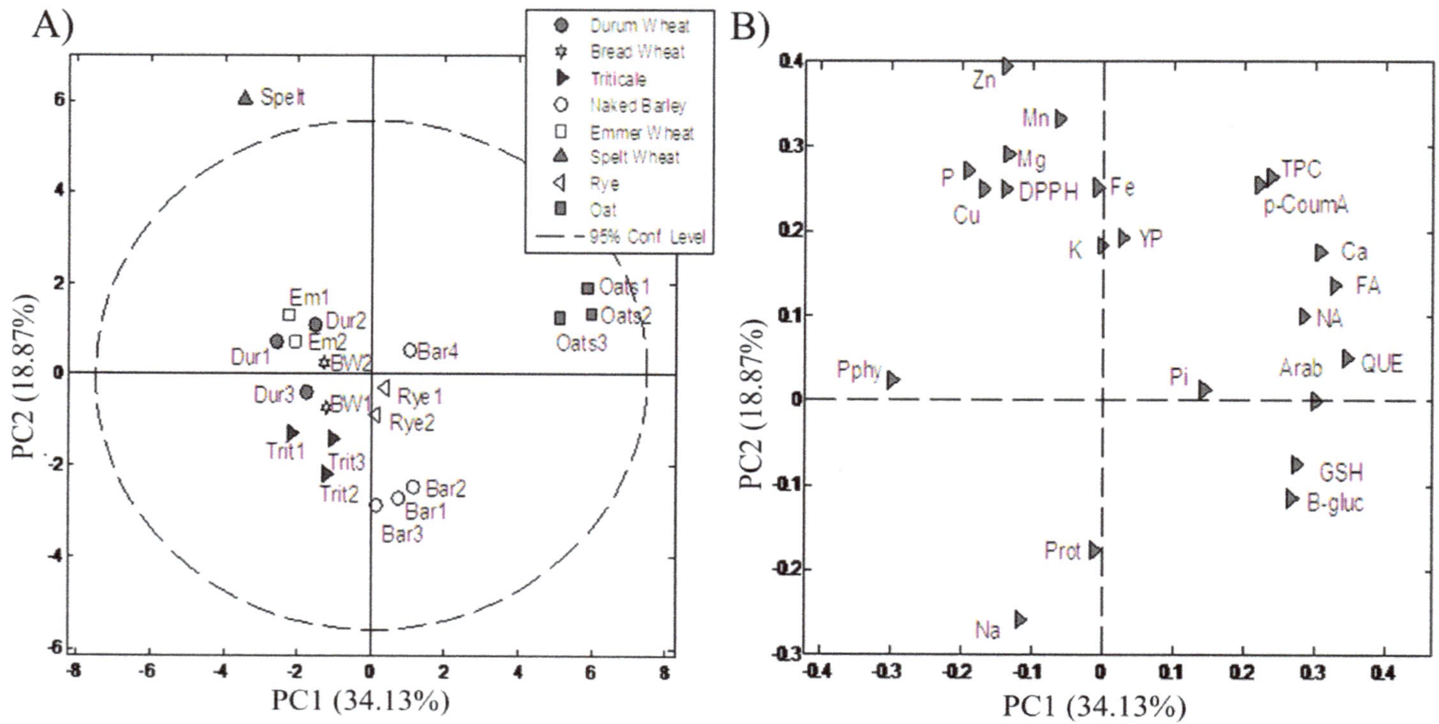

Figure 4. PCA for evaluated phytochemicals in tested genotypes: (**A**) score and (**B**) loading plot. Abbreviations: protein (Prot), inorganic phosphorus (Pi), phytic phosphorus (Pphy), yellow pigment (YP), total phenolic compounds (TPCs), total glutathione(GSH), β-glucan (β-glu), arabinoxylan (Arab), phosphorus (P), potassium (K), calcium (Ca), magnesium (Mg), sodium (Na), iron (Fe), manganese (Mn), zinc (Zn), copper (Cu), naringin (NA),p-coumaric acid(p-CoumA), ferulic acid (FA), and quercetin (QUE).

4. Discussion

Cereals are an important staple food worldwide, providing most of the macro- and micro-nutrients to humans. Owing to their high starch concentration, they are the foremost source of energy in the diet, but other nutrients, with functional properties, make them nutritious and desirable parts of any diet. From anutritional viewpoint, the diversity of species (wheat, rye, barley, oats, etc.) and species' genotypes could offer a whole range of bioactive compounds in nutrition. It is well known that most of the present-day genotypes are designed for high yields with high starch content; however, they are low in essential elements and bioactive compounds and poorly adaptable to stressful conditions. Thus, ancient and some older and colored grains are attracting attention nowadays as they are highly adaptable and tolerant to growth in limited conditions [13,14,44]. Importantly, they are rich in protein and various bioactive compounds, so they can therefore be considered nutraceuticals [9].

4.1. *Concentration of Protein and Dietary Fiber in the Standard and Ancient Grains*

The results showed that protein in samples of Bar2, Bar4, Trit3, Dur1, and spelt had concentrations $\geq$ 13%, indicating significant variability among analyzed genotypes in the phytochemical composition of the grains. Majzoobi et al. [10] obtained a wide range of protein content among various grains, including barley, spelt, and emmer wheat. They also accentuated that spelt and barley are the richest sources of dietary fiber, too. Our research revealed that the concentration of β-glucan, as a soluble fiber [4,45], was the highest in the samples of barley (particularly Bar2) and oat genotypes, while the samples of oats (mainly Oats1) and rye genotypes were found to be the best sources of arabinoxylan [2,45]. High concentrations of β-glucan and arabinoxylan in examined genotypes of barley, rye, and particularly oats can contribute to the beneficial microbiota in the human gut [46], and they can also be a fiber source for gluten-free bread production [47]. More importantly, examined triticale and barley genotypes could present a gene pool for the development of high-protein cereals.

4.2. *Essential Element Levels in the Standard and Ancient Grains*

Considering essential elements, examined genotypes of rye, oats, emmer, and spelt wheat had generally higher concentrations of essential elements compared to other genotypes. In contrast to barley genotypes (especially Bar2) with the highest Pi concentration, triticale and spelt revealed the highest concentration of Pphy, which is the most important factor that sequesters most of the essential elements [16,17]. Nevertheless, spelt was also the richest source of K, Mg, Mn, and Zn, which might suggest that it is capable of absorbing and accumulating higher amounts of minerals from the soil, making it a superior crop for growth in low-input systems [15]. Together with a high concentration of YP, the tested spelt genotype can be seen as a highly nutritious grain as it is unique and independent of other genotypes in terms of the potential bioavailability of essential elements. Similarities between spelt, durum wheat, and triticale genotypes in terms of the accumulation of essential elements indicate their increased ability to utilize environmental sources concerning the other cereals. The tested genotypes of emmer wheat and barley and, to a lesser extent, durum wheat are also capable of accumulating higher amounts of Fe, Mn, and Cu, while oats could be considered a good Ca source.

However, the lowest Pphy concentration refers to the higher potential availability of essential elements from the Oats3 grain, which ensures the enhanced absorption of Ca, Mg, Fe, and Mn during the digestion process [48,49]. A lower Phy/essential elements ratio was also present in rye and emmer wheat genotypes, but to a lesser degree.

4.3. *Antioxidants and Phenolic Acids in the Standard and Ancient Grains*

Plants synthesize antioxidants to combat oxidative stress, and, in seeds, they serve to protect viability during storage [50,51]. The concentration and antioxidant potential of phenolic compounds are important traits from a nutritional point of view, too [52]. In this research, some critical low-molecular-weight non-enzymatic antioxidants were analyzed in the grain of examined cereal genotypes, including the concentration of YP as a fat-soluble antioxidant, as well as the concentration of Pphy, TPC, and GSH as water-soluble antioxidants.

The antioxidant capacity to reduce DPPH radicals was shown to be the highest in Dur1, which was also a genotype with the highest GSH concentration. High DPPH radical scavenging activity was also noted in spelt (which had the greatest concentration of Pphy and YP), as well as Em2 and Rye2 (which were high in TPCs and GSH). The FRAP value was unsurprisingly the highest in Oats2, which showed the highest TPC value among all the tested samples. The increased variability of Pphy, YP, and GSH, along with pronounced similarity between examined durum wheat genotypes, Em2, spelt, and triticale genotypes, makes them a desirable source of antioxidants, as well as a gene pool for the further development of nutraceutical cereals.

In this research, the concentration of some common and rarely analyzed phenolic compounds in cereal grains has been shown. The results from this study revealed the presence of quercetin, daidzein, naringin, naringenin, catechin, and epicatechin in the tested whole-grain cereals. Such data represent novelty because, through literature review, it has been shown that the above-mentioned compounds can be mostly found in plant species, such as buckwheat [53], soybeans [54,55], quinoa [56], citrus fruit peels [57], apples, and Malus doumeri fruit [58,59]. Due to the lack of information on the phytochemical content in whole-grain cereals, and especially of ancient grains, such as spelt and emmer wheat, these results can be useful in enhancing data on this subject, especially nowadays when the quest for high nutritional quality in food is a world tendency.

Montevecchi et al. [11] showed in their study that ancient wheat varieties, with higher adaptability to local conditions and marginal soils, are a rich source of various phenolic compounds, containing conjugated flavones, such as vitexin and isovitexin. Due to the fact that p-coumaric acid and ferulic acid are the most present phenolics in cereal grains [29], they were observable in the grains of all examined genotypes with the highest concentration achieved in Oats3 and Oats2. While the greatest variability in the concentration of phenolic compounds was present in the grains of barley and oats [29,60], some specificity was present, especially when flavonoids were considered. Thus, samples of Oats1 and Oats2 showed the highest variability in naringin, quercetin, epicatechin, dihydro-*p*-coumaric, ferulic, and isoferulic acid, while the Trit2 samples showed the highest variability in dihydrocaffeic acid. The naringenin was mainly found in the grains of Em2 and oats. Since observed phenolic acids have profound health benefits [61], this research supports the statement that whole-grain products, particularly examined oat samples, are a valuable source of various micro- and trace-bioactives. Owing to the high level of dietary and prebiotic fiber [25] and prebiotic phenolic compounds [30], examined oat genotypes could be considered nutraceuticals.

4.4. PC Analysis

PCA was conducted to uncover the potential unique nutritional fingerprint of the tested whole-grain cereals. PCA revealed that the oats and barley genotypes were separated from all other analyzed species owing to their unique nutritive profile, while all the other species were similar due to their phylogenetic relationship. The association of oat genotypes with arabinoxylan, QUE, and NA, along with Ca, TPCs, FA, and pCoumA, to a minor extent, highlights their nutritional importance [10]. These compounds contribute significantly to antioxidant activity, play a prebiotic role [30], and serve as an important source of calcium. As previously mentioned, a possible association between the Bar1, Bar2, and Bar3 samples and the concentrations of protein, β-Glu, and GSH suggests the distinctiveness of barley genotypes as a valuable source of important bioactive compounds. The results obtained in our study are in agreement with those obtained in the study of Majzoobi et al. [10] who characterized barley as a highly nutritious grain, abundant in protein, vitamins, and minerals. These authors also emphasized the importance of barley's high β-glucan content, along with its phytosterols and polyphenols. Additionally, the strong association between triticale genotypes and protein content further supports the idea that triticale is a highly desirable and valuable nutritious crop. Consistent with our findings, Camerlengo and Kiszonas [62] recognized triticale as a nutrient-rich crop, offering higher levels of protein, health-promoting compounds, and a well-balanced amino acid composition. These findings make these cereals highly desirable and offer breeders the potential to develop genotypes with enhanced nutritional value.

5. Conclusions

While the importance of whole-grain cereals as a source of various nutrients has been confirmed, the comprehensive analysis conducted in this study shed new light on their nutritional uniqueness and importance as staple food and nutraceuticals at the same time. Some of the genotypes that were analyzed, including durum wheat, triticale, spelt, emmer wheat, and barley, emerge as significant and sustainable sources of protein. Additionally, their content of bioactive compounds—such as prebiotic fibers (β-glucan and arabinoxylan), essential minerals, and antioxidants—further enhances their value, positioning them as valuable ingredients for functional foods.

Owing to the high protein, and particularly GSH level, Cosmostar (durum wheat variety) could present a genetic pool for the development of high protein/GSH genotypes. While ancient grains (emmer and spelt wheat) are highly adaptable and sustainable crops, according to this study, Emmer FON, Emmer LP2-1-5, and Cimmyt spelt 1 are important sources of highly available essential elements. Special attention should be given to Cimmyt spelt 1, a genotype high in yellow pigment, which possesses high concentration of essential elements (particularly Zn) and a high capacity to reduce DPPH radicals. Thus, ancient grains from the MRIZP collection could serve as candidates for bio-fortification studies. The high concentration of phenolic compounds and the presence of phenolics, dihydro-*p*-coumaric acid, naringin, quercetin, epicatechin in grains of oats (Sopot), and catechin in barley grains (Apolon and Osvit) underline their unique nutritive profile, making them a desirable genetic pool for breeding cereal genotypes high in antioxidants and prebiotic compounds.

This study is the first report of the assessment the concentration of thirty-three parameters simultaneously in whole-grain cereals, including some ancient grains, such as spelt and emmer wheat. For the first time, the PC analysis revealed a distinctive nutritional profile for barley and oat samples. These findings emphasize the potential of these cereals as a source of important nutrients and phytochemicals, offering the possibility of developing genotypes with improved health benefits for humans.

Supplementary Materials: The following supporting information can be downloaded at https://www.mdpi.com/article/10.3390/foods13244116/s1. Table S1: Genotypes used in experiment; Table S2: Ratio between phytic acid (Phy) and essential elements in examined standard and ancient grains; Table S3: Correlation coefficients between PC axes and tested parameters (protein, inorganic phosphorus (Pi), phytic phosphorus (Pphy), yellow pigment (YP), total phenolics (TPCs), total glutathione (GSH), β-glucan (βGlu), arabinoxylan (Arab), naringin (NA), p-Coumaric acid (p-CoumA), ferulic acid (FA), and quercetin (QUE)).

Author Contributions: Conceptualization, V.D., V.K.R. and J.V.; methodology, V.D., J.V., M.D. and S.Đ.; software, V.K.R. and M.B.; validation, V.D., J.V., M.D. and S.Đ.; investigation, V.D., V.K.R. and J.V.; resources, M.S.; data curation, J.V. and M.B.; writing—original draft preparation, V.D. and J.V.; writing—review and editing, M.B.; visualization, V.D., V.K.R., J.V., M.D. and S.Đ.; supervision, M.S. All authors have read and agreed to the published version of the manuscript.

Funding: Research was supported by the Ministry of Education, Science, and Technological Development, Republic of Serbia (grant no. 451-03-66/2024-03/200040).

Institutional Review Board Statement: Not applicable.

Informed Consent Statement: Not applicable.

Data Availability Statement: The original contributions presented in the study are included in the article/Supplementary Materials, further inquiries can be directed to the corresponding author.

Acknowledgments: The authors are grateful to Boris Pisinov (Institute for Plant Protection and Environment) and Olgica Mandić (Institute of Public Health of Serbia "Dr. Milan Jovanović Batut") for their tremendous efforts in conducting the chemical analysis of the experimental material.

Conflicts of Interest: The authors declare no conflicts of interest. The funders had no role in the design of the study; in the collection, analyses, or interpretation of the data; in the writing of the manuscript; or in the decision to publish the results.

References

1. Bhardwaj, R.L.; Parashar, A.; Parewa, H.P.; Vyas, L. An Alarming Decline in the Nutritional Quality of Foods: The Biggest Challenge for Future Generations' Health. *Foods* **2024**, *13*, 877. [CrossRef] [PubMed]
2. Zannini, E.; Bravo Núñez, Á.; Sahin, A.W.; Arendt, E.K. Arabinoxylans as Functional Food Ingredients: A Review. *Foods* **2022**, *11*, 1026. [CrossRef] [PubMed]
3. Mitrevski, J.; Pantelic, N.Đ.; Dodevska, M.S.; Kojic, J.S.; Vulic, J.J.; Zlatanovic, S.; Gorjanović, S.; Lalicic-Petronijevic, J.; Marjanovic, S.; Antic, V.V. Effect of Beetroot Powder Incorporation on Functional Properties and Shelf Life of Biscuits. *Foods* **2023**, *12*, 322. [CrossRef] [PubMed]
4. Cassidy, Y.M.; McSorley, E.M.; Allsopp, P.J. Effect of soluble dietary fibre on postprandial blood glucose response and its potential as a functional food ingredient. *J. Funct. Foods* **2018**, *46*, 423–439. [CrossRef]
5. Jordanovska, S.; Jovović, Z.; Dolijanović, Ž.; Dragičević, V.; Branković, G.; Đekić, V. Nutritional properties of Macedonian landraces of small grain cereals as a source of new genetic variability. *Genetika* **2018**, *50*, 863–883. [CrossRef]
6. Rodríguez, J.P.; Rahman, H.; Thushar, S.; Singh, R.K. Healthy and Resilient Cereals and Pseudo-Cereals for Marginal Agriculture: Molecular Advances for Improving Nutrient Bioavailability. *Front. Genet.* **2020**, *11*, 49. [CrossRef] [PubMed]
7. FAO. *Statistical Yearbook 2023 Faostat Analytical Brief 79, Agricultural Production Statistics 2000–2022*; FAO: Rome, Italy, 2023; ISSN 2709-006X.
8. Garutti, M.; Nevola, G.; Mazzeo, R.; Cucciniello, L.; Totaro, F.; Bertuzzi, C.A.; Caccialanza, R.; Pedrazzoli, P.; Puglisi, F. The Impact of Cereal Grain Composition on the Health and Disease Outcomes. *Front. Nutr.* **2022**, *9*, 888974. [CrossRef] [PubMed]
9. Basile, G.; De Maio, A.C.; Catalano, A.; Ceramella, J.; Iacopetta, D.; Bonofiglio, D.; Saturnino, C.; Sinicropi, M.S. Ancient Wheat as Promising Nutraceuticals for the Prevention of Chronic and Degenerative Diseases. *Curr. Med. Chem.* **2023**, *30*, 3384–3403. [CrossRef]
10. Majzoobi, M.; Jafarzadeh, S.; Teimouri, S.; Ghasemlou, M.; Hadidi, M.; Brennan, C.S. The Role of Ancient Grains in Alleviating Hunger and Malnutrition. *Foods* **2023**, *12*, 2213. [CrossRef] [PubMed]
11. Montevecchi, G.; Setti, L.; Olmi, L.; Buti, M.; Laviano, L.; Antonelli, A.; Sgarbi, E. Determination of Free Soluble Phenolic Compounds in Grains of Ancient Wheat Varieties (*Triticum* sp. pl.) by Liquid Chromatography–Tandem Mass Spectrometry. *J. Agric. Food Chem.* **2019**, *67*, 201–212. [CrossRef] [PubMed]
12. Suchowilska, E.; Bieńkowska, T.; Stuper-Szablewska, K.; Wiwart, M. Concentrations of Phenolic Acids, Flavonoids and Carotenoids and the Antioxidant Activity of the Grain, Flour and Bran of *Triticum polonicum* as Compared with Three Cultivated Wheat Species. *Agriculture* **2020**, *10*, 591. [CrossRef]
13. Konvalina, P.; Suchý, K.; Stehno, Z.; Capouchová, I.; Moudrý, J. Drought Tolerance of Different Wheat Species (*Triticum* L.). In *Vulnerability of Agriculture, Water and Fisheries to Climate Change*; Behnassi, M., SyomitiMuteng'e, M., Ramachandran, G., Shelat, K., Eds.; Springer: Dordrecht, The Netherlands, 2014. [CrossRef]
14. Dragičević, V.; Brankov, M.; Stoiljković, M.; Tolimir, M.; Kanatas, P.; Travlos, I.; Simić, M. Kernel color and fertilization as factors of enhanced maize quality. *Front. Plant Sci.* **2022**, *13*, 1027618. [CrossRef]
15. Dragicevic, V.; Dolijanović, Ž.; Janosevic, B.; Brankov, M.; Stoiljkovic, M.; Dodevska, M.S.; Simić, M. Enhanced Nutritional Quality of Sweet Maize Kernel in Response to Cover Crops and Bio-Fertilizer. *Agronomy* **2021**, *11*, 981. [CrossRef]
16. Brouns, F. Phytic Acid and Whole Grains for Health Controversy. *Nutrients* **2022**, *14*, 25. [CrossRef] [PubMed]
17. Zhang, Y.Y.; Stockmann, R.; Ng, K.; Ajlouni, S. Revisiting phytate-element interactions: Implications for iron, zinc and calcium bioavailability, with emphasis on legumes. *Crit. Rev. Food Sci. Nutr.* **2020**, *62*, 1696–1712. [CrossRef]
18. Feizollahi, E.; Mirmahdi, R.S.; Zoghi, A.; Zijlstra, R.T.; Roopesh, M.S.; Vasanthan, T. Review of the beneficial and anti-nutritional qualities of phytic acid, and procedures for removing it from food products. *Food Res. Int.* **2021**, *143*, 110284. [CrossRef]
19. Brankovic, G.; Dragičević, V.; Dodig, D.; Zoric, M.; Knežević, D.; Žilić, S.; Denčić, S.; Šurlan, G. Genotype x Environment interaction for antioxidants and phytic acid contents in bread and durum wheat as influenced by climate. *Chil. J. Agric. Res.* **2015**, *75*, 139–146. [CrossRef]
20. Garg, M.; Sharma, N.; Sharma, S.; Kapoor, P.; Kumar, A.; Chunduri, V.; Arora, P. Biofortified Crops Generated by Breeding, Agronomy, and Transgenic Approaches Are Improving Lives of Millions of People around the World. *Front. Nutr.* **2018**, *5*, 12. [CrossRef] [PubMed]
21. Safdar, L.B.; Foulkes, M.J.; Kleiner, F.H.; Bhosale, R.A.; Fisk, I.D.; Boden, S.A. Challenges facing sustainable protein production: Opportunities for cereals. *Plant Commun.* **2023**, *4*, 100716. [CrossRef] [PubMed]

22. Barrett, E.M.; Batterham, M.J.; Ray, S.; Beck, E.J. Whole grain, bran and cereal fibre consumption and CVD: A systematic review. *Br. J. Nutr.* **2019**, *121*, 914–937. [CrossRef]
23. Prasadi, N.P.V.; Joye, I.J. Dietary Fibre from Whole Grains and Their Benefits on Metabolic Health. *Nutrients* **2020**, *12*, 3045. [CrossRef]
24. Dodevska, M.S.; Sobajic, S.S.; Dragicevic, V.D.; Stankovic, I.; Ivanovic, N.D.; Djordjevic, B.I. The Impact of Diet and Fibre Fractions on Plasma Adipocytokine Levels in Prediabetic Adults. *Nutrients* **2021**, *13*, 487. [CrossRef] [PubMed]
25. Schupfer, E.; Pak, S.C.; Wang, S.; Micalos, P.S.; Jeffries, T.; Ooi, S.L.; Golombick, T.; Harris, G.; El-Omar, E. The effects and benefits of arabinoxylans on human gut microbiota–A narrative review. *Food Biosci.* **2021**, *43*, 101267. [CrossRef]
26. Mio, K.; Ogawa, R.; Tadenuma, N.; Aoe, S. Arabinoxylan as well as β-glucan in barley promotes GLP-1 secretion by increasing short-chain fatty acids production. *Biochem. Biophys. Rep.* **2022**, *32*, 101343. [CrossRef]
27. Prins, A.; Kosik, O. Genetic Approaches to Increase Arabinoxylan and β-Glucan Content in Wheat. *Plants* **2023**, *12*, 3216. [CrossRef]
28. Syeunda, C.; Awika, J.M. Mechanisms of flavonoid inhibition of Maillard reaction product formation in relation to whole grains processing. *Food Chem.* **2024**, *449*, 139237. [CrossRef] [PubMed]
29. Shahidi, F.; Danielski, R.; Ikeda, C. Phenolic compounds in cereal grains and effects of processing on their composition and bioactivities: A review. *J. Food Bioact.* **2021**, *15*, 15281. [CrossRef]
30. Zhang, Y.; Li, Y.; Ren, X.; Zhang, X.; Wu, Z.; Liu, L. The positive correlation of antioxidant activity and prebiotic effect about oat phenolic compounds. *Food Chem.* **2023**, *402*, 134231. [CrossRef] [PubMed]
31. Zhu, F. Anthocyanins in cereals: Composition and health effects. *Food Res. Int.* **2018**, *109*, 232–249. [CrossRef]
32. Alves-Santos, A.M.; Araújo Sugizaki, C.S.; Lima, G.C.; Veloso Naves, M.M. Prebiotic effect of dietary polyphenols: A systematic review. *J. Funct. Foods* **2020**, *74*, 104169. [CrossRef]
33. Dragičević, V.; Sredojević, S.; Perić, V.; Nišavić, A.; Srebrić, M. Validation study of a rapid colorimetric method for the determination of phytic acid and norganic phosphorus from grains. *Acta Period. Technol.* **2011**, *42*, 11–21. [CrossRef]
34. Sari-Gorla, M.; Ferrario, S.; Rossini, L.; Frova, C.; Villa, M. Developmental expression of glutathione-S-transferase in maize and its possible connection with herbicide tolerance. *Euphytica* **1993**, *67*, 221–230. [CrossRef]
35. *ISO13730*; Meat and Meat Products—Determination of Total Phosphorus Content—Spectrometric Method. Institute for Standardization of Serbia: Belgrade, Serbia, 1996. Available online: https://iss.rs/en/project/show/iso:proj:22789 (accessed on 16 December 2024).
36. Vancetovic, J.; Zilic, S.; Bozinovic, S.; Ignjatovic-Micic, D. Simulating of Top-Cross system for enhancement of antioxidants in maize grain. *Span. J. Agric. Res.* **2014**, *12*, 467–476. [CrossRef]
37. Singleton, V.L.; Orthofer, R.; Lamuela-Raventos, R.M. Analysis of total phenols and other oxidation substrates and antioxidants by means of Folin-Ciocalteu reagent. *Methods Enzymol.* **1999**, *299*, 152–178.
38. Brand-Williams, W.; Cuvelier, M.E.; Berset, C. Use of a free radical method to evaluate antioxidant activity. *LWT-Food Sci. Technol.* **1995**, *28*, 25–30. [CrossRef]
39. Wong, S.P.; Leong, L.P.; Koh, J.H.W. Antioxidant activities of aqueous extracts of selected plants. *Food Chem.* **2006**, *99*, 775–783. [CrossRef]
40. Szőllősi, R.; Varga, I.S.I. Total antioxidant power in some species of Labiatae: Adaptation of FRAP method. *Acta Biol. Szeged.* **2002**, *46*, 125–127.
41. McCleary, B.V.; Codd, R. Measurement of (1 → 3),(1 → 4)-β-D-glucan in barley and oats: A streamlined enzymic pro-cedure. *J. Sci. Food Agric.* **1991**, *55*, 303–312. [CrossRef]
42. Đurović, S.; Nikolić, B.; Luković, N.; Jovanović, J.; Stefanović, A.; Šekuljica, N.; Miljin, D.; Knežević-Jugović, Z. The impact of high-power ultrasound and microwave on the phenolic acid profile and antioxidant activity of the extract from yellow soybean seeds. *Ind. Crops Prod.* **2018**, *122*, 223–231. [CrossRef]
43. Helrich, K. *AOAC Official Methods of Analysis of AOAC International*, 15th ed.; USA Association of Analytical Communities/AOAC Inc.: Washington, DC, USA, 1990.
44. Shen, S.; Huang, R.; Li, C.; Wu, W.; Chen, H.; Shi, J.; Chen, S.; Ye, X. Phenolic Compositions and Antioxidant Activities Differ Significantly among Sorghum Grains with Different Applications. *Molecules* **2018**, *23*, 1203. [CrossRef] [PubMed]
45. Torbica, A.; Radosavljević, M.; Belović, M.; Djukić, N.; Marković, S. Overview of nature, frequency and technological role of dietary fibre from cereals and pseudocereals from grain to bread. *Carbohydr. Polym.* **2022**, *290*, 119470. [CrossRef]
46. Nie, C.; Yan, X.; Xie, X.; Zhang, Z.; Zhu, J.; Wang, Y.; Wang, X.; Xu, N.; Luo, Y.; Sa, Z.; et al. Structure of β-glucan from Tibetan hull-less barley and its in vitro fermentation by human gut microbiota. *Chem. Biol. Technol. Agric.* **2021**, *8*, 12. [CrossRef]
47. Bender, D.; Regner, M.; D'Amico, S.; Jäger, H.; Tömösközi, S.; Schoenlechne, R. Effect of arabinoxylan extraction on gluten-freesourdough-breadproperties. *J. Food Qual.* **2018**, *2018*, 5719681. [CrossRef]
48. Johnson, C.R.; Thavarajah, D.; Thavarajah, P. The influence of phenolic and phytic acid food matrix factors on iron bioavailability potential in 10 commercial lentil genotypes (*Lens culinaris* L.). *J. Food Compos. Anal.* **2013**, *3*, 82–86. [CrossRef]
49. Wang, Z.; Liu, Q.; Pan, F.; Yuan, L.; Yin, X. Effects of increasing rates of zinc fertilization on phytic acid and phytic acid/zinc molar ratio in zinc bio-fortified wheat. *Field Crops Res.* **2015**, *184*, 58–64. [CrossRef]
50. Zandi, P.; Schnug, E. Reactive Oxygen Species, Antioxidant Responses and Implications from a Microbial Modulation Perspective. *Biology* **2022**, *11*, 155. [CrossRef]

51. Kurek, K.; Plitta-Michalak, B.; Ratajczak, E. Reactive Oxygen Species as Potential Drivers of the Seed Aging Process. *Plants* **2019**, *8*, 174. [CrossRef]
52. Liu, Z.; Ren, Z.; Zhang, J.; Chuang, C.-C.; Kandaswamy, E.; Zhou, T.; Zuo, L. Role of ROS and Nutritional Antioxidants in Human Diseases. *Front. Physiol.* **2018**, *9*, 477. [CrossRef] [PubMed]
53. Bai, C.Z.; Feng, M.L.; Hao, X.L.; Zhong, Q.M.; Tong, L.G.; Wang, Z.H. Rutin, quercetin, and free amino acid analysis in buckwheat (*Fagopyrum*) seeds from different locations. *Genet. Mol. Res.* **2015**, *14*, 19040–19048. [CrossRef]
54. Guzmán-Ortiz, F.A.; San Martín-Martínez, E.; Valverde, M.E.; Rodríguez-Aza, Y.; Berríos, J.D.J.; Mora-Escobedo, R. Profile analysis and correlation across phenolic compounds, isoflavones and antioxidant capacity during germination of soybeans (*Glycine max* L.). *CyTA-J. Food* **2017**, *15*, 516–524. [CrossRef]
55. Kim, M.Y.; Jang, G.Y.; Lee YLee, M.; Ji, Y.M.; Yoon, N.; Lee, S.H.; Kim, K.M.; Lee, J.; Jeong, H.S. Free and bound form bioactive compound profiles in germinated black soybean (*Glycine max* L.). *Food Sci. Biotechnol.* **2016**, *25*, 1551–1559. [CrossRef] [PubMed]
56. Bhinder, S.; Kumari, S.; Singh, B.; Kaur, A.; Singh, N. Impact of germination on phenolic composition, antioxidant properties, antinutritional factors, mineral content and Maillard reaction products of malted quinoa flour. *Food Chem.* **2021**, *346*, 128915. [CrossRef]
57. Andrade, M.A.; Barbosa, C.H.; Shah, M.A.; Ahmad, N.; Vilarinho, F.; Khwaldia, K.; Silva, A.S.; Ramos, F. *Citrus* By-Products: Valuable Source of Bioactive Compounds for Food Applications. *Antioxidants* **2023**, *12*, 38. [CrossRef]
58. Shelke, V.; Kale, A.; Kulkarni, Y.A.; Bhanudas Gaikwad, A. Phloretin: A comprehensive review of its potential against diabetes and associated complications. *J. Pharm. Pharmacol.* **2024**, *76*, 201–212. [CrossRef]
59. Sun, B.; Huo, H.-Z.; Cai, A.-H.; Xie, Y.-C.; Li, H.-Y.; Li, D.-P. Determination of contents of eight phenolic acids in *Malus doumeri* fruit by HPLC. *Guangxi Zhiwu/Guihaia* **2021**, *41*, 1135–1144. [CrossRef]
60. Ge, X.; Jing, L.; Zhao, K.; Su, C.; Zhang, B.; Zhang, Q.; Han, L.; Yu, X.; Li, W. The phenolic compounds profile, quantitative analysis and antioxidant activity of four naked barley grains with different color. *Food Chem.* **2021**, *335*, 127655. [CrossRef]
61. Vinayagam, R.; Xiao, J.; Xu, B. An insight into anti-diabetic properties of dietary phytochemicals. *Phytochem. Rev.* **2017**, *16*, 535–553. [CrossRef]
62. Camerlengo, F.; Kiszonas, A.M. Genetic factors influencing triticale quality for food. *J. Cereal Sci.* **2023**, *113*, 103744. [CrossRef]

foods

MDPI

Article

The Properties of Damaged Starch Granules: The Relationship Between Granule Structure and Water–Starch Polymer Interactions

Andrés Gustavo Teobaldi [1,2], Esteban Josué Carrillo Parra [1,2], Gabriela Noel Barrera [1,2] and Pablo Daniel Ribotta [1,2,*]

[1] Instituto de Ciencia y Tecnología de los Alimentos Córdoba (ICYTAC-CONICET), Universidad Nacional de Córdoba, Av. Filloy S/N, Ciudad Universitaria, Córdoba CP 5000, Argentina; ateobaldi@agro.unc.edu.ar (A.G.T.); ejocarrillo@mi.unc.edu.ar (E.J.C.P.); gbarrera@agro.unc.edu.ar (G.N.B.)
[2] Departamento de Química Industrial y Aplicada, Facultad de Ciencias Exactas, Físicas y Naturales (FCEFyN), Universidad Nacional de Córdoba (UNC), Av. Vélez Sarsfield 1611, Córdoba CP 5000, Argentina
* Correspondence: pdribotta@unc.edu.ar

Abstract: The morphology of wheat starch granules with different damaged starch (DS) content was analyzed using a particle size analyzer and scanning electron microscopy (SEM); the granular structure was studied using FT-IR spectroscopy and X-ray diffraction (XRD); and the granule–water interaction was evaluated by thermogravimetric analysis (TGA) and dynamic vapor sorption (DVS). The increase in the level of DS shifted the population of B-type granules towards larger particle diameters and shifted the population of A-type granules towards smaller particle diameters. The appearance of the surface of the starch-damaged granules was rough and flaky (SEM images). Crystallinity reductions were related to higher mechanical damage levels of the granular structure (FT-IR and XRD). Higher DS increased the liquid-water absorption capacity of the granules. Higher DS was associated with increments in less-bound water proportions and reductions in more strongly bound water proportions and related to reductions in the evaporation temperature of these water populations (TGA analyses). Concerning DVS data, the results suggested that the driving force for water–monolayer attachment to the starch granules decreased as DS increased. Therefore, it was suggested that the changes in granule structure led to a weaker water–starch polymer chain interactions due to the increase in DS. The results contribute to a better understanding of the influence of mechanical damage on the starch granular structure, which could be related to the rheological and thermal behavior of starch-based systems with different DS.

Keywords: starch; mechanical damage; granule–water interactions; granule structure; crystallinity

1. Introduction

Native wheat starch (NS) is a polymer macromolecule that consists primarily of linear amylose and highly branched amylopectin [1]. The polymer chains that constitute starch are organized into structures called granules. Native starch granules' crystallinity varies from 15% to 45%. The semicrystalline solid characteristic of starch granules is determined by the structural arrangement of amylose and amylopectin. These polymers are arranged with their longitudinal axes perpendicular to the surface of the granule, giving rise to the structure in the form of concentric rings or layers typical of starch granules. Two types of layers can be identified, crystalline and amorphous, which are arranged alternately. The crystalline layers are mainly composed of amylopectin, specifically by the double helices of the amylopectin side chains, and in this matrix, some amylose molecules are dispersed.

https://doi.org/10.3390/foods14010021

The amorphous layers originate at the branching points of amylopectin and are made up especially of amylose, which are arranged between the branches of amylopectin forming double helices. For most starches, the external surface of starch granules is the first barrier to processes such as granule hydration, enzyme hydrolysis, and chemical reaction with modifying agents. Consequently, it is becoming recognized that the nature of the granule surface, particularly the presence of surface proteins and lipids, may significantly affect the properties of the starch (Pérez et al., 2009) [2].

The process of grinding wheat until obtaining the desired particle size is achieved through applying compression, abrasion, shear and impact forces, and a combination of these. The granules that are damaged during milling constitute what is called damaged starch (DS). The degree of starch damage during milling is influenced by three main factors: the composition (amylose/amylopectin ratio) of the starch, the hardness of the grain, and the milling conditions. In hard wheat, damaged starch can constitute 8% or more of the total starch in the flour.

Damaged starch granules exhibit partial loss of birefringence, higher water absorption, and higher susceptibility to enzymatic hydrolysis. Damaged granules absorb between 200 and 430% of their weight in water at room temperature, while native granules absorb between 39 and 87%, increasing the flour's water absorption capacity [3,4]. The polymer chains in the crystalline regions of the granules interact strongly among them, affecting the water penetration and allowing only water access to the amorphous regions. After the partial interruption of the crystalline order due to mechanical damage, water access is less restricted to these regions of the granule, increasing the water absorption capacity [5]. This condition makes damaged starch granules more accessible to amylases hydrolysis, compared to native ones, which results in the production of dextrins [6,7]. The degradation of starch molecules by milling occurs mainly at the α-(1$\rightarrow$4) glycosidic bonds in the crystalline double helix, although it can also occur at the α-(1$\rightarrow$6) branch points in the amorphous lamellae of starch granules. The breakdown of α-(1$\rightarrow$4) glycosidic bonds in the crystalline region is denoted by the decrease in starch crystallinity after milling [8]. Molecules that form double helices in amorphous conformation are more flexible than crystalline ones, which helps reduce the impact of mechanical shear during milling [9,10]. Several authors [11,12] have shown that milling produces an alteration in the structure and functionality of starch granules. The damaged starch content influences the physicochemical and rheological properties of the flours, directly affecting the industrial performance of the flours [7,13,14]. Consequently, the damaged starch content of flour is of crucial importance in its characterization.

The objective of this work was to evaluate the influence of the milling damage level of wheat-starch granules' structure on their hydration and water-holding properties, focusing on the mechanical process and its impact on the interactions between granules and water in liquid and gaseous states.

2. Materials and Methods

2.1. Samples

Native wheat starch (NS: 4.8% DS) was milled in a Whisper Series Bench Top disk mill (Rocklabs, Auckland, New Zeeland) at different times to obtain samples with different damaged starch levels. The temperature was controlled during the milling; it was kept under 40 °C to avoid starch oxidation reactions. Three levels of damaged starch were obtained after milling: DS1 (1 min in disk mill, 14.7% DS); DS2 (2 min in disk mill, 21.4% DS); and DS4 (4 min in disk mill, 32.2% DS). The damaged starch content was determined using the AACC 76–30 A method [15]. Each measurement was made in triplicate.

2.2. Particle Size Distribution Analysis

The particle size distributions of the samples were evaluated using a Laser Scattering Particle Size Distribution Analyzer (HORIBA Scientific, LA-960, Kyoto, Japan). The starch samples were placed in the dispersion tank of the particle size analyzer, which contained distilled water. For the measurements, a starch refractive index of 1.33 was selected. The analysis of particle size distributions was performed using LA-950 Software, which generated a multimodal report. All the measurements were carried out in duplicate.

2.3. Fourier Transform Infrared Spectroscopy (FT-IR)

Starch samples were analyzed using an infrared spectrometer. FT-IR spectra between 650 and 4000 cm^{-1} were obtained with a Nicolet iN10 infrared microscope with a cooled detector (Thermo Scientific, Waltham, MA, USA) and a resolution of 8 cm^{-1}. The spectra obtained were analyzed with OMNICTM 8.3 software (Thermo Scientific, Waltham, MA, USA).

2.4. X-Ray Diffraction (XRD)

The degree of crystallinity of the samples was determined using X-ray diffraction with a diffractometer (Philips PW1800, San Jose, CA, USA). The X-ray patterns were obtained using Cu Kα radiation (λ = 0.154 nm) under the following operating conditions: 30 mA, 40 kV, a scanning range of 5–40° (2θ), and a scanning speed of 0.01 °/s. Both the crystalline and amorphous fractions were quantified, and the percentage of crystallinity was calculated following the procedure outlined by Barrera et al. [8]. Peakfit v4.12 software (Peakfit, Jandel Scientific, San Rafael, CA, USA) was used to deconvolute the diffractograms in order to quantify the crystalline peaks and amorphous phases.

2.5. Scanning Electron Microscopy (SEM)

Scanning electron microscopy (SEM) was used to evaluate the surface characteristics of the starch granules. The samples were mounted on a sample holder and coated with a 30 nm thick layer of gold using a sputter coating system. To carry out the observations, a Supra 55 VP scanning electron microscope (Carl Zeiss Co., Oberkochen, Germany) was used at an acceleration potential of 1 Kv. The observations were conducted using a secondary electron detector "SE" and an "In Lens". Photographs were taken using automatic image capture software. Images were obtained with magnifications of 7000×–48,000×.

2.6. Water Absorption Capacity (WAC)

The water retention capacity (WAC) of the samples was determined using a method proposed by Palavecino et al. [16]. For the test, 0.5 g of sample (W1) was weighed in centrifuge tubes and 6 mL of distilled water was added. Each tube was incubated at 25 °C for 30 min and shacked at 0, 10, 20, and 30 min. The tubes were centrifuged for 20 min at 3000× g before draining the water excess. Each tube with wet sample (W2) was weighed and WAC was calculated as the coefficient between W1 and W2. Each measurement was made in duplicate.

$$\mathrm{WAC} = \frac{\mathrm{W2}}{\mathrm{W1}} \tag{1}$$

2.7. Thermogravimetric Analysis (TGA)

The thermogravimetric analysis (TGA) was carried out to analyze the interaction between water and starch granules. To standardize the moisture content of the samples, 2 g of the solid was placed in an airtight container along with a 15% v/v sulfuric acid solution. The containers were stored in a temperature-controlled chamber at 25 °C for 14 days until the equilibrium between ambient humidity and each sample was achieved.

Then, hydrated starches (~5 mg) were weighed into open aluminum capsules and analyzed over a temperature range of 25–120 °C at a rate of 4 °C/min. The temperature range used was selected (25–150 °C) to avoid starch decomposition. The graphs of the weight loss (considered as water loss) vs. temperature of each sample were obtained, as was the first derivative of these curves, that is, the rate of water loss (Derivative Thermo-Gravimetry (DTGA) (%/°C)). These curves were analyzed using TRIOS 5.1 software (TA Instruments, New Castle, DE, USA). The DTGA curves were analyzed by deconvolution to separate the overlapping peaks corresponding to different water populations through Gaussian functions and quantification areas (Peakfit 4.12, Jandel Scientific, San Rafael, CA, USA) [17]. The size of the peaks and the average temperature at which they occurred were determined. The size of the different peaks was calculated as the peak area (%) relative to the total peaks. Adjustments with R^2 greater than 0.99 were considered. All the measurements were performed at least in duplicate.

2.8. Dynamic Vapor Sorption (DVS)

Dynamic vapor adsorption equipment, Advantage I model (Surface Measurement Systems Ltd., London, UK) was used. Approximately 25 mg of starch samples were weighed into the DVS pans and placed inside a chamber at a constant temperature (25 °C) and relative humidity (0% to 90%) controlled by the nitrogen flow. The equilibration moisture content at each humidity value was determined by a change in dm/dt of 0.002 (%/min). The determination of the adsorption and desorption curves was carried out in the range of 0 to 0.9 of P/P0 (partial pressure of water vapor inside the chamber/saturation pressure of pure water vapor at the same temperature). The pressure relationship that determines the aqueous activity of the gas inside the chamber was equal to the definition of the water activity (aw) of food at equilibrium [18]. Measurements were carried out in duplicate.

The theoretical models of Guggenheim, Anderson, and de Boer (GAB) and Brunauer–Emmett–Teller (BET) were adjusted to the experimental data (regression coefficient: 0.99) using SigmaPlot 10.0 software (Systat Software, Inc., Surrey, Germany). Nonlinear regression analysis was applied over the range 0–0.9 aw in the GAB model and 0–0.5 aw in the BET model to determine the apparent monolayer coverage. In those models, W was the equilibrium moisture content (g H_2O/100 g dry solids), Wm was the monolayer content (g H_2O/100 g dry solids), C was the model constant, and aw was the water activity. In the GAB model, Kg was the model constant. The monolayer content Wm was a measure of the availability of active sites for water sorption [19]. It represented the moisture content where dehydrated food materials showed their maximum shelf life since degradation reactions were limited [20]. Parameter C was related to the binding force of water to the primary binding sites. Regarding the Kg parameter, when this approached 1, there were practically no differences between the multilayer water molecules and those of free water [19].

2.9. Statistical Analysis

Statistical analysis was conducted using analysis of variance (ANOVA). Means were compared with an LSD Fisher test at a significance level of 0.05. INFOSTAT statistical software v2020 (*Facultad de Ciencias Agropecuarias*, UNC, Córdoba, Argentina) was utilized for the analysis.

3. Results and Discussion

3.1. Particle Size Distribution Analysis

Starches extracted from different plant sources display unique granular shapes and structures. Starch granules can vary in shape, including spherical, oval, polygonal, lenticular (disk-shaped), and elongated forms. Their sizes can range from less than 1 μm to 100 μm in diameter. Wheat starch displays a bimodal particle size distribution: the larger granules (A-type) are disk-shaped or lenticular, with diameters between 18 and 33 μm, while the smaller granules (B-type) are spherical, with diameters ranging from 2 to 5 μm [21].

The particle size distribution of the starch granules is shown in Figure 1a. Increased mechanical damage to the starch granules caused a shift in the population of B-type granules toward larger particle diameters (Figure 1b). This shift can be attributed to a higher swelling in cold water or at ambient temperature. As a result, the peak corresponding to the B-type granule population began to overlap with that of the A-type granule population, making the bimodal distribution nearly indistinguishable. The rapid hydration and swelling of starch granules may occur due to alterations in their surface structure. The milling process can modify the granular surface, impacting the hydration and swelling of starch granules [8]. Previous studies have shown that the B-type granules in wheat starch are more susceptible to damage than A-type granules [22].

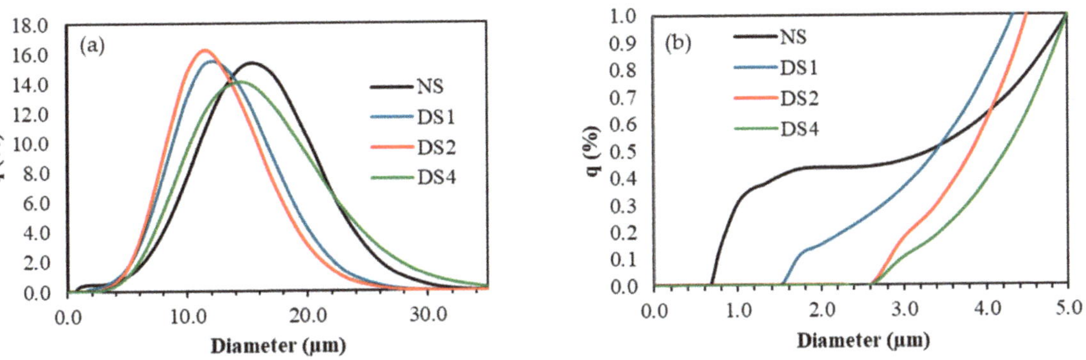

Figure 1. Particle size distributions of starch samples. (**a**) Complete particle size distributions, (**b**) B-type starch granules particle size distributions. q: volume (%). NS: 4.8% DS, DS1: 14.7% DS, DS2: 21.4% DS, DS4: 32.2% DS.

As for the A-type granule population, increased mechanical damage caused a shift toward smaller particle diameters, likely due to granular rupture. However, in the sample with the highest level of damage (DS4), the A-type granule population shifted toward larger diameters, approaching the distribution observed in the native starch sample. This suggests that A-type granules may become more prone to damage at higher grinding times, leading to their hydration, which was previously insignificant compared to B-type granules at lower grinding times. Hong et al. [23] reported that A-type wheat starch granules increase in size with prolonged regrinding (1, 3, and 5 h).

3.2. Fourier Transform Infrared Spectroscopy (FT-IR)

The FTIR spectra from 4000 to 700 cm^{-1} of the starch samples are shown in Figure S1. The spectrum region located between 4000 cm^{-1} and 1500 cm^{-1} showed the same characteristic peaks for all the analyzed samples. All four samples exhibited the same characteristic peaks at 3500 cm^{-1}, 2900 cm^{-1}, 2100 cm^{-1}, and 1650 cm^{-1}. The peak at 1650 cm^{-1} was associated with the water molecules absorbed in the amorphous region [24]. The peak at 2100 cm^{-1} is attributed to free water content [25]. The peaks at 2900 and 3500 cm^{-1} correspond to CH_2 deformation and OH bonds, respectively [26].

The most important structural differences could be observed in the region located between 1500 and 700 cm^{-1} (Figure 2), considered the "fingerprint". These are related to information about changes in the polymer structure and starch conformation. The peaks at approximately 1412 cm^{-1} were assigned to the bending vibrations of the -CH_2 bonds and the stretching of the O-C-O bonds. The bands observed near 1340 cm^{-1} correspond to the bending vibrations of the C-O-H bonds and the twisting vibrations of the -CH_2 bond. Around 1150 cm^{-1}, a band was observed to be assigned to the vibrations of the C-O-C glycosidic bonds and the complete glucose ring that may be present in different vibrational modes and bending conformations [27]. At 1070 cm^{-1} and 1005 cm^{-1}, the spectra presented two bands assigned to the crystalline and amorphous regions of starch, respectively. The peak observed at 930 cm^{-1} was assigned to the vibrational modes of the α-(1→4) glycoside bonds of the starch glucose chain backbone. Finally, the signals observed at 850 cm^{-1} and 760 cm^{-1} correspond to the deformation of the C bonds (C1 of the glucose structure)-H and CH_2, and the stretching vibrations of the C-C bonds [27].

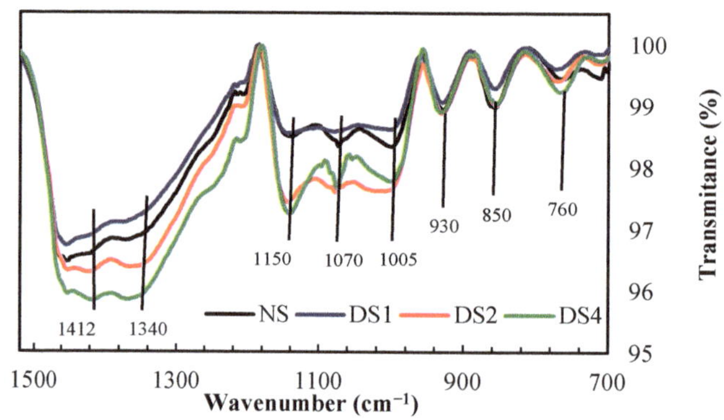

Figure 2. FT-IR spectrum "fingerprint" region of starch samples. q: volume (%). NS: 4.8% DS; DS1: 14.7% DS; DS2: 21.4% DS; DS4: 32.2% DS.

Some authors have suggested that the infrared spectroscopy technique is sensitive to the so-called "short-range order", defined as the order of the double helix of the starch structure, as opposed to the "long-range order" related to the packing of the double helices [28].

An increase in the intensity of the band located at around 1005 cm^{-1} has been associated with a decrease in the crystallinity of starch. On the other hand, an increase in the intensity of the band located at around 1070 cm^{-1} has been associated with an increase in the crystallinity of starch. The ratio (R) between the intensities (I) of the bands (R = I1070/I1005) has been analyzed to study changes in the crystallinity of starch granules because of different processes [26,29]. In this work, a decrease in the intensity of the band to 1070 cm^{-1} and an increase in the intensity of the band to 1005 cm^{-1} were observed, and therefore a decrease in the R ratio with the increase in the content of damaged starch. This result suggests that, by increasing the level of mechanical damage of the granules, the amorphous proportion of the polymeric structures that make up the granular arrangement increases.

3.3. X-Ray Diffraction (XRD)

Radial confirmation of amylopectin is the basis of the structure of starch granules [2]. Taking this into account, the short chains of amylopectin are what largely determine the level of crystallinity of the starch granules.

The diffraction patterns of starch samples with different damaged starch content showed the characteristic profiles of starch from cereals (A-type) (Figure 3). The main refraction peaks of this type of pattern are observed at 2°, 15° and 23°, and a doublet at approximately 17° and 18° [30]. The type of crystalline polymorph is determined by how the double helices of the amylopectin chains are organized and packed in the starch granule and by the average length of amylopectin chains. Wheat starch exhibits this diffraction pattern because, on average, it has short chain lengths [31].

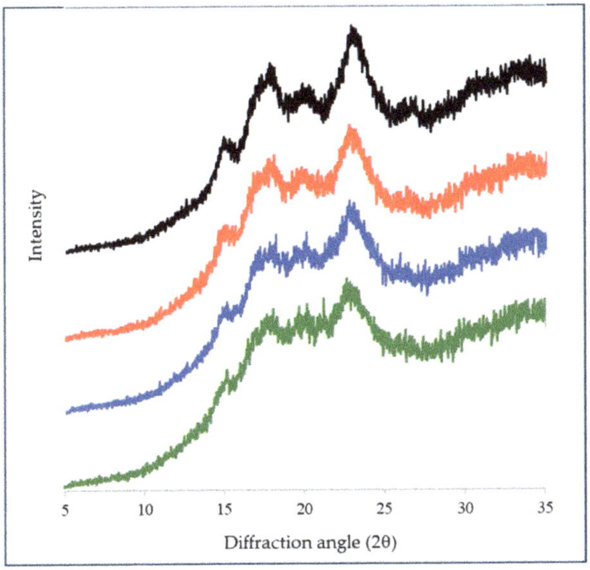

Figure 3. X-ray diffraction patterns of starch samples. NS: 4.8% DS (black); DS1: 14.7% DS (red); DS2: 21.4% DS (blue); DS4: 32.2% DS (green).

The intensity of the peaks in the diffractograms decreased with the amount of damaged starch (Figure 3). These findings suggest a loss of structural order in the starch granules. In agreement with these results, it has been shown that starch crystallinity, measured by X-ray diffraction, differential scanning calorimetry, and nuclear magnetic resonance, depends on the severity of the milling conditions and time [32].

The degree of total crystallinity of the starch was determined from the deconvolution of the peaks obtained from the diffraction patterns, and then calculating the integrals of the areas of the amorphous and crystalline peaks. The degree of total crystallinity for the starch samples with different damaged starch content gradually decreased ($p < 0.05$) from 34.1% (NS) to 32.3%; and 30.4% and 28.9% for the DS1, DS2, and DS4 samples, respectively. These results confirm that there was a loss of crystallinity of the damaged starch granules due to mechanical action. Franzoni et al. [33] found similar crystallinity values for starch samples with different contents of damaged starch, using low-field and single-sided NMR analysis.

Morrison et al. [34] observed a similar behavior and proposed that mechanical damage to starch granules causes a gradual loss of the molecular order of the polymers that form it. These authors noted that the crystallinity determined by X-ray diffraction and the double helix content of amylopectin decrease as the damaged starch content increases, leading to the idea that mechanical damage could produce a new amorphous material, which would be differentiated from the amorphous part of the native starch granules. The gelatinization enthalpy values correlated with the decrease in the crystallinity of the starch granules, as observed in previous studies [3,35]. The enthalpy of gelatinization of starch granules involves different thermodynamic events that include the swelling of the granules, fusion of crystals (endothermic processes), and hydration and rearrangement of the polymer chains (exothermic processes) [32].

Various authors have reported that mechanically treating starch samples results in defects accumulating in the crystal structure. Dome et al. [36] stated that the grinding process disrupts the double helices of amylopectin and the single helices of amylose. This disruption occurs due to the breaking of hydrogen bonds or glycosidic bonds within the crystalline regions. Therefore, milling causes the release of intercalated amylose molecules from the crystalline regions into the amorphous regions [36].

3.4. Scanning Electron Microscopy (SEM)

Scanning electron microscopy (SEM) is a widely used tool to study the structural characteristics and architecture of starch granules. Although image acquisition requires a high vacuum and sample preparation techniques that tend to damage or alter the structure of the granules [2], this microscopic technique has proven to be very useful for studying the surface and the internal structure of starch granules. A disadvantage of the SEM technique is that the morphological analysis obtained from the images is only qualitative. The images obtained using the SEM showed the presence of large and lenticular and small and spherical starch granules, which correspond to A-type and B-type granules, respectively (Figure 4).

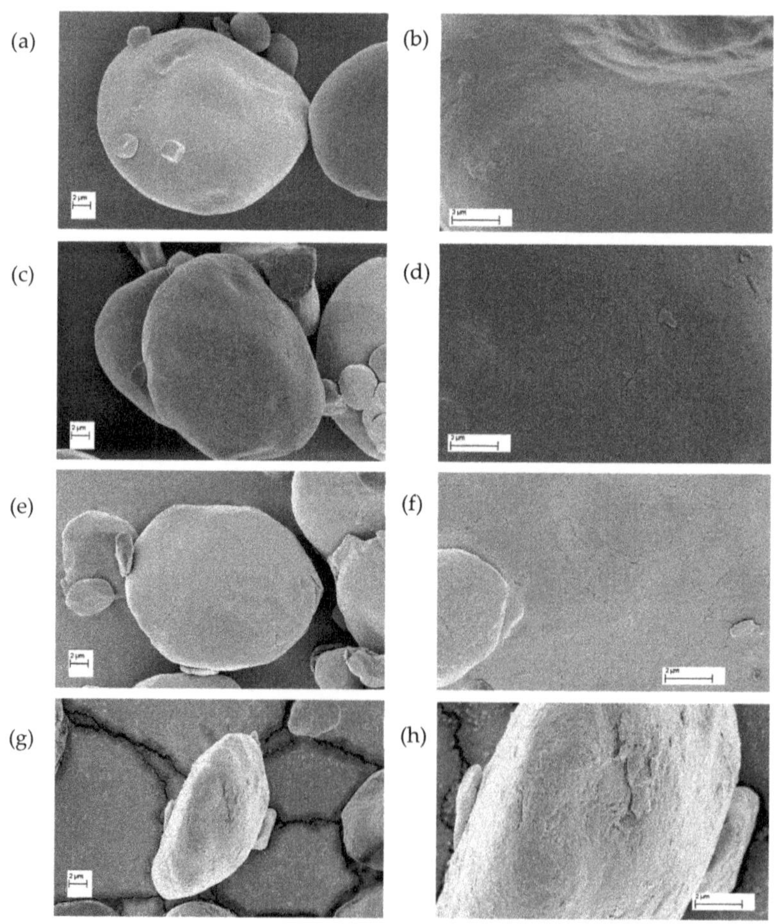

Figure 4. SEM images of starch samples. On the left: 3.00 K X magnification. On the right: 8.00 K X magnification. (**a**,**b**) NS: 4.8% DS, (**c**,**d**) DS1: 14.7% DS, (**e**,**f**) DS2: 21.4% DS, (**g**,**h**) DS4: 32.2% DS. (**a**,**c**,**e**,**g**) images: 3 kV, 3000×. (**b**,**d**,**f**,**h**) images: 3 kV, 8000×.

The analysis of the images of the native starch (NS) sample revealed that the surface of the granules was smooth and flat, although there were reliefs or marks associated with points where the granules could have been in close contact with another one [8,22]. However, in the images of samples DS1, DS2, and DS4, it was observed that the surface of the granules went from having a smooth and polished appearance to a rough and flaky appearance, because of the grinding process (Figure 4). Similarly, Tester et al. [22] reported that, as a consequence of grinding, the granular surfaces became distorted, rough, and with numerous clumps, and in this sense, few intact granules after this process were observed.

3.5. Water Absorption Capacity (WAC)

Damage to starch granules during milling increased their water absorption capacity (Table 1). However, the swelling power decreases when damaged starch granules are generated under extremely severe conditions [37]. Since milling can promote an alteration in the proportion of hydrogen bonding interactions, it is possible that, because of this phenomenon, more hydroxyl groups within the double helices become available to interact with the surrounding water molecules [38], thus increasing the water absorption capacity (and, at the same time, increasing its swelling power). Likewise, damaged starch granules have a structurally altered surface, which facilitates the access of water molecules, which consequently promotes their absorption [39]. However, excessive granular damage during milling may cause the collapse of the semi-crystalline layers and contribute to the decrease in the swelling power of the granules.

Table 1. Effect of DS on water absorption capacity of starch.

Sample	WAC
NS	1.93 ± 0.01 [a]
DS1	2.28 ± 0.05 [b]
DS2	2.53 ± 0.01 [c]
DS4	2.76 ± 0.04 [d]

WAC: water absorption capacity. NS: 4.8% DS; DS1: 14.7% DS; DS2: 21.4% DS; DS4: 32.2% DS. Values followed by different letters are significantly different ($p < 0.05$).

Wang and Copeland [40] suggested that the integrity of amylopectin dominates the swelling power of starch granules. Presumably, milling depolymerized amylose chains, which could result in a decrease in the swelling power of the granules, with the increase in the degree of milling [39]. From this, it is possible to infer that damaged starch may have higher solubility compared to native starch. The greater the severity of milling (and, consequently, the higher the damaged starch content), the greater the starch solubility [37,41]. Some authors have observed that solubility is positively correlated with damaged starch content [42]. The depolymerization of starch granules caused by the mechanical milling process increases the proportion of molecules of lower molecular weight and consequently increases their solubility. González et al. [43] observed a negative exponential correlation between the solubility and the degree of crystallinity of rice starch. According to this, the structural disorder (associated with the increase in the amorphous regions of the starch granules) caused by the reductions in the crystalline regions also contributed to the increase in solubility. Furthermore, the solubility of damaged starch increases noticeably in hot water (T° > 50 °C) in contrast to water at room temperature (~25 °C) [44]. The reason for this behavior could be that the molecules move faster at higher temperatures, which accelerates the dispersion of amylose from the amorphous region and other water-soluble factions.

3.6. Thermogravimetric Analysis (TGA)

Thermogravimetric analysis (TGA) is a thermal analysis method based on the measurement of the mass loss of a material as a function of temperature. The observed changes in thermal behavior can be attributed specifically to the evaporation of water from the system, due to heating. The curves of the percentage of weight loss (%) as a function of the temperature of the samples are shown in Figure 5. The percentage of weight loss at different temperatures is shown in Table 2.

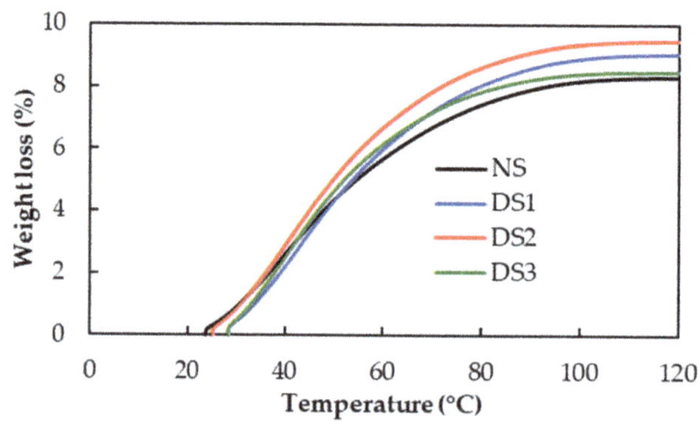

Figure 5. Curves of the percentage of weight loss as a function of temperature for starch samples, obtained by TGA. NS: 4.8% DS; DS1: 14.7% DS; DS2: 21.4% DS; DS4: 32.2% DS.

Table 2. Effect of DS on weight loss at 30, 50, 70, 90, 100, and 110 °C measured by TGA.

Sample	Weight Loss (%)					
	30 °C	50 °C	70 °C	90 °C	100 °C	150 °C
NS	4.1 ± 0.1 [a]	43.1 ± 0.5 [a]	77.1 ± 1.0 [a]	94.3 ± 1.0 [a]	97.8 ± 0.3 [a]	99.7 ± 0.2 [a]
DS1	4.8 ± 0.2 [ab]	47.7 ± 0.9 [b]	79.7 ± 0.5 [b]	94.7 ± 0.1 [ab]	98.0 ± 0.1 [ab]	99.6 ± 0.0 [a]
DS2	9.0 ± 0.6 [c]	53.3 ± 1.3 [c]	82.6 ± 0.6 [c]	95.7 ± 0.3 [bc]	98.5 ± 0.2 [bc]	99.6 ± 0.1 [a]
DS4	5.8 ± 0.6 [b]	55.5 ± 1.9 [c]	84.7 ± 0.4 [d]	96.7 ± 0.1 [c]	98.9 ± 0.2 [c]	99.5 ± 0.1 [a]

NS: 4.8% DS; DS1: 14.7% DS; DS2: 21.4% DS; DS4: 32.2% DS. Values followed by different letters in the same column are significantly different ($p < 0.05$).

According to the results, the percentage of water lost at 50 °C and 70 °C increased gradually and significantly ($p < 0.05$) with the increase in the damaged starch content. These results suggest that, as damage to the starch granule structures increased, the water bound to them was released at a higher rate. In addition, the temperature at which different percentages of water were lost (30, 50, 75, 80, 90, and 100%) was determined, which are detailed in Table 3. As a general trend, it can be observed that the temperature at which a certain percentage of water was lost from the samples decreased with the increase in the damaged starch content. This effect was more significant at water percentages of 75, 80, and 90%. As in the analysis previously carried out, the results indicated that, by increasing the content of damaged starch, water was released faster, that is, at lower temperatures.

The DTGAs curves (Figure 6) seem to reflect the presence of two overlapping events, a main peak at around 45 °C and a second one (a small shoulder of the first one) at around 60 °C. Two peaks were obtained for each starch sample, indicating the existence of two populations of water: one population that presents weaker interactions with the starch granules (population 1) and another population that shows stronger interactions with them (population 2). Table 4 shows the temperatures at which the main peak occurs for each sample and the areas of each peak relative to the total area of the curves. Water in the presence of heat establishes a high-pressure vapor gradient within the granular structure. Starch retains water through the hydrogen bonds between amylose and amylopectin branches and inter-amylopectin helices [45].

Table 3. Effect of DS on temperature at 30, 50, 75, 80, 90, and 100% of weight loss measured by TGA.

Sample	Temperature (°C)					
	30%	50%	75%	80%	90%	100%
NS	40.1 ± 0.9 [a]	51.2 ± 2.9 [a]	67.1 ± 2.5 [b]	71.5 ± 2.4 [c]	82.8 ± 2.1 [c]	165.3 ± 43.9 [a]
DS1	42.3 ± 0.4 [a]	51.1 ± 0.5 [a]	66.2 ± 0.4 [b]	70.3 ± 0.4 [bc]	81.8 ± 0.3 [bc]	195.8 ± 0.3 [a]
DS2	39.7 ± 0.6 [a]	48.5 ± 0.6 [a]	63.3 ± 0.7 [ab]	67.5 ± 0.6 [ab]	78.9 ± 0.7 [ab]	196.7 ± 0.8 [a]
DS4	39.8 ± 0.8 [a]	47.5 ± 1.0 [a]	61.5 ± 0.7 [a]	65.6 ± 0.5 [a]	76.5 ± 0.3 [a]	192.9 ± 5.2 [a]

NS: 4.8% DS; DS1: 14.7% DS; DS2: 21.4% DS; DS4: 32.2% DS. Values followed by different letters in the same column are significantly different ($p < 0.05$).

Table 4. Effect of DS on peak temperature (T) and relative peak area percentages (A) for the deconvolved peaks measured by TGA.

Sample	T1 (°C)	A1 (%)	T2 (°C)	A2 (%)
NS	41.7 ± 0.5 [b]	37.5 ± 0.3 [a]	61.5 ± 0.3 [c]	62.5 ± 0.3 [c]
DS1	40.4 ± 0.4 [b]	38.7 ± 0.5 [b]	58.2 ± 0.3 [b]	61.3 ± 0.5 [b]
DS2	37.9 ± 0.7 [a]	42.5 ± 0.2 [c]	56.4 ± 0.8 [b]	57.5 ± 0.2 [a]
DS4	37.7 ± 0.5 [a]	43.0 ± 0.3 [c]	51.7 ± 0.9 [a]	57.0 ± 0.3 [a]

T1 and A1 correspond to peak 1. T2 and A2 correspond to the peak. NS: 4.8% DS; DS1: 14.7% DS; DS2: 21.4% DS; DS4: 32.2% DS. Values followed by different letters in the same column are significantly different ($p < 0.05$).

After the elimination of excess water, drying speeds always follow three well-defined stages: the first stage is related to the evaporation of interstitial water between the granules, which evaporates almost as pure water; the second stage is associated with the surface water of the granules; and finally, the third stage relates to the most strongly bound water (or structural water). The water molecules most tightly bound to starch can be interpreted as water in the crystalline regions and the amorphous regions of the starch granules.

Water population 1 presented maximum peak temperatures between 41.7 °C (NS) and 37.7 °C (DS4), with a significant decrease observed as the damaged starch content increased. On the other hand, water population 2 also showed a gradual and significant ($p < 0.05$) decrease in peak maximum temperatures from 61.5 °C (NS) to 51.7 °C (DS4). Because water population 1 appeared at lower temperatures, it can be related to excess water, the interstitial (located between the granules), and the water found on the surface of the starch granules. The water population 2 can be attributed to those molecules that are more closely bound to the starch in the pores and channels of the granules, associated with the amorphous and crystalline regions.

By increasing the damaged starch content from 4.8 to 32.2%, a decrease in the peak temperature of both populations was detected. However, this decrease was greater for the peak of water population 2 (10.2 °C for population 2 and 3 °C for population 1). In contrast, the increase in damaged starch content caused a gradual and significant increase ($p < 0.05$) in the relative area of population 1, and a gradual and significant ($p < 0.05$) decrease in the relative area of population 2. The area under the curve gives us information about the number of water molecules associated with that population.

In summary, the increase in the content of damaged starch caused different effects: (a) an increase in the proportion of less bound water and a decrease in its evaporation temperature (fraction of water found on the surface of the granules and in the interstitial regions) and (b) a decrease in the amount of more strongly bound and in its evaporation temperature (fraction of water found in the crystalline and amorphous regions of the granules). Consequently, the greater the degree of damage to the starch granules, the more easily water was released, and the proportion of less bound water increased.

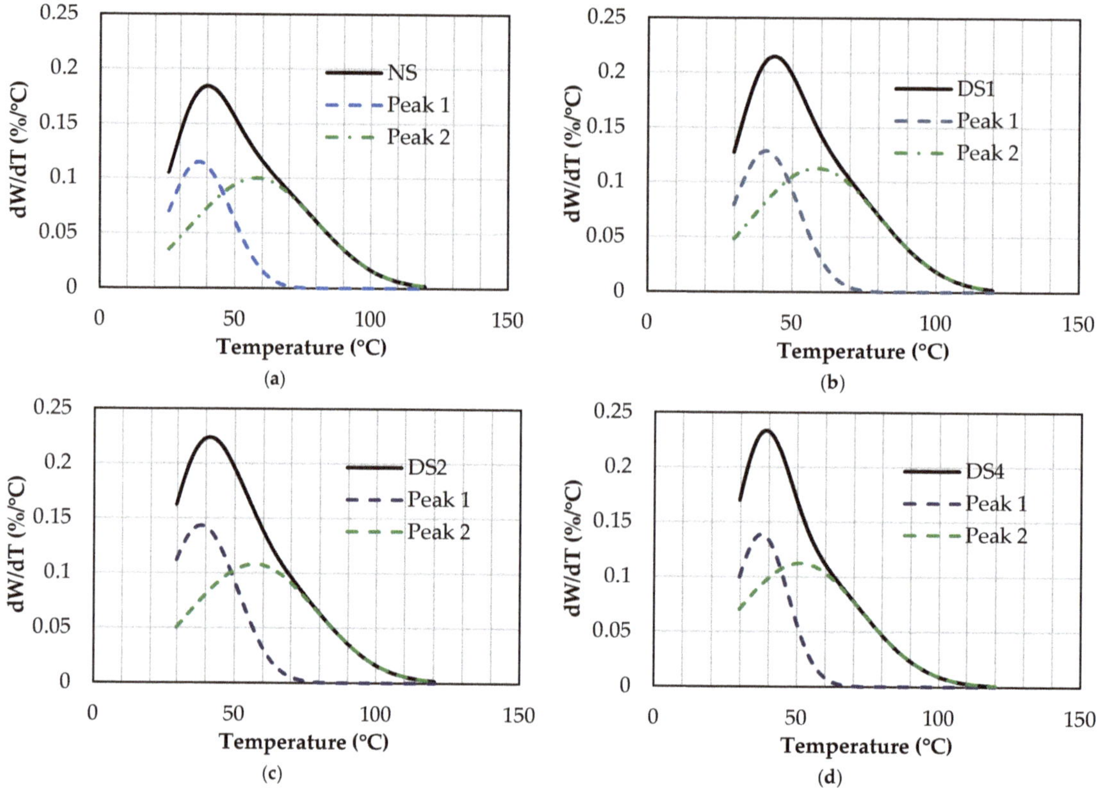

Figure 6. Curves of derivate of weight loss as a function of temperature (dW/dT) for starch samples, obtained by TGA. (**a**) NS: 4.8% DS; (**b**) DS1: 14.7% DS; (**c**) DS2: 21.4% DS; (**d**) DS4: 32.2% DS.

3.7. Dynamic Vapor Sorption (DVS)

The relationship between the total moisture content and the corresponding water activity of a product at a constant temperature generates what is called a sorption isotherm [46].

All the curves showed an increase in the equilibrium moisture content with increasing water activity (aw) (Figure 7 shows those obtained for NS and DS4). The sigmoidal shape found in all isotherms of the samples is typical of type II isotherms based on Brunauer's classification [47]. This type of isotherm is due to the monolayer–multilayer sorption theory.

The DVS curves obtained for the samples with different damaged starch content were similar, although, a change in the increasing slope of the curve was observed from the aw values of 0.7. This upward curvature can result from the condensation of vapor molecules within the interstitial spaces of materials or polymers. The sorption of water by starch occurs through the interaction of water molecules with hydroxyl groups, which are primarily found in the amorphous regions and on the surface of starch crystals. In contrast, the crystalline regions of starch typically resist moisture penetration and have energetically less favorable sites for water sorption [48]. However, in the aw region above 0.7, the sorption patterns showed different slopes of increasing moisture content with increasing water activity. Comparing the adsorption process of the NS and DS4 samples, above an aw of 0.7, a higher water adsorption was observed when the damaged starch content was 32.2% (DS4) (Figure 7). This was inversely related to the crystallinity percentages of the starch samples, since the DS4 sample presented a lower crystallinity (determined by XRD), and therefore, an increase in the proportion of amorphous regions.

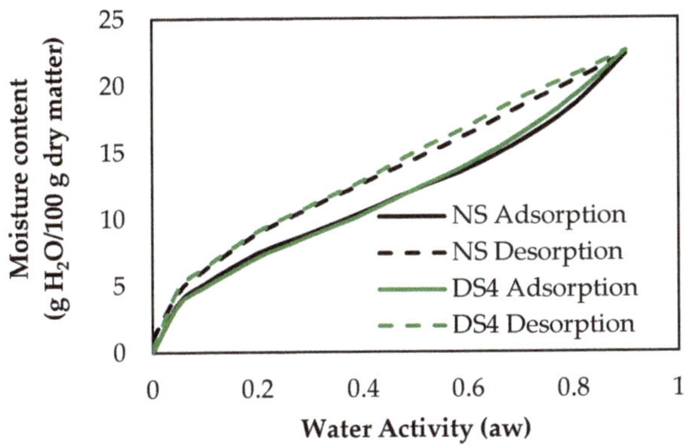

Figure 7. Sorption isotherms for NS and DS4 starch samples. NS: 4.8% DS; DS4: 32.2% DS.

In the wet state prior to desorption, nearly all the hydroxyl groups in the molecular structure of the starches were occupied by absorbed water molecules. During the desorption process, these hydroxyl groups were gradually released, allowing them to come closer together and form hydrogen bonds with each other. Through this interaction, the interstitial spaces of the polymers that formed starch were irreversibly reduced, so that the trapping of capillary water molecules decreased. Some researchers have suggested that irreversible polymer shrinkage may create kinetic barriers to moisture migration, which could account for the observed "non-zero" moisture contents at aw = 0 during desorption [48].

The two analyzed models fitted the experimental data (correlation coefficients > 0.99) related to the water adsorption process. Figure 8 shows the curves obtained experimentally and those obtained through mathematical adjustments for the GAB and BET models. The adjustment parameters of the isotherms from the models applied for water adsorption are shown in Table 5. The monolayer value (associated with the parameter Wm) for the damaged starch samples was estimated using the GAB and BET models. The monolayer values were similar for all the starch samples, ranging from 0.086 to 0.090 g water/g solids and 0.068 to 0.070 g water/g solids for GAB and BET, respectively. However, a decrease in the value of parameter C was observed when DS increased from 4.8 to 32.2%. The C value is related to temperature and isosteric heat (also known as differential enthalpy of sorption, which provides a measure of the energy of the union between the adsorbed water and the molecules of food) during the adsorption of water in food. In the GAB model, this decrease was observed gradually, from 18.1 for NS to 14.5 for DS4. While in the BET model, the decrease was from 20.1 to 18.6 for AN and DS4, respectively. These modifications suggested that the driving force for the monolayer attachment also decreased. Therefore, the energy involved in the unions of the starch polymer chains with the adsorbed water was lower with increasing mechanical damage.

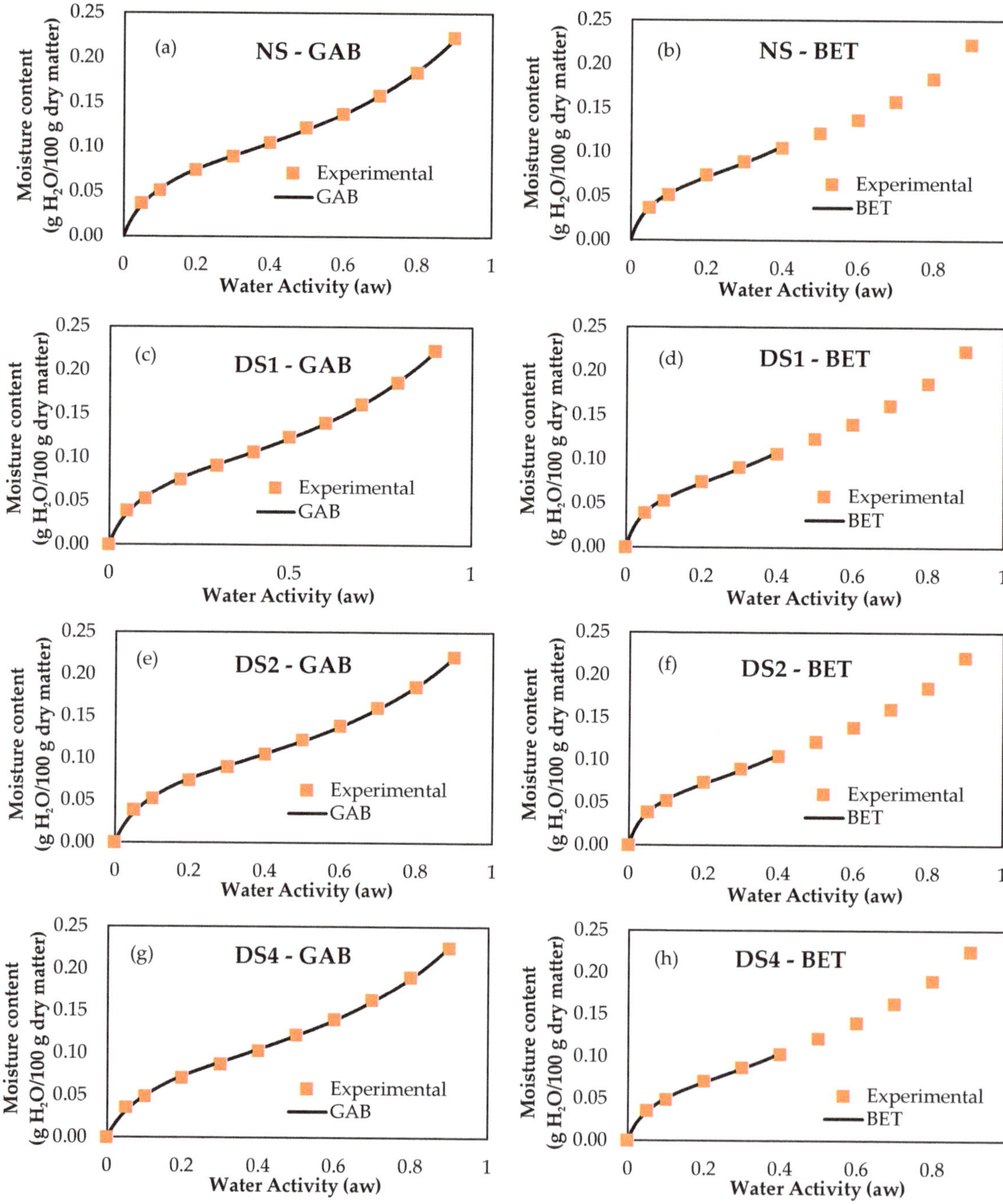

Figure 8. The sorption curve fits with the GAB and BET models for the starch samples. (**a**,**b**) NS: 4.8% DS; (**c**,**d**) DS1: 14.7% DS; (**e**,**f**) DS2: 21.4% DS; (**g**,**h**) DS4: 32.2% DS.

Table 5. Effect of DS on parameters determined from the fit of the GAB and BET models to the experimental data for starches measured by DVS.

Model	Sample	W_m	K_G	C	R^2	s
GAB	NS	0.086	0.693	18.1	0.99	0.0013
	DS1	0.089	0.685	18.4	0.99	0.0015
	DS2	0.088	0.684	18.0	0.99	0.0016
	DS4	0.090	0.690	14.5	0.99	0.0022
BET	NS	0.069		20.1	0.99	0.0017
	DS1	0.070		21.5	0.99	0.0016
	DS2	0.068		21.3	0.99	0.0016
	DS4	0.068		18.6	0.99	0.0014

W_m: monolayer moisture content; Kg and C: model constants; s is the standard deviation of the fit. NS: 4.8% DS; DS1: 14.7% DS; DS2: 21.4% DS; DS4: 32.2% DS.

4. Conclusions

The damage caused by the milling process of the starch granules produced physical changes in their structures. These modifications could be observed qualitatively by scanning electron microscopy, where more irregular, rough, and less uniform granule surfaces were distinguished compared to those of the native starch granules. Analyzing the particle size distribution, it was concluded that the B-type and A-type granules were susceptible to the effects of the mechanical damage produced during the milling times analyzed.

Different techniques used during this stage of the work (X-ray diffraction and infrared spectroscopy) allowed us to confirm that there was a decrease in the crystallinity of the starch granules by the milling process. This loss of the degree of crystallinity led to modifications in the interactions with water molecules.

TGA and WAC analyses suggested that the increase in the content of damaged starch generated a greater absorption of water and that the interactions established between water and the polymers of the damaged starch samples were weaker than those established with native starch. Consequently, the greater the degree of damaged starch, the higher the proportion of the less bound water and the more easily water was released. Additionally, DVS analysis showed a higher absorption of water at high aw values and a lower energy involved in the unions of the starch polymer chains with the adsorbed water when increased the mechanical damage.

The main findings of this research provide a deeper understanding of the changes associated with the mechanical damage of the starch granular structure and its interaction with water. Further studies are needed to relate these results to the hydration properties of flours and their relationship with the rheological and thermal characteristics of starch-based systems.

Supplementary Materials: The following supporting information can be downloaded at https://www.mdpi.com/article/10.3390/foods14010021/s1, Figure S1. FT-IR spectrum of starch samples.

Author Contributions: A.G.T. led the investigation, design, formal analysis, and writing of the original draft. E.J.C.P. and G.N.B. assisted with the investigation and formal analysis, participated in the analysis and discussion of the results, and provided review and editing of the draft. P.D.R. organized the conceptualization of the project, participated in the analysis and discussion of the results, and provided review and editing of the draft. All authors have read and agreed to the published version of the manuscript.

Funding: This research was supported by the "Agencia Nacional de Promoción de la Investigación, el Desarrollo Tecnológico y la Innovación" (PICT-2016-1150 AND PICT-2019-1122), the "Consejo Nacional de Ciencia y Técnica" (CONICET) from Argentina and the "Secretaría de Ciencia y Tecnología" of the "Universidad Nacional de Córdoba" (UNC).

Institutional Review Board Statement: Not applicable.

Informed Consent Statement: Not applicable.

Data Availability Statement: The original contributions presented in this study are included in the article/Supplementary Materials. Further inquiries can be directed to the corresponding author.

Acknowledgments: The authors would like to thank Fernanda Quiroga (CPA-ICYTAC, CONICET-UNC) for technical support.

Conflicts of Interest: The authors declare that no conflicts of interest exist.

References

1. Apriyanto, A.; Compart, J.; Fettke, J. A review of starch, a unique biopolymer—Structure, metabolism and in planta modifications. *Plant Sci.* **2022**, *318*, 111223. [CrossRef] [PubMed]
2. Pérez, S.; Baldwin, P.M.; Gallant, D.J. Structural features of starch granules I. In *Starch: Chemistry and Technology*, 3rd ed.; BeMiller, J.N., Whistler, R.L., Eds.; Academic Press, Elsevier Inc.: New York, NY, USA, 2009; pp. 149–187.
3. Barrera, G.N.; León, A.E.; Ribotta, P.D. Effect of damaged starch on wheat starch thermal behavior. *Starch/Staërke* **2012**, *64*, 786–793. [CrossRef]
4. Berton, B.; Scher, J.; Villieras, F.; Hardy, J. Measurement of hydration capacity of wheat flour: Influence of composition and physical characteristics. *Powder Technol.* **2002**, *128*, 326–331. [CrossRef]
5. Lv, Y.; Zhang, L.; Li, M.; He, X.; Hao, L.; Dai, Y. Physicochemical properties and digestibility of potato starch treated by ball milling with tea polyphenols. *Int. J. Biol. Macromol.* **2019**, *129*, 207–213. [CrossRef]
6. Arya, S.S.; Prajakta, D.S.; Ashish, G.W. Importance of damaged starch in bakery products-a review. *Starch/Staërke* **2015**, *1*, 2019.
7. Barrera, G.N.; León, A.E.; Ribotta, P.D. Use of enzymes to minimize the rheological dough problems caused by high levels of damaged starch in starch–gluten systems. *J. Sci. Food Agric.* **2016**, *96*, 2539–2546. [CrossRef]
8. Barrera, G.N.; Calderón-Domínguez, G.; Chanona-Pérez, J.; Gutiérrez-López, G.F.; León, A.E.; Ribotta, P.D. Evaluation of the mechanical damage on wheat starch granules by SEM, ESEM, AFM and texture image analysis. *Carbohydr. Polym.* **2013**, *98*, 1449–1457. [CrossRef]
9. Lu, X.; Wang, Y.; Li, Y.; Huang, Q. Assembly of Pickering emulsions using milled starch particles with different amylose/amylopectin ratios. *Food Hydrocoll.* **2018**, *84*, 47–57. [CrossRef]
10. Liu, X.; Li, E.; Jiang, Y.; Tian, Y. Milling process of Starch. In *Physical Modifications of Starch*, 3rd ed.; Sui, Z., Kong, X., Eds.; Springer: Singapore, 2018; pp. 155–174.
11. Lee, Y.T.; Shim, M.J.; Goh, H.K.; Mok, C.; Puligundla, P. Effect of jet milling on the physicochemical properties, pasting properties, and in vitro starch digestibility of germinated brown rice flour. *Food Chem.* **2019**, *282*, 164–168. [CrossRef]
12. Yu, J.; Wang, S.; Wang, J.; Li, C.; Xin, Q.; Huang, W.; Zhang, Y.; He, Z.; Wang, S. Effect of laboratory milling on properties of starches isolated from different flour millstreams of hard and soft wheat. *Food Chem.* **2015**, *172*, 504–514. [CrossRef]
13. Liu, C.; Li, L.; Hong, J.; Zheng, X.; Bian, K.; Sun, Y.; Zhang, J. Effect of mechanically damaged starch on wheat flour, noodle and steamed bread making quality. *Int. J. Food Sci. Technol.* **2014**, *49*, 253–260. [CrossRef]
14. Wang, S.; Yu, J.; Xin, Q.; Wang, S.; Copeland, L. Effects of starch damage and yeast fermentation on acrylamide formation in bread. *Food Control.* **2017**, *73*, 230–236. [CrossRef]
15. American Association of Cereal Chemists (AACC). *Approved Methods of the AACC*, 9th ed.; The Association: St. Paul, MN, USA, 2000.
16. Palavecino, P.M.; Penci, M.C.; Calderón-Domínguez, G.; Ribotta, P.D. Chemical composition and physical properties of sorghum flour prepared from different sorghum hybrids grown in Argentina. *Starch/Staërke* **2016**, *68*, 1055–1064. [CrossRef]
17. Blanco Canalis, M.S.; León, A.E.; Ribotta, P.D. Incorporation of dietary fiber on the cookie dough. Effects on thermal properties and water availability. *Food Chem.* **2019**, *271*, 309–317. [CrossRef]
18. Gely, M.C.; Giner, S.A. Water-corn equilibrium: Temperature dependency of the GAB model parameters and calculation of the heat of sorption. *Dry. Technol.* **2000**, *18*, 1449–1464. [CrossRef]
19. Quirijns, E.J.; Van Boxtel, A.J.B.; Van Loon, W.K.P.; Van Straten, G. Sorption isotherms, GAB parameters and isosteric heat of sorption. *J. Sci. Food Agric.* **2005**, *85*, 1805–1814. [CrossRef]
20. Escalona-García, L.A.; Pedroza-Islas, R.; Natividad, R.; Rodríguez-Huezo, M.E.; Carrillo-Navas, H.; Pérez-Alonso, C. Oxidation kinetics and thermodynamic analysis of chia oil microencapsulated in a whey protein concentrate-polysaccharide matrix. *J. Food Eng.* **2016**, *175*, 93–103. [CrossRef]

21. Jay-line, J. Structural features of starch granules II. In *Starch: Chemistry and Technology*, 3rd ed.; BeMiller, J.N., Whistler, R.L., Eds.; Academic Press, Elsevier Inc.: New York, NY, USA, 2009; pp. 192–236.
22. Tester, R.F.; Morrison, W.R.; Gidley, M.J.; Kirkland, M.; Karkalas, J. Properties of damaged starch granules. III. Microscopy and particle size analysis of undamaged granules and remnants. *J. Cereal Sci.* **1994**, *19*, 59–67.
23. Hong, J.; Li, L.; Li, C.; Liu, C.; Zheng, X.; Bian, K. Effect of Heat–Moisture Treatment on Physicochemical, Thermal, Morphological, and Structural Properties of Mechanically Activated Large A- and Small B-Wheat Starch Granules. *J. Food Sci.* **2019**, *84*, 2795–2804. [CrossRef]
24. Xu, J.; Kuang, Q.; Wang, K.; Zhou, S.; Wang, S.; Liu, X.; Wang, S. Insights into molecular structure and digestion rate of oat starch. *Food Chem.* **2017**, *220*, 25–30. [CrossRef]
25. Olsson, A.; Salm, L. The association of water to cellulose and hemicellulose in paper examined by FTIR spectroscopy. *Carbohydr. Res.* **2004**, *339*, 813–818. [CrossRef] [PubMed]
26. Wang, S.; Wang, J.; Zhang, W.; Li, C.; Yu, J.; Wang, S. Molecular order and functional properties of starches from three waxy wheat varieties grown in China. *Food Chem.* **2015**, *181*, 43–50. [CrossRef] [PubMed]
27. Dankar, I.; Haddarah, A.; Omar, F.E.L.; Pujolà, M.; Sepulcre, F. Characterization of food additive-potato starch complexes by FTIR and X-ray diffraction. *Food Chem.* **2018**, *260*, 7–12. [CrossRef] [PubMed]
28. Sevenou, O.; Hill, S.E.; Farhat, I.A.; Mitchell, J.R. Organisation of the external region of the starch granule as determined by infrared spectroscopy. *Int. J. Biol. Macromol.* **2002**, *31*, 79–85. [CrossRef]
29. Warren, F.J.; Gidley, M.J.; Flanagan, B.M. Infrared spectroscopy as a tool to characterise starch ordered structure—A joint FTIR—ATR, NMR, XRD and DSC study. *Carbohydr. Polym.* **2016**, *139*, 35–42. [CrossRef]
30. Zeng, J.; Li, G.; Gao, H.; Ru, Z. Comparison of A and B Starch Granules from Three Wheat Varieties. *Molecules* **2001**, *16*, 10570–10591. [CrossRef]
31. Maningat, C.; Seib, P.; Bassi, S.; Woo, K. Wheat Starch: Production, Properties, Modification and Uses. In *Starch: Chemistry and Technology*, 3rd ed.; BeMiller, J.N., Whistler, R.L., Eds.; Academic Press, Elsevier Inc.: New York, NY, USA, 2009; pp. 441–510.
32. Biliaderis, C.G. Structural Transitions and Related Physical Properties of Starch. In *Starch: Chemistry and Technology*, 3rd ed.; BeMiller, J.N., Whistler, R.L., Eds.; Academic Press, Elsevier Inc.: New York, NY, USA, 2009; pp. 292–372.
33. Franzoni, M.B.; Teobaldi, A.G.; Barrera, G.N.; Acosta, R.H.; Ribotta, P.D.; Velasco, M.I. NMR characterization of structure and moisture sorption dynamics of damaged starch granules. *Carbohydr. Polym.* **2022**, *285*, 119220. [CrossRef]
34. Morrison, W.R.; Tester, R.F.; Gidley, M.J. Properties of damaged starch granules II. Crystallinity, molecular order, and gelatinisation of ball milled starches. *J. Cereal Sci.* **1994**, *19*, 209–217. [CrossRef]
35. Teobaldi, A.G.; Barrera, G.N.; Severini, H.; Ribotta, P.D. Influence of damaged starch on thermal and rheological properties of wheat starch and wheat starch–gluten systems in water and sucrose. *J. Sci. Food Agric.* **2023**, *103*, 1377–1384. [CrossRef]
36. Dome, K.; Podgorbunskikh, E.; Bychkov, A.; Lomovsky, O. Changes in the crystallinity degree of starch having different types of crystal structure after mechanical pretreatment. *Polymers* **2020**, *12*, 641. [CrossRef]
37. Moraes, J.; Alves, F.S.; Franco, C.M.L. Effect of ball milling on structural and physicochemical characteristics of cassava and Peruvian carrot starches. *Starch/Staërke* **2013**, *65*, 200–209. [CrossRef]
38. Hackenberg, S.; Verheyen, C.; Jekle, M.; Becker, T. Effect of mechanically modified wheat flour on dough fermentation properties and bread quality. *Eur. Food Res. Technol.* **2017**, *243*, 287–296. [CrossRef]
39. Wang, Q.; Li, L.; Zheng, X. A review of milling damaged starch: Generation, measurement, functionality and its effect on starch-based food systems. *Food Chem.* **2020**, *315*, 126267. [CrossRef] [PubMed]
40. Wang, S.; Copeland, L. Phase transitions of pea starch over a wide range of water content. *J. Agric. Food Chem.* **2012**, *60*, 6439–6446. [CrossRef] [PubMed]
41. Li, C.; Dhital, S.; Gilbert, R.G.; Gidley, M.J. High-amylose wheat starch: Structural basis for water absorption and pasting properties. *Carbohydr. Polym.* **2020**, *245*, 116557. [CrossRef]
42. Liu, C.; Hong, J.; Zheng, X. Effect of heat-moisture treatment on morphological, structural and functional characteristics of ball-milled wheat starches. *Starch/Staërke* **2017**, *69*, 1500141. [CrossRef]
43. González, L.C.; Loubes, M.A.; Tolaba, M.P. Incidence of milling energy on dry-milling attributes of rice starch modified by planetary ball milling. *Food Hydrocoll.* **2018**, *82*, 155–163. [CrossRef]
44. Wu, F.; Li, J.; Yang, N.; Chen, Y.; Jin, Y.; Xu, X. The Roles of Starch Structures in the Pasting Properties of Wheat Starch with Different Degrees of Damage. *Starch/Staërke* **2018**, *70*, 1700190. [CrossRef]
45. Orlowska, M.; Utzig, E.; Randzio, S.L. Thermogravimetric study of water state in wheat starch gels obtained under high pressures. *Ann. N. Y. Acad. Sci.* **2010**, *1189*, 55–61. [CrossRef]
46. Barbosa-Cánovas, G.V.; Fontana, A.J.; Schmidt, S.J.; Labuza, T.P. *Water Activity in Foods. Fundamentals and Applications*, 2nd ed.; Blackwell Publishing: Hoboken, NJ, USA, 2020; pp. 207–226.

47. Aviara, N.A. Moisture sorption isotherms and isotherm model performance. Evaluation for food and agricultural products. In *Sorption in 2020s*, 3rd ed.; Kyzas, G., Lazaridis, N., Eds.; IntechOpen: London, UK, 2020; p. 143.
48. Mutungi, C.; Schuldt, S.; Onyango, C.; Schneider, Y.; Jaros, D.; Rohm, H. Dynamic Moisture Sorption Characteristics of Enzyme-Resistant Recrystallized Cassava Starch. *Biomacromolecules* **2011**, *12*, 660–671. [CrossRef]

foods

MDPI

Article

The Wheat Starchy Endosperm Protein Gradient as a Function of Cultivar and N-fertilization Is Reflected in Mill Stream Protein Content and Composition

Wisse Hermans, Justine Busschaert, Yamina De Bondt, Niels A. Langenaeken and Christophe M. Courtin *

Laboratory of Food Chemistry and Biochemistry & Leuven Food Science and Nutrition Research Centre (LFoRCe), KU Leuven, Kasteelpark Arenberg 20, B-3001 Heverlee, Belgium; wisse.hermans@kuleuven.be (W.H.); yamina.debondt@kuleuven.be (Y.D.B.); niels.langenaeken@kuleuven.be (N.A.L.)
* Correspondence: christophe.courtin@kuleuven.be; Tel.: +32-1632-1917

Abstract: Within the wheat starchy endosperm, the protein content increases biexponentially from the inner to outer endosperm. Here, we studied how this protein gradient is reflected in mill fractions using three cultivars (Claire, Apache, and Akteur) grown without and with N-fertilization (300 kg N ha^{-1}). The increasing protein content in successive break fractions was shown to reflect the protein gradient within the starchy endosperm. The increasing protein content in successive reduction fractions was primarily due to more aleurone contamination and protein-rich material being harder to reduce in particle size. The miller's bran fractions had the highest protein content because of their high sub-aleurone and aleurone content. Additionally, the break fractions were used to deepen our understanding of the protein composition gradient. The gradient in relative gluten content, increasing from inner to outer endosperm, was more pronounced without N-fertilization than with and reached levels up to 87.3%. Regarding the gluten composition gradient, no consistent trends were observed over cultivars when N-fertilization was applied. This could, at least partly, explain why there is no consensus on the gluten composition gradient in the literature. This study aids millers in managing fluctuations in the functionality of specific flour streams, producing specialized flours, and coping with lower-quality wheat.

Keywords: milling; protein distribution; gluten; break fractions; sub-aleurone; N-fertilization; wheat varieties; flour quality

1. Introduction

Wheat (*Triticum aestivum* L.) accounts for 19% of global protein intake, making it the number one protein source worldwide [1]. A wheat kernel comprises four main parts: the botanical bran (pericarp, testa, and nucellar epidermis) (6–8% *w*/*w*), aleurone layer (6–7% *w*/*w*), starchy endosperm (80–84% *w*/*w*) and embryo (3% *w*/*w*) [2,3]. The starchy endosperm is spatially heterogeneous in architecture and composition [4,5]. Small, protein-rich sub-aleurone cells can be found at its periphery [5–7]. The prismatic endosperm cells are located beneath the sub-aleurone cells, extend towards the center of the cheeks, and measure 128–200 μm × 40–60 μm × 40–60 μm [5]. The central endosperm cells lie in the center of the cheeks, are round or polygonal, low in protein, and have dimensions of 72–144 μm × 69–120 μm [5]. The compositional heterogeneity within the starchy endosperm includes a gradient in protein content and protein composition from the center of the cheeks towards the boundary with the aleurone layer [4,6–9]. The gradient in protein content can be described as an ascending biexponential curve with protein contents as high as 32% beneath the aleurone layer due to the presence of protein-rich sub-aleurone cells [7]. The extent of this gradient highly depends on the cultivar and the level of nitrogen fertilization (N-fertilization) [7].

Millers aim to separate the starchy endosperm from the aleurone layer and botanical bran and grind this into white flour using roller milling, a dry fractionation process. Using multiple sets of corrugated break rolls, smooth reduction rolls, and plansifters or sieves, more than 50 white flour fractions [10] and 10 bran-rich side streams are produced in an industrial mill. Fluctuations in the functionality of specific flour streams, stemming from differences in the starchy endosperm protein gradient, may introduce variability in the miller's end products. At the same time, these fluctuations also create opportunities for the development of customized flour formulations. Insight into how the protein gradient in the endosperm is reflected in mill streams is, therefore, of utmost importance.

In the scientific literature, laboratory roller mills are often used. They operate on a similar principle as industrial mills but vary in terms of the number of flour- and bran-rich fractions they yield, as well as their efficiency (flour yield). Pairs of corrugated break rolls open the kernels and scrape off the resulting bran particles to remove the adhering starchy endosperm [11]. The pairs of smooth reduction rolls reduce the starchy endosperm particles to fine flour [11]. It was reported that the last break (BFs) and reduction fractions (RFs) are higher in protein content compared to the other flour fractions [12,13]. This is often ascribed to aleurone and botanical bran contamination [12,14]. However, Tosi et al. [14] hypothesized that the lower protein content of the first BF and RF could be attributed not only to lower levels of bran contamination but also to the fact that these could originate from the inner endosperm, while the later BFs and RFs would be derived from the outer endosperm. Simmons and Sutton [15] and Wang et al. [16] observed that the increase in protein content from the first to last BF was more outspoken than throughout the RFs. To the best of our knowledge, there is currently no literature available on the protein content of mill fractions derived from different cultivars grown at different levels of N-fertilization, and, therefore, also displaying a high variability in gradient in protein content within the starchy endosperm.

In addition to the protein content, the protein composition plays a crucial role in determining the functionality of flour streams [17]. Two important aspects concerning protein composition are the relative gluten content (the proportion of proteins being gluten proteins) and the gluten composition (the relative proportion of gluten protein types) [4].

Sequential pearling of kernels is not very well suited for studying the gradient in relative gluten content due to its limited spatial specificity [4,18]. The non-spherical shape of wheat kernels and the crease cause outer endosperm fractions to be contaminated with aleurone material, which does not contain gluten [19]. Recently, Hermans et al. [9] used laser microdissection to separate sub-aleurone tissue from inner endosperm tissue and showed with proteomics that the sub-aleurone has a significantly higher relative gluten content than the inner endosperm. Although this study effectively detected significant differences in relative gluten content, it could not provide accurate mass-based relative gluten contents [9]. However, it did give a very strong indication that the gradient in relative gluten content increases from the inner to outer endosperm. Unlike the gradient in relative gluten content, sequential pearling of kernels can be used to investigate the gradient in gluten composition. However, the results reported by Tosi et al. [6] and He et al. [8] were inconclusive. For example, for various wheat cultivars and varying levels of N-fertilization, the proportion of high molecular weight glutenin subunits (HMW-GS) (expressed relative to total gluten content) was either observed to be higher, lower, or not significantly different in the sub-aleurone-enriched fraction compared to the inner endosperm-enriched fractions [6,8]. In both pearling studies, the presence of aleurone contamination could also interfere with the SDS-PAGE and SE-HPLC analysis performed in these studies, as the molecular weight of aleurone proteins ranges from <10 kDa to near 100 kDa [20]. Hermans et al. [9] did not detect a significant difference in HMW-GS protein proportion between the sub-aleurone and inner endosperm. Hence, there is no consensus on the gradient in gluten composition, while there are strong indications that the gradient in relative gluten content increases from the inner endosperm toward the sub-aleurone, although the magnitude of this increase remains unclear.

Limited research has been conducted regarding the protein composition of mill fractions. However, there are indications that the relative gluten content changes throughout both the BFs and RFs [15,16]. Regarding gluten composition, no differences were noted across the BFs and RFs [15,16]. However, Wang et al. [16] did document an increase in the proportion of gliadins across the RFs.

Because of the industrial relevance and the identified gaps in the existing literature, the main goal of this study was to investigate how the protein gradient within the starchy endosperm, affected by cultivar and N-fertilization, is translated to the protein content and composition of mill fractions. To investigate this, three wheat cultivars, grown with and without N-fertilization, of which the gradient in protein content was analyzed by Hermans et al. [7], were milled. The resulting mill fractions were subsequently investigated in detail. The findings of this study are relevant for both fundamental research and the milling industry. Given the trend toward less N-fertilization (imposed by governments) for environmental reasons, investigating the effect of N-fertilization on the protein content and composition of mill streams is extremely relevant. Due to reduced N-fertilization, industrial millers are required to produce functional ingredients from lower-quality raw materials. This capability, in turn, could accelerate the transition to reduced N-fertilization.

2. Materials and Methods

2.1. Materials

Sodium dihydrogen phosphate dihydrate, sodium phosphate dibasic dodecahydrate, TRIS, hydrochloric acid, 1-propanol, dithiothreitol (DTT), and acetonitrile were procured from Sigma-Aldrich (Bornem, Belgium), trifluoroacetic acid (TFA) and absolute ethanol (99.8% v/v) from Thermo Fisher Scientific (Aalst, Belgium), and sodium chloride and urea from VWR International (Oud-Heverlee, Belgium). All chemicals were at least of analytical grade.

2.2. Sample Set

Wheat (*Triticum aestivum* L., Poaceae) of three cultivars (Claire (Cl), Apache (Ap), Akteur (Akt)) was cultivated at the experimental site Brakumdreef (Lubbeek, Belgium). Each cultivar was cultivated without (0 kg N ha^{-1}) or with N-fertilization (a total amount of 300 kg N ha^{-1}). The N, which was applied in solid form as calcium ammonium nitrate (27% N), was equally distributed over tillering and the beginning of stem elongation. All information regarding the cultivation of the wheat is described in Hermans et al. [7]. The wheat was harvested in August 2017 and subsequently cleaned to remove impurities such as stones and residual husks.

These three wheat cultivars were chosen because they differ in protein characteristics and, therefore, suitability for bread making. Claire is a class C cultivar, which is considered unsuitable for bread making. Apache is a class B cultivar suitable for bread making when mixed with higher classes. Akteur is classified in class E, the most suitable for bread making. Moreover, for the six wheat samples used in this study, the gradient in protein content within the starchy endosperm has been determined by Hermans et al. [7]. Finally, the difference in protein composition between the sub-aleurone and inner endosperm of Akteur, grown at 300 kg N ha^{-1}, was studied by Hermans et al. [9] using laser microdissection and nanoLC-MS/MS.

2.3. Wheat Milling

Wheat kernels were milled using a Bühler MLU-202 laboratory mill (Bühler group, Uzwil, Switzerland) [21]. Kernels of the cultivars Akteur and Apache were tempered to a moisture content of 16.5%, whereas kernels of the cultivar Claire were tempered to a moisture content of 16.0%. A schematic representation of the Bühler MLU-202 laboratory mill is given in Figure S1 (Supplementary Materials). The mill produces six flour fractions, a miller's bran fraction, and a shorts fraction. The first pair of break rolls exert a shearing action to open the kernels and separate inner endosperm portions from the remainder of the kernel rather than crushing them. Subsequently, the second and third pair of break rolls scrape off the bran particles to remove the adhering starchy endosperm [11]. After each pair of break rolls, there is a set of sieves, eventually leading to three break flour fractions (BFs). BF1 only passed the first pair of break rolls, while BF3 passed all three pairs of break rolls. Miller's bran particles are sieved out (>530 μm) after the third set of break rolls. The gaps between the first, second, and third pair of break rolls were set at 0.35 mm, 0.18 mm, and 0.07 mm, respectively. Particles that are too large to end up in the BFs (>150 μm) but do not end up in the miller's bran fraction are sent to three pairs of smooth reduction rolls, which, after sieving, lead to three reduction flour fractions (RFs). The shorts are sieved out after the third pair of reduction rolls. The gaps between the first, second, and third pair of reduction rolls were set at 0.08 mm, 0.06 mm, and 0.04 mm, respectively. All milling trials were performed in duplicate. After the determination of the yields, the duplicate fractions were mixed.

2.4. Chemical Analysis

The protein content of the kernels and mill fractions was determined according to an adaptation of the AOAC Official Method 990-03 (AOAC International, 1995) to an automated micro-DUMAS protein analysis system (Vario Max Cube, Elementar, Hanau, Germany). The number 5.70 was chosen as the N-to-protein conversion factor for all samples (kernels, flour, and miller's bran). That choice disregards small differences that could derive from the milling fractions exhibiting variations in protein composition due to differences in their starchy endo-sperm protein-to-aleurone and botanical bran protein ratios, but also due to differences in cultivar, level of N-fertilization, and the part of the starchy endosperm from which they originate. For the kernels, shorts fractions, and miller's bran fractions, the N determination was preceded by a milling step with a CryoMill (RETSCH, Haan, Germany; grinding mode: dry at room temperature) to obtain a fine, homogeneous sample because only 1 mg of sample is needed for the micro-DUMAS system. The ash content was determined following an adaptation of the AACC International method 08-01.01 (AACC International, 1999) [22]. The moisture content was determined according to the AACC International method 44-15-02 (AACC International, 1999). The ash and moisture content measurements were performed in triplicate. The protein content was analyzed in duplicate at least.

2.5. Light Microscopy

Mill streams were embedded in Historesin, as described by Dornez et al. [23]. Cross sections of 1 μm thickness were prepared with a rotary microtome HM355 (MicromLabergerate GmbH, Walldorf, Germany). These sections were transferred to microscopy slides (Thermo Scientific Cel-line, SSG Braunschweig, Germany), and proteins were stained with 0.1% (*w*/*v*) Naphthol Blue Black in 7% (*v*/*v*) acetic acid in water for 2 min. Finally, bright-field images were captured at 100× magnification with a Nikon Eclipse 80i microscope (Nikon Inc., New York, NY, USA).

2.6. Modified Osborne Fractionation and Analytical RP-HPLC

Wheat proteins were separated into albumins and globulins, gliadins, and glutenins using a modified Osborne fractionation. Afterward, RP-HPLC was performed to determine the relative amounts of these fractions and the relative amounts of different gluten types (ω-, α- and γ-gliadins, D-low-molecular-weight glutenin subunits (LMW-GS) (sulfur-poor), B/C-LMW-GS (sulfur-rich) and HMW-GS) [18,24,25]. Albumins and globulins were separated, starting from an amount of sample corresponding to 100 mg protein, by conducting two sequential extractions with a 3.0 mL salt solution (sodium phosphate buffer in water (0.05 M; pH 7.6) with 0.4 M NaCl) and a third extraction with 3.0 mL water. Each extraction (10 min) was followed by a centrifugation step (10 min, 4500× *g*). The three supernatants were combined, diluted to 10 mL with the extraction buffer, and filtered (0.45 mm). Gliadins were extracted from the remaining pellet by performing three sequential extractions with ethanol/water (60/40, *v/v*) according to the same procedure. Subsequently, glutenins were extracted similarly using 50% aqueous 1-propanol containing Tris-HCl (0.05 M, pH 7.5), urea (2 M), and 1% DTT at 60 °C for 20 min under a nitrogen atmosphere. All extracts were loaded (2 μL) on an Aeris C18 column (3.6 μm, 150 × 2.1 mm, Phenomenex, Torrance, CA, USA) using an HPLC system (Shimadzu, Kyoto, Japan) consisting of an LC-20AT pump, a DGU-20A5 degasser, a SIL-20ACHT autosampler, a CTO-20AC column oven, and an SPD-10AVP UV/VIS detector. The oven was set at 60 °C and UV/VIS detector at 214 nm. The mobile phases MilliQ (Milli-Q Plus, Merck Millipore, Darmstadt, Germany) water (A) and acetonitrile containing 0.1% (*v/v*) TFA (B) were used at a flow rate of 0.3 mL/min, following a linear gradient from 24% B to 90% B in 70 min. The proportions of albumins and globulins, gliadins, and glutenins were calculated from their respective area compared to the total area of all protein fractions. The proportion of gluten types, ω-, α- and γ-gliadins, HMW-GS, and LMW- GS were deduced from the corresponding areas relative to the total area of the gliadin or glutenin fraction. Examples of the chromatograms obtained by modified Osborne fractionation and RP-HPLC are given in Figure S2 (Supplementary Materials). Samples were analyzed in triplicate.

2.7. Statistical Analysis

A Tukey's honest significant difference (HSD) test was applied to detect statistically significant differences between means. Principal component analysis (PCA) was performed to transform a high-dimensional dataset into a new coordinate system. Regression models were fitted using the standard least squares method. The effects (main, two-way interaction, and quadratic effects) included in the models were determined using backward elimination and F-tests. The significance level was set at 5%. The coefficients of determination (R^2 and $R^2_{adjusted}$ (R^2_{adj})) of the predicted models were calculated. The statistical analysis was performed using JMP Pro 17.0.0 software (SAS Institute, Cary, NC, USA).

3. Results and Discussion

3.1. Effect of Kernel Protein Content on the Wheat Milling Yields

The cultivar and level of N-fertilization are the major factors determining the protein characteristics of wheat. Hence, we chose to work with three cultivars from different baking classes, each grown at 0 and 300 kg N ha^{-1}. The high variability in protein characteristics is already reflected in the kernel protein content, which varies from 7.2% to 14.2% (Figure S3 (Supplementary Materials)). The cultivar Claire has the lowest protein content, while Akteur has the highest. This was expected as Claire was chosen because of its low suitability and Akteur for its high suitability for bread making. Note that the differences between cultivars increase if N-fertilization is applied, confirming their different N-use efficiency.

The wheat kernels of the cultivars Claire, Apache, and Akteur, each cultivated at 0 and 300 kg N ha^{-1}, were milled using a laboratory roller mill. The mill produces six flour fractions, a miller's bran fraction, and a shorts fraction. The six flour fractions include three BFs and three RFs. The yields of these eight fractions are displayed in Table 1. The flour yields vary from 62.8% to 69.5%. As expected, these are significantly lower than the 72% to 80% yield achieved by industrial millers [4,11]. However, the kernel protein content appears to determine the milling behavior of the kernels. To establish the relationships between the kernel protein content and milling yields, a PCA was performed (Figure 1). In a PCA loading plot, positive relations between variables are indicated by arrows in the same direction, and inverse relations by opposing arrows (Figure 1B). The total flour and miller's bran yield (varying from 26.8% to 32.3%) are not significantly correlated with the kernel protein content. The yield of the three break fractions combined (BF_{total}), varying from 20.3% to 34.6%, is inversely correlated with the kernel protein content (R^2 = 0.96). This means that 96% of the variation in the BF_{total} yield can be explained by the variation in kernel protein content, and more specifically, the yield of BF_{total} decreases by 2.1% for each 1% increase in kernel protein content. The variation in BF_{total} yield is mainly caused by the variation in BF_1 yield (Figure 1B). The yields of the reduction fractions combined

(RF_{total}) (varying from 33.5% to 48.6%) and the shorts fractions (varying from 1.8% to 4.5%) are positively correlated with the kernel protein content, having an R^2 of 0.84 and 0.75, respectively. Hence, higher protein contents seem to make the starchy endosperm harder. If the starchy endosperm, separated from the bran by the corrugated break rolls, is harder to reduce in particle size, there will be more large particles. As these cannot pass the sieves leading to the BFs, the yield of the BFs decreases, and more material is transferred to the reduction rolls, increasing the yields of the RFs. Additionally, the yields of the shorts increase because the material transferred to the reduction rolls is harder to reduce in particle size. Hourston et al. [26] and Kent [27] previously addressed the relationship between starchy endosperm hardness and protein content.

Table 1. Wheat milling yields of the cultivars Claire (Cl), Apache (Ap), and Akteur (Akt), grown at 0 or 300 kg N ha^{-1} addition levels. BF: break fraction, RF: reduction fraction. Labeling with different letters points out significant differences (Tukey's HSD test, $p \leq 0.05$) between the milling yields of different wheat samples.

Mill Fraction	Yield (%)					
	Cl0	Cl300	Ap0	Ap300	Akt0	Akt300
Flour	66.7 ± 1.7^{AB}	62.8 ± 1.7^{B}	68.3 ± 0.2^{A}	69.5 ± 0.4^{A}	69.5 ± 0.0^{A}	69.0 ± 0.4^{A}
BF_{total}	33.1 ± 1.5^{A}	24.7 ± 0.4^{B}	34.4 ± 1.1^{A}	22.2 ± 0.4^{BC}	34.6 ± 0.1^{A}	20.3 ± 0.1^{C}
BF1	22.0 ± 1.7^{A}	14.4 ± 0.1^{B}	24.4 ± 0.9^{A}	14.0 ± 0.4^{B}	25.1 ± 0.3^{A}	12.8 ± 0.1^{B}
BF2	6.7 ± 0.6^{A}	5.9 ± 0.1^{AB}	6.0 ± 0.1^{AB}	5.1 ± 0.1^{BC}	5.6 ± 0.2^{ABC}	4.7 ± 0.1^{C}
BF3	4.5 ± 0.4^{A}	4.5 ± 0.2^{A}	4.1 ± 0.1^{A}	3.1 ± 0.0^{B}	3.9 ± 0.0^{A}	2.9 ± 0.2^{B}
RF_{total}	33.5 ± 0.2^{C}	38.6 ± 2.0^{B}	33.9 ± 1.2^{C}	47.3 ± 0.0^{A}	34.9 ± 0.2^{BC}	48.6 ± 0.5^{A}
RF1	26.4 ± 3.1^{D}	33.0 ± 1.3^{BC}	29.8 ± 0.8^{CD}	38.0 ± 1.0^{AB}	29.3 ± 0.4^{CD}	39.3 ± 0.3^{A}
RF2	6.4 ± 2.6^{A}	5.0 ± 0.7^{A}	3.6 ± 0.5^{A}	8.0 ± 0.4^{A}	4.7 ± 0.6^{A}	8.0 ± 0.1^{A}
RF3	0.8 ± 0.3^{A}	0.5 ± 0.0^{A}	0.5 ± 0.0^{A}	1.3 ± 0.6^{A}	0.9 ± 0.0^{A}	1.3 ± 0.2^{A}
Miller's bran	31.5 ± 2.1^{AB}	32.3 ± 0.3^{A}	29.5 ± 0.4^{ABC}	26.7 ± 1.0^{C}	28.0 ± 0.2^{BC}	26.8 ± 0.5^{C}
Shorts	1.8 ± 0.5^{B}	4.5 ± 1.4^{A}	2.2 ± 0.3^{AB}	3.8 ± 0.6^{AB}	2.5 ± 0.1^{AB}	4.2 ± 0.1^{AB}

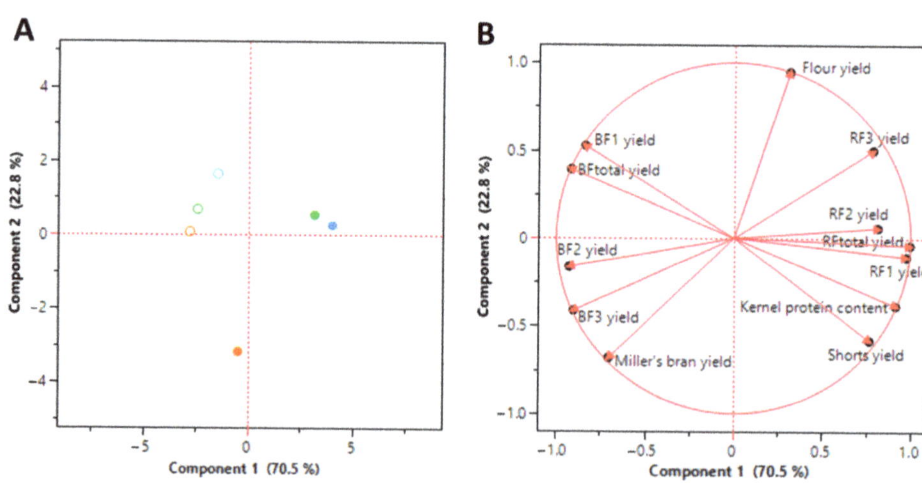

Figure 1. Principal component analysis (**A**) score and (**B**) loading plot of the kernel protein content and milling yields. BF: break fraction, RF: reduction fraction. Principle component 1 explains 70.5% of the variability and principle component 2 22.8%. Orange: Claire, Green: Apache, Blue: Akteur. ○: 0 kg N ha^{-1}, •: 300 kg N ha^{-1}.

3.2. How Is the Protein Gradient within the Starchy Endosperm Reflected in the Mill Streams?

The protein contents of the eight mill streams of the three cultivars, grown without and with N-fertilization (300 kg N ha^{-1}), are displayed in Figure 2. If no N-fertilization is applied, an increase in protein content is observed from BF1 to BF3, on the one hand, and RF1 to RF3, on the other hand. Both increases are of comparable magnitude. For example, for Cl0, the protein content rises 1.3% (or a relative increase of 23.5%) in the BFs, while 2.0% (or a relative increase of 32.8%) in the RFs. This is very interesting as in the absence of N-fertilization, Claire, Apache, and Akteur exhibit very similar and gentle gradients in protein content [7]. From the inner endosperm to the sub-aleurone, the protein content of Cl0 increased from 6.3 ± 1.0% to 14.0 ± 3.7%, of Ap0 from 6.7 ± 0.7% to 18.2 ± 3.8% and of Akt0 from 5.9 ± 1.0% to 19.1 ± 4.5% [7].

If N-fertilization (300 kg N ha^{-1}) is applied, the gradients in protein content within the starchy endosperm are notably steeper and also differ between the cultivars, with Akteur having the steepest gradient [7]. From the inner endosperm to the sub-aleurone, the protein content of Akt300 increased from 9.8 ± 1.6% to 31.7 ± 3.3%, compared to 8.7 ± 0.4% to 24.4 ± 2.3% for Cl300 [7]. Analyzing the mill streams when N-fertilization is applied, similar to when no N-fertilization is applied, an increase in protein content is also observed across both the BFs and RFs. However, in this case, the increase in the protein content in the BFs is substantially higher than in the RFs. For example, for Akt300, the protein content in the BFs rises by 5.2% (or a relative increase of 42.0%), whereas in the RFs, it increases by only 1.7% (or a relative increase of 13.6%). In addition, we observed that, with N-fertilization, the increases in protein content in the RFs are lower or comparable to the increases in protein content across the RFs without N-fertilization. Hence, these increases in the reduction fractions are independent of the gradient in protein content and are hypothesized to be solely the result of protein-rich material being harder to reduce in particle size. The rise in protein content in the BFs is significantly higher for the 300 kg N ha^{-1} samples compared to the samples without N-fertilization, with the rise being the lowest for Cl300 and highest for Akt300. This is highly comparable with their gradients in protein content [7]. Given the mechanism of action of the break rolls, this is a strong indication that the gradient in protein content is reflected in the BFs. However, the increasing protein content in the BFs could also be caused by increasing levels of contamination with aleurone, which contains approximately 30% protein [28]. Given the aleurone's high ash 315 content of approximately 12%, ash content can be used as a marker for aleurone contamination [29,30]. The ash contents of the mill fractions are displayed in Figure 3. As expected, the miller's bran and shorts fractions, which contain elevated aleurone levels, exhibit significantly greater ash content than the flour fractions. Across the RFs, the ash content increases significantly for each of the six samples. This indicates that the rises in protein content in the RFs do not solely stem from the protein-rich material being harder to reduce in particle size but also from increased levels of aleurone contamination. Except for Cl300, none of the samples show a significant increase in ash content across the BFs. The values of 0.4 to 0.6% are also representative of commercial flours. This demonstrates that the increase in protein content in the BFs is not caused by increasing levels of aleurone contamination and, therefore, confirms our hypothesis that the gradient in protein content is reflected in the BFs.

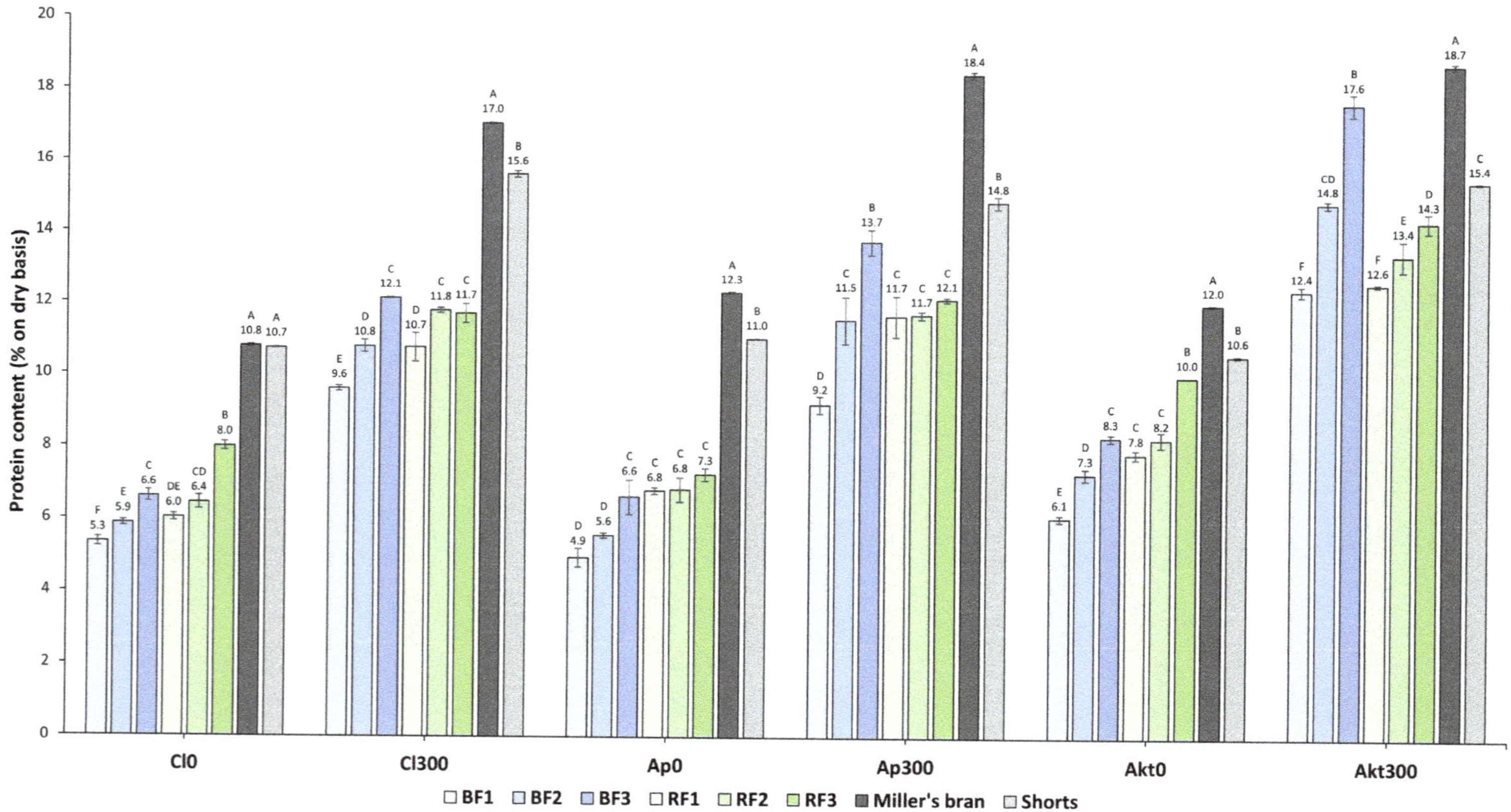

Figure 2. Protein contents of the mill streams of the cultivars Claire (Cl), Apache (Ap), and Akteur (Akt) grown with and without N-fertilization (300 kg N ha^{-1}). BF: break fraction, RF: reduction fraction. Labeling with different letters points out significant differences (Tukey's HSD test, $p \leq 0.05$) between mill fractions of one cultivar grown at one N-fertilization level.

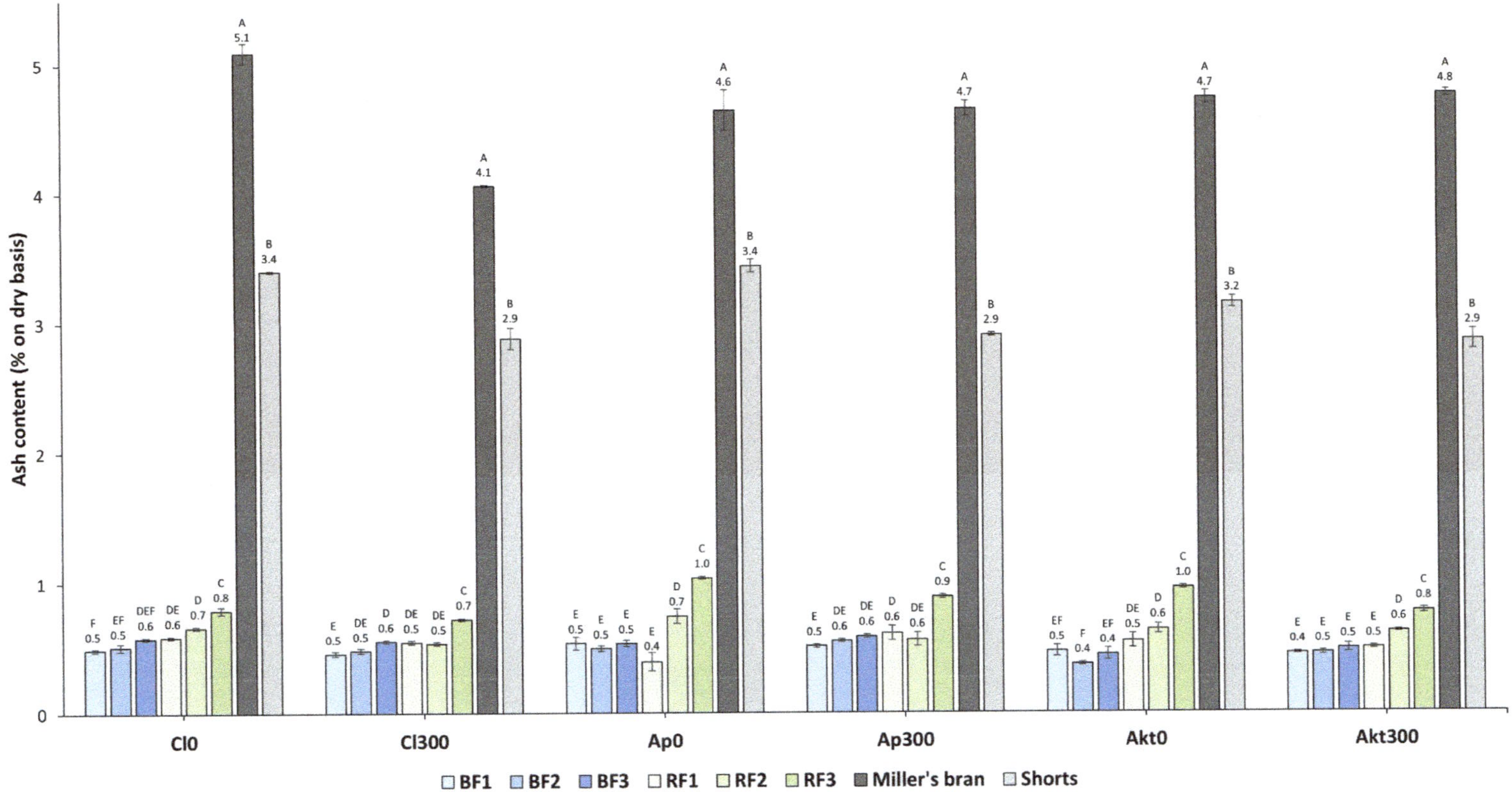

Figure 3. Ash contents of the mill streams of the cultivars Claire (Cl), Apache (Ap), and Akteur (Akt) grown with and without N-fertilization (300 kg N ha^{-1}). BF: break fraction, RF: reduction fraction. Labeling with different letters points out significant differences (Tukey's HSD test, $p \leq 0.05$) between mill fractions of one cultivar grown at one N-fertilization level.

This is also demonstrated by microscopy. In Figure 4, microscopy images of the eight mill fractions of Akt300 are shown. Proteins are stained blue with Naphthol Blue Black. In the BFs and RFs, one can observe the typical starchy endosperm material, high in starch granules, embedded in a protein matrix. Compared to the other flour fractions, BF3 also contains a notable amount of large protein aggregates. These are characteristic of the subaleurone cells, lying at the periphery of the starchy endosperm. However, the significant amount of sub-aleurone material still present in the miller's bran fraction shows the incomplete recovery of the starchy endosperm in the flour fractions. This is also revealed by analyzing the obtained flour yields, ranging from 62.8% to 69.5% (Table 1), because the starchy endosperm constitutes 80–84% of a wheat grain [2,3]. This means that only a part of the periphery of the starchy endosperm is recovered in the flour fractions, mainly in BF3, and the remaining part remains attached to the bran. This, together with the high protein content of the aleurone, also explains why the miller's bran fraction has the highest protein content of all the mill fractions (Figure 2). Between samples, the miller's bran protein content follows a similar trend as the kernel protein content. For the shorts, no clear cultivar-dependent effects can be distinguished.

To conclude, there are strong indications suggesting that BF1 is enriched in the inner endosperm, BF2 in the more distant starchy endosperm, and BF3 in the peripheral starchy endosperm. However, it is not possible to precisely delineate the specific starchy endosperm region to which each BF corresponds, and there is likely an overlap between the starchy endosperm regions present in the three BFs. Furthermore, a part of the sub-aleurone remains adhered to the miller's bran fraction, preventing it from being included in BF3.

This reflection of the gradient in protein content in the BFs provides opportunities. The BFs obtained in this study could be used to gain insight into the gradient in protein composition and its dependence on cultivar and level of N-fertilization. However, there are some limitations. The incomplete recovery of the starchy endosperm, coupled with the vague and intertwined boundaries of the starchy endosperm regions corresponding to each BF, leads to an underestimation of the gradient in protein content within the starchy endosperm. This is shown by comparing the gradient in protein content of Akt300, measured in detail by Hermans et al., to the rise in protein content in the BFs, which, respectively, increase from 9.8 to 31.7% [7] and 12.4 to 17.6% (Figure 3). Consequently, this will similarly affect the estimation of the gradient in protein composition.

3.3. Analysis of the Protein Composition Gradient within the Starchy Endosperm Using the Break Fractions

The protein composition of the BFs of the cultivars Claire, Apache, and Akteur, grown with and without N-fertilization (300 kg N ha^{-1}), is shown in Table 2. Over the different BFs, cultivars, and N-fertilization levels, the relative gluten content varies from 76.61% to 87.29%. Statistical modeling revealed that the factors 'break fraction', 'cultivar', and 'N-fertilization', as well as their interactions, significantly affect the relative gluten content ($R^2{}_{adj}$ = 0.95). A visual representation of the estimated model is given in Figure S4A (Supplementary Materials). Here, the parameter estimates are presented in descending order of absolute value, from highest to lowest. The relative gluten content is the lowest for the cultivar Claire and the highest for Akteur, and for each cultivar, N-fertilization increases the relative gluten content. Akteur, having the highest relative gluten content, reaffirms its high suitability for baking purposes. Furthermore, the relative gluten content significantly increases from BF1 to BF3, and hence, from the inner to outer endosperm, across all cultivars at each level of N-fertilization, except for Cl300 (Table 2). This confirms the results of Hermans et al. [9] and also of Simmons and Sutton [15] and Wang et al. [16]. For Akt300, the relative gluten content even increased to 87.29% in BF3 (Table 2), but, taking into account the underestimation (described in Section 3.2), this is expected to be substantially higher in the sub-aleurone cells.

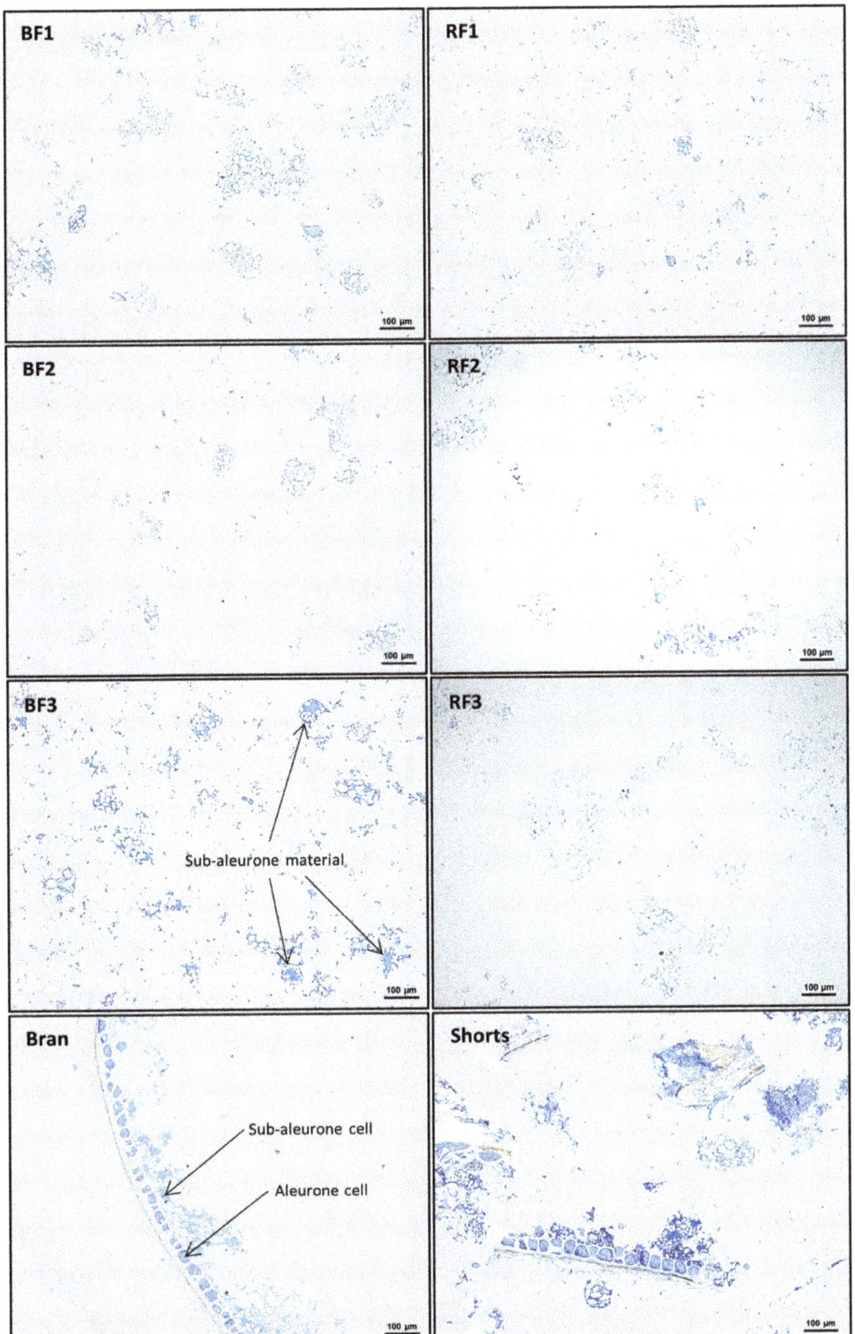

Figure 4. Microscopy images of the mill streams of Akteur, grown at 300 kg/ha, in which proteins are stained with Naphthol Blue Black. BF: break fraction, RF: reduction fraction.

Table 2. Protein composition of the break fractions (BFs) of the cultivars Claire (Cl), Apache (Ap), and Akteur (Akt), without or with N-fertilization (300 kg N ha^{-1}). GLIA/GLU: gliadin-over-glutenin ratio, HMW-GS: high molecular weight glutenin subunits, LMW-GS: low molecular weight glutenin subunits (D-type: sulfur-poor, B- and C-type: sulfur-rich). Different letters indicate significant differences (Tukey's HSD test, $p \leq 0.05$) in the relative content of a specific protein type between the break fractions of a specific cultivar grown at one level of N-fertilization.

Sample	Mill Fraction	Relative Gluten Content (%)	Gluten Composition (%)							
			ω-Gliadin	α-Gliadin	γ-Gliadin	D-LMW-GS	B/C-LMW-GS	HMW-GS	GLIA/GLU	LMW-GS/HMW-GS
Cl0	BF1	76.61 ± 0.51^{C}	5.74 ± 0.05^{B}	31.47 ± 0.36^{AB}	30.51 ± 0.82^{B}	0.87 ± 0.04^{A}	23.61 ± 0.70^{A}	7.79 ± 0.23^{A}	2.10 ± 0.09^{B}	3.14 ± 0.01^{B}
	BF2	78.54 ± 0.39^{B}	5.79 ± 0.05^{B}	31.88 ± 0.08^{A}	31.24 ± 0.19^{B}	0.86 ± 0.02^{A}	22.83 ± 0.22^{A}	7.41 ± 0.08^{AB}	2.22 ± 0.03^{B}	3.20 ± 0.01^{A}
	BF3	81.15 ± 0.16^{A}	6.09 ± 0.04^{A}	31.12 ± 0.06^{B}	33.20 ± 0.42^{A}	0.85 ± 0.01^{A}	21.64 ± 0.28^{B}	7.10 ± 0.12^{B}	2.38 ± 0.05^{A}	3.17 ± 0.01^{AB}
Cl300	BF1	83.64 ± 0.87^{A}	7.46 ± 0.19^{B}	35.73 ± 0.54^{A}	28.88 ± 1.23^{A}	0.84 ± 0.04^{B}	19.51 ± 1.01^{AB}	7.56 ± 0.39^{AB}	2.59 ± 0.19^{AB}	2.69 ± 0.01^{A}
	BF2	85.53 ± 0.52^{A}	7.81 ± 0.07^{A}	36.16 ± 0.20^{A}	30.53 ± 0.95^{A}	0.82 ± 0.06^{B}	17.80 ± 0.74^{B}	6.88 ± 0.28^{B}	2.93 ± 0.16^{A}	2.71 ± 0.01^{A}
	BF3	84.37 ± 1.12^{A}	7.77 ± 0.11^{AB}	34.03 ± 0.08^{B}	28.13 ± 1.50^{A}	1.01 ± 0.04^{A}	21.15 ± 0.96^{A}	7.90 ± 0.45^{A}	2.33 ± 0.16^{B}	2.81 ± 0.09^{A}
Ap0	BF1	77.45 ± 0.37^{B}	7.45 ± 0.01^{B}	33.66 ± 0.32^{B}	25.14 ± 0.60^{B}	0.91 ± 0.03^{A}	25.03 ± 0.66^{A}	7.82 ± 0.20^{B}	1.96 ± 0.08^{B}	3.32 ± 0.01^{A}
	BF2	81.20 ± 0.60^{A}	7.63 ± 0.10^{A}	34.99 ± 0.44^{A}	27.22 ± 0.13^{A}	0.82 ± 0.03^{A}	22.22 ± 0.46^{B}	7.13 ± 0.19^{A}	2.32 ± 0.07^{A}	3.23 ± 0.02^{B}
	BF3	82.69 ± 0.70^{A}	7.59 ± 0.04^{A}	33.49 ± 0.44^{B}	28.91 ± 0.84^{A}	0.84 ± 0.04^{A}	21.96 ± 0.96^{B}	7.21 ± 0.31^{A}	2.34 ± 0.14^{A}	3.16 ± 0.01^{C}
Ap300	BF1	81.05 ± 0.34^{C}	8.01 ± 0.16^{A}	29.85 ± 0.46^{A}	26.64 ± 0.46^{A}	0.94 ± 0.02^{B}	25.64 ± 0.72^{B}	8.92 ± 0.35^{B}	1.82 ± 0.08^{A}	2.98 ± 0.04^{A}
	BF2	83.11 ± 0.33^{B}	8.01 ± 0.18^{A}	28.01 ± 0.73^{B}	25.09 ± 0.50^{B}	1.19 ± 0.08^{A}	27.53 ± 1.02^{A}	10.17 ± 0.30^{A}	1.57 ± 0.09^{B}	2.82 ± 0.03^{B}
	BF3	84.46 ± 0.13^{A}	7.79 ± 0.04^{A}	26.29 ± 0.20^{C}	24.74 ± 0.24^{B}	1.32 ± 0.07^{A}	29.19 ± 0.34^{A}	10.67 ± 0.04^{A}	1.43 ± 0.02^{B}	2.86 ± 0.04^{B}
Akt0	BF1	78.50 ± 0.21^{B}	6.10 ± 0.05^{A}	27.88 ± 0.41^{A}	21.30 ± 0.76^{A}	1.20 ± 0.03^{A}	32.51 ± 0.93^{A}	11.02 ± 0.24^{A}	1.24 ± 0.06^{A}	3.06 ± 0.02^{A}
	BF2	80.90 ± 1.00^{A}	6.52 ± 0.26^{A}	29.73 ± 1.34^{A}	23.11 ± 1.41^{A}	1.08 ± 0.08^{A}	29.36 ± 2.20^{A}	10.20 ± 0.72^{A}	1.47 ± 0.18^{A}	2.98 ± 0.02^{B}
	BF3	82.19 ± 0.29^{A}	6.44 ± 0.02^{A}	28.46 ± 1.04^{A}	23.37 ± 1.84^{A}	1.06 ± 0.07^{A}	29.96 ± 1.99^{A}	10.72 ± 0.84^{A}	1.40 ± 0.17^{A}	2.90 ± 0.04^{C}
Akt300	BF1	84.50 ± 0.35^{C}	10.09 ± 0.35^{A}	28.76 ± 0.86^{A}	21.54 ± 0.89^{B}	1.35 ± 0.05^{A}	26.41 ± 1.38^{A}	11.85 ± 0.66^{A}	1.53 ± 0.13^{A}	2.34 ± 0.02^{B}
	BF2	85.78 ± 0.34^{B}	10.22 ± 0.29^{A}	28.44 ± 0.70^{A}	21.92 ± 0.55^{AB}	1.33 ± 0.05^{A}	26.70 ± 1.05^{A}	11.38 ± 0.44^{A}	1.54 ± 0.10^{A}	2.46 ± 0.03^{A}
	BF3	87.29 ± 0.18^{A}	10.17 ± 0.25^{A}	28.26 ± 0.51^{A}	23.45 ± 0.22^{A}	1.36 ± 0.05^{A}	25.71 ± 0.58^{A}	11.05 ± 0.35^{A}	1.62 ± 0.06^{A}	2.45 ± 0.04^{A}

The increase in relative gluten content throughout the BFs, and hence, from the inner to outer endosperm, appears to be independent of the cultivar but is affected by the level of N-fertilization. The increase in relative gluten content throughout the BFs is of a higher magnitude without N-fertilization compared to with N-fertilization. This was according to expectations because a basic machinery, i.e., a basic set of albumins and globulins, is hypothesized to be present throughout the entire starchy endosperm [9]. In other words, once the basic set of albumins and globulins is built up, the excess of N is predominantly accumulated as gluten proteins. For example, for Akt0, BF1 and BF3 have a protein content of 6.07% and 8.29%, respectively. In BF1, the absolute gluten content (percentage of gluten proteins on total dry weight) is 4.76%, and the non-gluten protein content is 1.30%, while for BF3, those are 6.81% and 1.48%, respectively. Hence, the difference in absolute gluten content between BF1 and BF3 is 2.05%, whereas it is only 0.17% in non-gluten protein content. This shows that the 2.22% extra accumulated proteins in BF3 compared to BF1 are almost all gluten proteins, which confirms the hypothesis that once the basic set of albumins and globulins is built up, the excess of N is predominantly accumulated as gluten proteins. Increasing the level of N-fertilization will, therefore, mainly lead to additional accumulation of gluten proteins in both the inner and outer endosperm. This will level out the gradient in relative gluten content as in the inner endosperm, with the lower relative gluten content, there is more potential for an increase compared to the outer endosperm with the higher initial relative gluten content. Assuming that increasing the level of N-fertilization from 0 to 300 kg N ha^{-1} only leads to extra accumulation of proteins, we can calculate that 89.5%, 89.8%, and 91.0% of the extra accumulated proteins are gluten in BF1, BF2, and BF3 of Akteur, respectively. Considering that this study provides underestimations of the magnitudes of the gradients in protein composition, we can hypothesize that, for the subaleurone cells, more than 91.0% of the proteins additionally accumulated due to N-fertilization are gluten proteins. Taking this further, one could expect that the additional protein accumulation in the sub-aleurone, going from a high level of N-fertilization, e.g., 150 kg N/ha, to an even higher level, e.g., 300 kg N/ha, is almost purely due to the additional accumulation of gluten proteins for the cultivar Akteur. Important to note here is that not only relatively speaking, more gluten proteins are being accumulated in the outer endosperm due to higher levels of N-fertilization, but also in absolute numbers. All this confirms the hypothesis that the sub-aleurone cells act as storage facilities for "excess" N.

Regarding gluten composition, Table 2 shows that the three cultivars vary in their proportion of gluten types. Claire exhibits a relatively high amount of γ-gliadins, whereas Akteur is characterized by a high proportion of HMW-GS. Consequently, the LMW-GS/HMW-GS ratio is the lowest for the cultivar Akteur. The gliadin-over-glutenin (GLIA/GLU) ratio, on the other hand, decreases from Claire, over Apache, to Akteur (Table 2, Figure S4B (Supplementary Materials)). For both bread-making indicators, the LMW-GS/HMW-GS and GLIA/GLU ratio, low values are desired, and this once again highlights the high baking quality of the cultivar Akteur. Note that for GLIA/GLU and LMW-GS/HMW-GS, 'cultivar' is the most determining factor, while for the relative gluten content, 'break fraction' is (Figure S4B (Supplementary Materials)). In general, the effect of N-fertilization seems to be dependent on the cultivar. However, more N-fertilization consistently results in a higher proportion of ω-gliadin and a decrease in the LMW-GS/HMW-GS ratio (Figure S4C (Supplementary Materials)).

From BF1 over BF2 to BF3, the trends in gluten composition appear to depend on the cultivar–N-fertilization combination. However, some general trends can be observed. In the absence of N-fertilization, there is a trend of increasing ω- and γ-gliadin and decreasing B/C-LMW-GS and HMW-GS proportions from BF1 to BF3, and therefore, from the inner to outer endosperm (Table 2). As a result, the GLIA/GLU ratio significantly increased from BF1 to BF3 for Cl0 and Ap0. This can also be observed in Figure S4B (Supplementary Materials), displaying a visual representation of the model estimating the GLIA/GLU ratio in function of the cultivar, level of N-fertilization, and BF (R^2_{adj} = 0.89). The decrease in the proportion of HMW-GS from the inner to outer endosperm was also reported by He et al. [8] at low N levels (100 kg N ha^{-1}) and Tosi et al. [6]. The increase in the proportion of ω-gliadin was only observed at 350 kg N ha^{-1} by He et al. [8]. He et al. [8] did not individually quantify the proportions of α-gliadin, γ-gliadin, and LMW-GS. At 300 kg N ha^{-1}, no consistent trends were observed. The gradient in proportion of certain gluten types decreased, increased, or remained constant depending on the cultivar. He et al. [8] conducted similar observations for the proportion of LMW-GS and gliadins combined and HMW-GS at 350 kg N ha^{-1} of N-fertilization. Therefore, it seems that different cultivars exhibit similar gradients in gluten composition in the absence of N-fertilization, while there is more variation at high levels of N-fertilization. This may partially explain the (seemingly) contradictory literature on the gradient in gluten composition.

Finally, in the current study, the differences in gluten composition between the inner endosperm and sub-aleurone, as observed in Akt300 by Hermans et al. [9] through the utilization of cryosectioning and nanoLC-MS/MS, did not manifest between BF1 (enriched in inner endosperm) and BF3 (enriched in peripheral endosperm). Hermans et al. [9] demonstrated that gluten of the sub-aleurone contains a higher proportion of ω-gliadins but a lower proportion of LMW-GS and γ-gliadins compared to the inner endosperm. The inability to detect this here may be attributed to two factors: firstly, analyzing the BFs provides an underestimation of the actual gradients (described in Section 3.2), and secondly, this study analyzes gluten proteins differently. The current study classifies gluten proteins according to extractability and hydrophobicity into gluten types like γ-gliadins, HMW-GS, etc., while Hermans et al. [9] use protein sequences for classification into gluten types. These methods do not lead to perfect one-to-one matches [31].

3.4. Relevance of Sub-Aleurone in the Milling Industry

Knowledge of how the protein gradient within the starchy endosperm is reflected in the mill streams can be very relevant for the milling industry and lead to new opportunities. From previous research, we know that for Cl0, the protein content increased from about 6.3% in the inner endosperm to 14.0% in the sub-aleurone. For Akt300, this rise was from about 9.8% to 31.7% [7]. Integrating this with the current study on the composition of mill streams can lead to new insights. Making the conservative assumption that the relative gluten content of BF1 corresponds to that of the inner endosperm and that of BF3 to that of the sub-aleurone, the relative gluten content increases from 76.6% to 81.2% across the starchy endosperm of Cl0 and from 84.5% to 87.3% for Akt300 (Table 2). Combining the results on the gradient in protein content and protein composition, the absolute gluten content (relative to total dry weight) increases from 4.8% in the inner endosperm to 11.4% in the sub-aleurone for Cl0 and 8.3% to 27.7% for Akt300. While this is an underestimation, it demonstrates the enormous potential of the sub-aleurone as a gluten source. Given the dependence of both the gradient in protein content and protein composition on the wheat cultivar and level of N-fertilization, even higher values are possible for other cultivar-N-fertilization combinations. However, knowing that in this study, BF3 of Akt300 contains 17.6% protein, of which 87.3% gluten proteins, we can hypothesize that at least 1 of the more than 50 flour streams of an industrial mill contains more than 20% protein of which more than 90% is gluten, due to the presence of sub-aleurone material. This fraction(s) could be used in specialized flours.

Due to the trend towards less N-fertilization (imposed by governments), the production of functional ingredients from low-quality raw materials is an immense challenge. However, this study shows that we can lower the level of N-fertilization while maintaining a high protein content and favorable protein composition just by selecting mill streams. For example, when lowering the level of N-fertilization, and hence, also the quality of the wheat, the high-protein flour streams could be collected to produce flour for bread making, whereas the flour streams with a lower protein content are well-suited for the production of, e.g., cookies, cakes, and coatings. This can be one of the components contributing to the transition to reduced N-fertilization, along with increasing wheat's nitrogen use efficiency

(NUE) through breeding, soil health management, implementing precision agriculture practices, etc.

Industrial mills never achieve full recovery of the starchy endosperm. This means that at least some sub-aleurone material, which is the part of the kernel that is most enriched in gluten, remains adhered to the miller's bran. Despite this, most miller's bran (about 90%) ends up in animal feed [32]. This creates big opportunities. Nowadays, there is no selection procedure prior to using miller's bran for whole-meal bread making. Hence, one opportunity lies in selecting miller's bran based on its sub-aleurone protein properties before its incorporation into whole-meal products like bread. We hypothesize that, for example, Akt300 bran will perform better in whole-meal bread making than Cl0 bran. Additionally, further research is required to explore the separation of sub-aleurone material adhered to the bran from the bran itself. This is in line with the goal of promoting sustainable N management, where both increasing wheat's NUE and incorporating protein from side streams into human food are crucial. Both are essential for alleviating the environmental impact of N-fertilization. Finally, while this study provides insight into the effect of cultivar (genetics) and N-fertilization (element of environment) on the gradient in protein composition, more research is needed to map the impact of environmental factors.

4. Conclusions

This study is the first to systematically investigate how the protein gradient within the starchy endosperm, influenced by cultivar and N-fertilization, translates into the protein content and composition of mill fractions. Both within the BFs and RFs, an increase in protein content was observed. The increase in protein content within the RFs could be attributed to increased levels of aleurone contamination and protein-rich material being harder to reduce in particle size. The rise in protein content within the BFs was shown to be a reflection of the gradient in protein content within the starchy endosperm. This implies that BF1 is enriched in the inner endosperm, BF2 in the more distant starchy endosperm, and BF3 in the peripheral starchy endosperm. Additionally, the miller's bran fraction contained a significant amount of peripheral starchy endosperm (sub-aleurone). Using the BFs, this is the first study that was able to investigate the magnitude of the gradient in relative gluten content within the starchy endosperm and how this is affected by cultivar and N-fertilization level. The gradient in relative gluten content, increasing from the inner to outer endosperm (up to 87.3%), was more pronounced without N-fertilization than with N-fertilization. Since 87.3% is an underestimation of the actual relative gluten content in the sub-aleurone, this confirms the hypothesis that the sub-aleurone cells are storage facilities for gluten proteins. Regarding the gradient in gluten composition, without N-fertilization, higher ω- and γ-gliadin proportions and lower B/C-LMW-GS and HMW-GS proportions were observed in the outer than inner endosperm. With N-fertilization, no consistent trends were observed over the different cultivars. This could, at least partly, explain why there is no consensus on the gradient in gluten composition in the literature. Knowledge of the reflection of the protein gradient in the mill streams can be highly valuable for the milling industry. It will aid in comprehending and managing fluctuations in the functionality of specific flour streams, producing specialized flours, coping with low-quality raw materials, and upcycling side streams. This study, along with research on enhancing wheat's NUE, contributes to a more efficient utilization of N fertilizer. Finally, further research is necessary to understand how environmental factors influence protein gradients and, hence, the composition in mill streams.

Supplementary Materials: The following supporting information can be downloaded at https://www.mdpi.com/article/10.3390/foods12234192/s1, Figure S1: Schematic representation of the Bühler MLU-202 laboratory mill. First, three pairs of corrugated break rolls break up the kernels and scrape off the resulting bran particles to remove the adhering starchy endosperm. Second, three pairs of smooth reduction rollers reduce the particle size of the starchy endosperm particles not ending up in the break fractions. BR: break fraction, RF: reduction fraction.; Figure S2: Chromatograms of (A) albumins and globulins, (B) gliadins, and (C) glutenins separated by modified

Osborne fractionation from the first break fraction of Akteur grown at 300 kg N ha^{-1} and analyzed using RP-HPLC. HMW-GS: high molecular weight glutenin subunits, LMW-GS: low molecular weight glutenin subunits.; Figure S3: Wheat kernel protein content on dry base of the cultivars Claire (Cl), Apache (Ap), and Akteur (Akt), grown at 0 or 300 kg N ha^{-1}. Different letters indicate significant differences.; Figure S4: Visual representations of models estimating (A) the relative gluten content, (B) GLIA/GLU ratio, and (C) the LMW-GS/HMW-GS ratio in function of the level of N-fertilization, the wheat cultivar and mill fraction. Abbreviations: Cl: Claire, Ap: Apache, Akt: Akteur, GLIA: gliadin, GLU: glutenin, HMW-GS: high molecular weight glutenin subunits, LMW-GS: low molecular weight glutenin subunits, BF: break fraction.

Author Contributions: W.H.: Methodology, Conceptualization, Investigation, Formal analysis, Visualization, Writing—original draft preparation. J.B.: Methodology, Conceptualization, Investigation, Writing—Review and editing. Y.D.B.: Methodology, Writing—Review and editing. N.A.L.: Methodology, Conceptualization, Writing—Review and editing, Supervision. C.M.C.: Methodology, Conceptualization, Resources, Funding acquisition, Writing—Review and editing, Supervision. All authors have read and agreed to the published version of the manuscript.

Funding: This work was funded by the Research Foundation—Flanders (FWO-Vlaanderen, Brussels, Belgium) [grant number G.0561.17N]. Wisse Hermans acknowledges the Research Foundation—Flanders (FWO-Vlaanderen, Brussels, Belgium) [grant number 1S06224N]. Yamina De Bondt acknowledges the Research Foundation—Flanders (FWO-Vlaanderen, Brussels, Belgium) for a position as a postdoctoral research fellow [grant number 12B3723N]. Niels Langenaeken acknowledges the Research Foundation—Flanders (FWO-Vlaanderen, Brussels, Belgium) for a position as a postdoctoral research fellow [grant number 12A8723N].

Data Availability Statement: The data used to support the findings of this study can be made available by the corresponding author upon request.

Conflicts of Interest: The authors declare no conflict of interest.

References

1. FAOSTAT. 2020. Available online: https://www.fao.org/faostat/en/#data/FBS (accessed on 29 August 2023).
2. Barron, C.; Surget, A.; Rouau, X. Relative amounts of tissues in mature wheat (*Triticum aestivum* L.) grain and their carbohydrate and phenolic acid composition. *J. Cereal Sci.* **2007**, *45*, 88–96. [CrossRef]
3. Marion, D.; Saulnier, L. Minor components and wheat quality: Perspectives on climate changes. *J. Cereal Sci.* **2020**, *94*, 103001. [CrossRef]
4. Shewry, P.R.; Wan, Y.; Hawkesford, M.J.; Tosi, P. Spatial distribution of functional components in the starchy endosperm of wheat grains. *J. Cereal Sci.* **2020**, *91*, 102869. [CrossRef] [PubMed]
5. Evers, T.; Millar, S. Cereal Grain Structure and Development: Some Implications for Quality. *J. Cereal Sci.* **2002**, *36*, 261–284. [CrossRef]
6. Tosi, P.; Gritsch, C.S.; He, J.; Shewry, P.R. Distribution of gluten proteins in bread wheat (*Triticum aestivum*) grain. *Ann. Bot.* **2011**, *108*, 23–35. [CrossRef]
7. Hermans, W.; Mutlu, S.; Michalski, A.; Langenaeken, N.A.; Courtin, C.M. The Contribution of Sub-Aleurone Cells to Wheat Endosperm Protein Content and Gradient Is Dependent on Cultivar and N-Fertilization Level. *JAFC* **2021**, *69*, 6444–6454. [CrossRef]
8. He, J.; Penson, S.; Powers, S.J.; Hawes, C.; Shewry, P.R.; Tosi, P. Spatial Patterns of Gluten Protein and Polymer Distribution in Wheat Grain. *JAFC* **2013**, *61*, 6207–6215. [CrossRef]
9. Hermans, W.; Geisslitz, S.; De Bondt, Y.; Langenaeken, N.A.; Scherf, K.A.; Courtin, C.M. NanoLC-MS/MS protein analysis on laser-microdissected wheat endosperm tissues: A comparison between aleurone, sub-aleurone and inner endosperm. *Food Chem.* **2024**, *437*, 137735. [CrossRef]
10. Dornez, E.; Gebruers, K.; Wiame, S.; Delcour, J.A.; Courtin, C.M. Insight into the Distribution of Arabinoxylans, Endoxylanases, and Endoxylanase Inhibitors in Industrial Wheat Roller Mill Streams. *JAFC* **2006**, *54*, 8521–8529. [CrossRef]
11. Delcour, J.A.; Hoseney, R.C. *Principles of Cereal Science and Technology*, 3rd ed.; AACC International Press: St. Paul, MN, USA, 2010.
12. Iqbal, Z.; Pasha, I.; Abrar, M.; Hanif, M.S.; Arif, A.T.; Masih, S. Protein concentration, composition and distribution in wheat flour mill streams. *Annals. Food Sci. Technol.* **2015**, *16*, 104–114.
13. Pojić, M.M.; Spasojević, N.B.; Atlas, M.Đ. Chemometric Approach to Characterization of Flour Mill Streams: Chemical and Rheological Properties. *Food Bioprocess Technol.* **2014**, *7*, 1298–1309. [CrossRef]
14. Tosi, P.; He, J.; Lovegrove, A.; Gonzáles-Thuillier, I.; Penson, S.; Shewry, P.R. Gradients in compositions in the starchy endo- sperm of wheat have implications for milling and processing. *Trends Food Sci. Technol.* **2018**, *82*, 1–7. [CrossRef] [PubMed]

15. Sutton, K.H.; Simmons, L.D. Molecular Level Protein Composition of Flour Mill Streams from a Pilot-Scale Flour Mill and Its Relationship to Product Quality. *Cereal Chem.* **2006**, *83*, 52–56. [CrossRef]
16. Wang, Y.G.; Khan, K.; Hareland, G.; Nygard, G. Distribution of Protein Composition in Bread Wheat Flour Mill Streams and Relationship to Breadmaking Quality. *Cereal Chem.* **2007**, *84*, 271–275. [CrossRef]
17. Goesaert, H.; Brijs, K.; Veraverbeke, W.S.; Courtin, C.M.; Gebruers, K.; Delcour, J.A. Wheat flour constituents: How they impact bread quality, and how to impact their functionality. *Trends Food Sci.* **2005**, *16*, 12–30. [CrossRef]
18. Jacobs, P.J.; Hemdane, S.; Claes, S.; Mulders, L.; Langenaeken, N.A.; Dewettinck, K.; Courtin, C.M. Wheat bran-associated subaleurone and endosperm proteins and their impact on bran-rich bread-making. *J. Cereal Sci.* **2018**, *81*, 99–107. [CrossRef]
19. Shewry, P. What Is Gluten—Why Is It Special? *Front. Nutr.* **2019**, *6*, 101. [CrossRef]
20. Janssen, F.; Courtin, C.M.; Wouters, A.G.B. Aqueous phase extractable protein of wheat bran and germ for the production of liquid and semi-solid foods. *Crit. Rev. Food Sci. Nutr.* **2023**, 1–19. [CrossRef]
21. Delcour, J.A.; Vanhamel, S.; De Geest, C. Physico-Chemical and Functional Properties of Rye Nonstarch Polysaccharides. I. Colorimetric Analysis of Pentosans and Their Relative Monosaccharide Compositions in Fractionated (Milled) Rye Products. *Cereal Chem.* **1989**, *66*, 107–111.
22. Van de Vondel, J.; Janssen, F.; Wouters, A.G.B.; Delcour, J.A. Air-water interfacial and foaming properties of native protein in aqueous quinoa (*Chenopodium quinoa Willd.*) extracts: Impact of pH- and heat-induced aggregation. *Food Hydrocoll.* **2023**, *144*, 108945. [CrossRef]
23. Dornez, E.; Cuyvers, S.; Holopainen, U.E.; Poutanen, K.; Delcour, J.A.; Courtin, C.M. Inactive Fluorescently Labeled Xylanase as a Novel Probe for Microscopic Analysis of Arabinoxylan Containing Cereal Cell Walls. *JAFC* **2011**, *59*, 6369–6375. [CrossRef] [PubMed]
24. Schalk, K.; Lexhaller, B.; Koehler, P.; Scherf, K.A. Isolation and characterization of gluten protein types from wheat, rye, barley and oats for use as reference materials. *PLoS ONE* **2017**, *12*, e0172819. [CrossRef] [PubMed]
25. Wieser, H.; Antes, S.; Seilmeier, W. Quantitative Determination of Gluten Protein Types in Wheat Flour by Reversed-Phase High-Performance Liquid Chromatography. *Cereal Chem.* **1998**, *75*, 644–650. [CrossRef]
26. Hourston, J.E.; Ignatz, M.; Reith, M.; Leubner-Metzger, G.; Steinbrecher, T. Biomechanical properties of wheat grains: The im-plications on milling. *J. R. Soc. Interface* **2017**, *14*, 20160828. [CrossRef] [PubMed]
27. Kent, N.L. Subaleurone endosperm cells of high protein content. *Cereal Chem.* **1966**, *43*, 585–601.
28. Brouns, F.; Hemery, Y.; Price, R.; Anson, N.M. Wheat Aleurone: Separation, Composition, Health Aspects, and Potential Food Use. *Crit. Rev. Food Sci. Nutr.* **2012**, *52*, 553–568. [CrossRef] [PubMed]
29. Bilge, G.; Sezer, B.; Eseller, K.E.; Berberoglu, H.; Koksel, H.; Boyaci, I.H. Ash analysis of flour sample by using laser- i nduced breakdown spectroscopy. *Spectrochim. Acta B At. Spectrosc.* **2016**, *124*, 74–78. [CrossRef]
30. Meziani, S.; Nadaud, I.; Tasleem-Tahir, A.; Nurit, E.; Benguella, R.; Branlard, G. Wheat aleurone layer: A site enriched with nutrients and bioactive molecules with potential nutritional opportunities for breeding. *J. Cereal Sci.* **2021**, *100*, 103225. [CrossRef]
31. Lexhaller, B.; Colgrave, M.L.; Scherf, K.A. Characterization and Relative Quantitation of Wheat, Rye, and Barley Gluten Protein Types by Liquid Chromatography–Tandem Mass Spectrometry. *Front. Plant Sci.* **2019**, *10*, 1530. [CrossRef]
32. Hossain, K.; Ulven, C.; Glover, K.; Ghavami, F.; Simsek, S.; Alamri, M.S.; Kumar, A.; Mergoum, M. Interdependence of Cultivar and Environment on Fiber Composition in Wheat Bran. *Aust. J. Crop Sci.* **2013**, *7*, 525–531.

foods

MDPI

Article

Mechanical and Thermal Properties and Moisture Sorption of Puffed Cereals Made from Brown Rice, Barley, Adlay, and Amaranth

Atsuko Takahashi and Keiko Fujii *

Department of Food Science and Nutrition, Faculty of Human Sciences and Design, Japan Women's University, 2-8-1 Mejirodai, Bunkyo-ku, Tokyo 112-8681, Japan; jwu00010722@fc.jwu.ac.jp
* Correspondence: kfujii@fc.jwu.ac.jp; Tel./Fax: +81-03-5981-3439

Abstract: The moisture sorption, rheological, and glass transition properties of puffed cereals, such as brown rice, barley, adlay, and amaranth, were assessed. The puffed cereals were stored in desiccators until their moisture content reached equilibrium. Moisture sorption isotherms were measured, and monomolecular adsorption moisture content was calculated through Brunauer−Emmett−Teller (BET) analysis. The glass transition temperature (Tg) was determined, and the internal structure was observed using a scanning electron microscope. The rupture force and apparent elastic modulus of puffed cereals decreased with increasing relative humidity (RH). The puffed cereals exhibited ductile fracture ,when the moisture content was >8%. The Tg of puffed cereals with 8% moisture content was approximately 40 °C. It was inferred that puffed cereals demonstrated a crispy texture in the glassy state when stored at <40 °C, but transitioned to a rubbery state at >40 °C, resulting in the loss of crispy texture.

Keywords: puff; cereal; moisture sorption; glass transition; rupture properties

1. Introduction

Because food is a multi-component system, microbial spoilage and deterioration owing to reactions can occur during transportation and storage. Therefore, various attempts have been made to protect food from deterioration. In food engineering, the development of methods to reduce temperature and moisture and improve shelf life is vital [1].

The curve plotting the equilibrium moisture content against various relative humidities at a certain temperature is called a moisture sorption isotherm. It is useful for analyzing storage and processing properties by predicting the moisture adsorption and drying of foods in a certain environment. Additionally, they can be used to classify food groups based on similarities in their behavior. Numerous foods are known to exhibit an inverted S-shape moisture sorption isotherm.

By determining the monomolecular adsorption moisture content from the region of strong moisture sorption that occurs at the rising edge of the moisture sorption isotherm curve, it was observed that microbial growth was inhibited, enzymatic and browning reactions were minimized, and numerous degradation reactions of food products were inhibited [1–7]. Previous studies using moisture sorption isotherms include those that examined the moisture sorption characteristics of biscuits [8] and wafers [9] and those that examined the relationship between moisture sorption and texture characteristics of various baked confections [10–17].

https://doi.org/10.3390/foods14020189

Additionally, numerous dried foods are in a glassy state, resulting in research on quality control that considers glass transition [18,19]. The focus on glass transition stems from its ability to explain the phenomena and rupture properties associated with food deterioration and shelf life.

Previous studies have revealed the moisture sorption and plasticity in moisture-sorbed snacks, corn cakes [20], and wafers [9] by exploring their glass transition points. However, despite 30 years of research on glass transition in food, few studies have been conducted on puffed cereals.

Therefore, this study aimed to clarify the moisture sorption and rheological properties of puffed cereals by examining the effects of moisture absorption on their rupture properties. This includes assessing alterations in the breaking behavior, monomolecular adsorption moisture content, and glass transition point of the sorbed puffed cereals. These aspects are characteristic of swollen foods.

2. Materials and Methods

2.1. Experimental Materials

Four types of cereals with characteristic nutritional functions and varying grain sizes were selected as samples for various puffed cereals: brown rice (domestic), barley (American), adlay (Iwate), and amaranth (Iwate). The long and short diameters of the four types of cereals (raw) are as follows. Brown rice: 4.88 mm, 2.85 mm; barley: 4.06 mm, 2.98 mm; adlay: 5.15 mm, 4.72 mm; amaranth: 1.16 mm, 1.16 mm.

Of the four cereals, brown rice, adlay, and amaranth were purchased from the Japan Millet Association, and barley was provided by Mitake Co. (Saitama, Japan). The cereals were swelled at the Sekizawa Shouten Ltd. (Saitama, Japan) swelling and processing plant in the presence of the authors.

The nutritional components of the four cereals in 100 g are shown in Table 1 [21].

Table 1. Nutritional composition of the four types of cereals.

	Energy (kcal)	Protein (g)	Fat (g)	Carbo-Hydrate (g)	Potassium (mg)	Calcium (mg)	Iron (mg)	Dietary-Fiber (g)
Brown rice	356	6.5	3.3	74.3	160	13	1	3.1
Barley	343	7	2.1	76.2	170	17	1.2	8.7
Adlay	360	13.3	1.3	72.2	85	6	0.4	0.6
Amaranth	358	12.7	6	64.9	600	160	9.4	7.4

2.2. Preparation of Puffed Cereals

The swelling of various cereals was investigated using a cereal puffing machine (Koyo Machine Manufacturing Co., Ltd. Hiroshima, Japan). A pressure kiln was preheated to 230 °C, and 2 kg of grain was placed inside the kiln. The pressure was increased to 0.1, 0.7, 0.9, and 1.1 MPa and was reduced rapidly to induce swelling. Briefly, raw samples (grains) were placed in a pressure kiln and heated at high pressure while rotating, and when a certain pressure was reached, the valve of the pressure kiln was struck with a hammer, the lid was opened, and the internal pressure was released at once. The samples swelled while retaining the shape of the grain. The heating temperature was 240 ± 20 °C, and the heating time was approximately 1 min for the 0.1 MPa treatment and 5–7 min for the 0.7–1.1 MPa treatment. The samples were stored in eight humidity-controlled containers with relative humidity (RH) ranging from 6 to 94% for 5–12 days until their moisture sorption reached equilibrium. The humidity control temperatures were 15 °C, 25 °C, and 35 °C. For humidity control, eight different saturated aqueous solutions of salts were used:

lithium bromide for 6.0–6.9% RH; lithium chloride for 11.3% RH; magnesium chloride for 32.1–33.3% RH; potassium carbonate for 43.2% RH; sodium bromide for 54.6–60.7% RH; sodium chloride for 74.9–75.6% RH; potassium chloride for 83.0–85.9% RH; and potassium nitrate for 90.8–95.4% RH.

2.3. Measurement of Moisture Content and Equilibrium Moisture Content

The moisture content of the moisture-sorbed puffed cereals (1.5–3.0 g) was measured. A moisture analyzer (MB45; OHAUS Co., Ltd. Tokyo, Japan) was used to measure the moisture content. The samples were heated at 105 °C until the rate of weight change reached 1 mg/120 s.

The equilibrium moisture content was calculated using the following equation:

$$\text{Equilibrium moisture content (\% dry basis [D.B.])} = \{\text{Weight at equilibrium moisture sorption} \times \text{Moisture content at moisture sorption equilibrium} / \text{Weight at equilibrium moisture sorption} \times (1 - \text{moisture content before moisture sorption})\} \quad (1)$$

2.4. Moisture Sorption Isotherm and Measurement of Monomolecular Adsorption Moisture Content

In many foods, the moisture sorption isotherm exhibits an inverted S-shape. A typical moisture sorption isotherm can be divided into three regions, as shown in Figure 1. In the region with the lowest water activity, moisture molecules are adsorbed on the food surface, forming a monomolecular layer. In this zone, moisture molecules are restricted, minimizing physical and chemical alterations and significantly reducing the deterioration rate of the food. Because the BET formula is often used to determine the monomolecular adsorption moisture content, it was calculated by fitting the values of each moisture sorption isotherm in the 5–35% RH range to the BET formula [4].

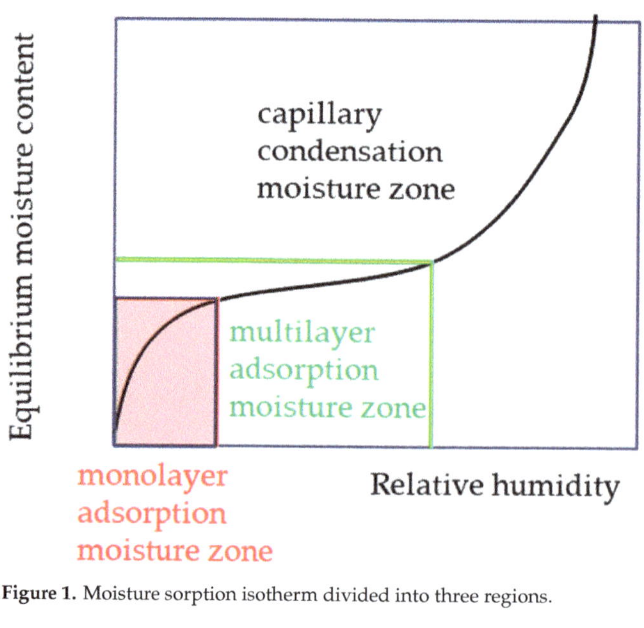

Figure 1. Moisture sorption isotherm divided into three regions.

Monomolecular adsorption moisture content was calculated based on the parameter analysis of the BET equation, as follows:

$$\text{BET formula (Brunauer–Emmett–Teller)} \quad A/(V(1-A)) = (C-1)/CVm \cdot A + 1/(CVm) \tag{2}$$

A: Water activity of food sample
V: Moisture content of food sample (g/100 g dry matter)
Vm: Monomolecular adsorption moisture content (g/100 g dry matter)
C: Constant determined by the properties of the food sample and temperature
$A/V(1-A) = A \cdot (C-1)/C \cdot Vm + 1/C \cdot Vm$
If $A/V(1-A) = y$, $A = x$, $(C-1)/C \cdot Vm = a$, and $1/C \cdot Vm = b$
This can be expressed by the linear equation $y = ax + b$.
$a = (C-1)/C \cdot Vm$... Slope of the straight line
$b = 1/(C-1)$...y–intercept
$C = (b + a)/b$
$Vm = 1/(b + a)$

∴ Vm: monomolecular adsorption (g/100 g dry matter) was calculated
∗ The BET formula reportedly fits well in the 0.05–0.35 water activity range.

2.5. Rupture Properties of Puffed Cereals Following Moisture Sorption

The rupture properties of various moisture-sorbed puffed cereals were measured with a rheometer (RE-3305; Yamaden Co., Tokyo, Japan) using the single-grain method [22,23]. An acrylic resin plunger with a diameter of 40 mm was used. The compression ratio was 99%, and the compression speeds were 0.05, 0.1, 1, and 10 mm/s at the four levels. The rupture force and apparent elastic modulus were calculated from the resulting force-deformation curves. For the amaranth specimens, automatic creep meter analysis device software Rupture Strength Analysis (HS Windows Ver. 2.5; BAS-3305H; Yamaden Co., Ltd.) was used at a compression rate of 10 mm/s. The measurement temperature was 25 °C and the measurements were taken three times.

2.6. Measurement of Glass Transition Temperature (Tg)

The Tg of the puffed cereals was measured using a differential scanning calorimeter (Diamond DSC; Perkin Elmer Japan Co., Kanagawa, Japan). The samples were ground in a mortar, and 7–42 mg in weight was placed in a large stainless steel pan. The measurement temperature ranged from −50 °C to 100 °C, and the scan speed was 10 °C/min. Two scans were obtained for each sample.

2.7. Observation of Tissue Structure

Various puffed cereals were sliced into approximately 1 mm–thick slices, and platinum–palladium deposition was performed. The tissue structure was observed using a scanning electron microscope (S-800; Hitachi, Ltd., Tokyo, Japan) [24] at 40×, 100×, and 500× magnifications.

2.8. Statistical Processing

SPSS (Ver. 25.0; IBM Japan, Ltd., Tokyo, Japan) was used for statistical processing and analysis. Tukey's multiple comparisons were performed after one-way analysis of variance (ANOVA) to test variations in puffed cereals between samples. In all cases, a risk rate of <5% was set as the significance level.

3. Results and Discussion

3.1. Moisture Sorption Characteristics

Various types of puffed cereals were humidified, and their equilibrium moisture content was determined. Subsequently, the moisture sorption isotherms for each cereal puff were plotted, with RH on the horizontal axis and equilibrium moisture content on the vertical axis (Figure 2). In this study, the majority of the moisture sorption isotherms exhibited an inverted S-shape, except for a few isotherms.

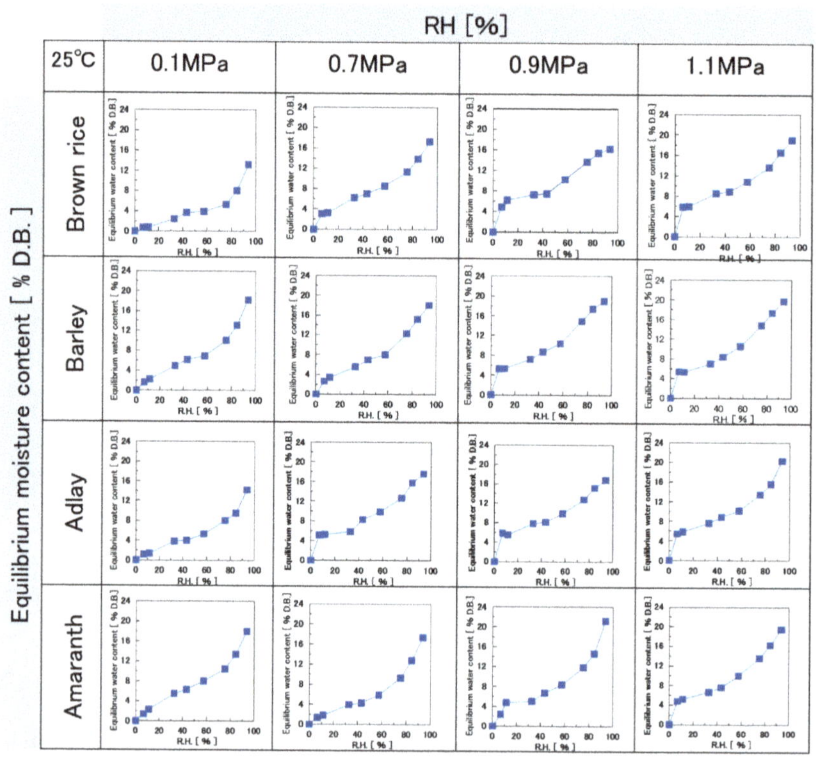

Figure 2. Moisture sorption isotherms of each puffed cereal at 25 °C.

The moisture sorption isotherms of various puffed cereals at 0.1 MPa and amaranth puffs at 0.7 MPa indicated that all samples exhibited moisture content < 3% D.B. at 15% RH. None of the samples were swollen. However, the other swollen samples exhibited a higher moisture content of approximately 6% D.B. at 15% RH. This indicates that moisture sorption increases significantly in the region where the isotherm curve rises. Thus, it was inferred that sorption was stronger in puffed cereals with advanced swelling.

Figure 3 shows the moisture sorption isotherms at different temperatures for various puffed cereals swelled at 0.9 MPa. Up to 80% RH, there was no difference in moisture content between the puffed cereal samples. However, at 15 °C and approximately 90% RH, the moisture content increased, specifically for barley and adlay, reaching approximately 32% D.B. At 25 °C and 35 °C, the moisture contents of brown rice, barley, and adlay at approximately 90% RH were similar. Conversely, for amaranth, the moisture content increased as the humidity conditioning temperature decreased (35 °C < 25 °C < 15 °C). This aligns with the general inference that the moisture content decreases at higher temperatures [1], likely owing to the temperature dependence of the adsorption potential function [2] and alterations in intermolecular forces between the solid and the adsorbent.

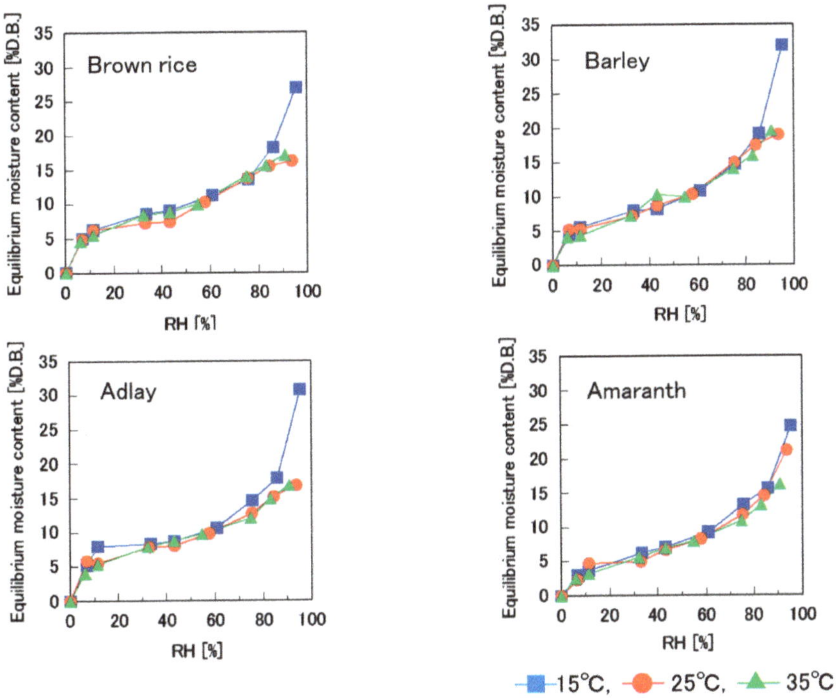

Figure 3. Moisture sorption isotherm of each puffed cereal of 0.9 MPa at 15 °C, 25 °C, and 35 °C.

3.2. Monomolecular Adsorption Moisture Content

The moisture sorption isotherms of various puffed cereals humidified at 15 °C, 25 °C, and 35 °C exhibited an almost inverse S-shape.

Table 2 presents the results of the amount of monomolecular adsorption for all the samples at 15 °C, 25 °C, and 35 °C. As shown, the monomolecular adsorption moisture content of puffed cereals was approximately 10–18 g/100 g dry solid, which is higher than that of the 3.7–4.7 g/100 g dry solid observed in similar dried foods, such as cookies, biscuits, and snacks [25]. Foods with relatively similar values were 11 g/100 g dry solid for stick agar and 14 g/100 g dry solid [25] for kanpyo. Wada et al. [25] studied the effect of absorbed moisture on the textural properties of low-moisture foods and observed that samples with high monomolecular adsorption had high starch content and low oil and sugar content. In the present study, the results in Table 2 show that amaranth has the lowest value at all temperatures. Looking at the nutritional compositions of the four types of puffed cereals in Table 1, amaranth had the lowest value for carbohydrates and the highest value for fat among the four species. This could be the reason for the lower monomolecular adsorption of amaranth.

3.3. Rupture Properties of Puffs After Moisture Sorption (Rupture Force)

Figure 4 shows the relationship between the equilibrium moisture content and rupture force of various puffed cereals swollen at 0.9 MPa. In all samples, the rupture force increased to approximately 8% of the equilibrium moisture content. However, at higher equilibrium moisture contents, the rupture force decreased as the equilibrium moisture content increased. This aligns with numerous studies [10,20,25–33] that have demonstrated that stress in moisture-sorbed foods increases or decreases with increasing equilibrium moisture content. This is attributed to the fact that starch binds and becomes more viscous with moisture absorption, which increases its rupture properties. However, in the higher humidity range, the adsorbed moisture molecules cause flow deformation, which reduces the rupture properties [32]. In the present study, the reduction in the rupture force at high equilibrium moisture content in the 15 °C humidity-controlled samples was attributed to the moisture sorption isotherm (Figure 3). This demonstrates that the equilibrium moisture content in all four types of puffed cereals was significantly higher at 15 °C humidity, resulting in water acting as a plasticizer and causing flow deformation.

Table 2. Monomolecular adsorption moisture content of each puffed cereal [g water/100 g dry solid].

		0.7 MPa	0.9 MPa	1.1 MPa
15 °C	Brown rice	14.6	17.3	16.6
	Barley	14.7	16.3	17.2
	Adley	15.6	15.3	15.8
	Amaranth	14.4	14.6	13.6
25 °C	Brown rice	15.1	13.2	16.0
	Barley	12.7	13.2	12.5
	Adley	9.72	14.4	13.7
	Amaranth	13.0	10.2	17.8
35 °C	Brown rice	12.3	16.8	13.6
	Barley	16.5	14.5	13.9
	Adley	16.1	16.2	13.2
	Amaranth	10.6	12.7	13.8

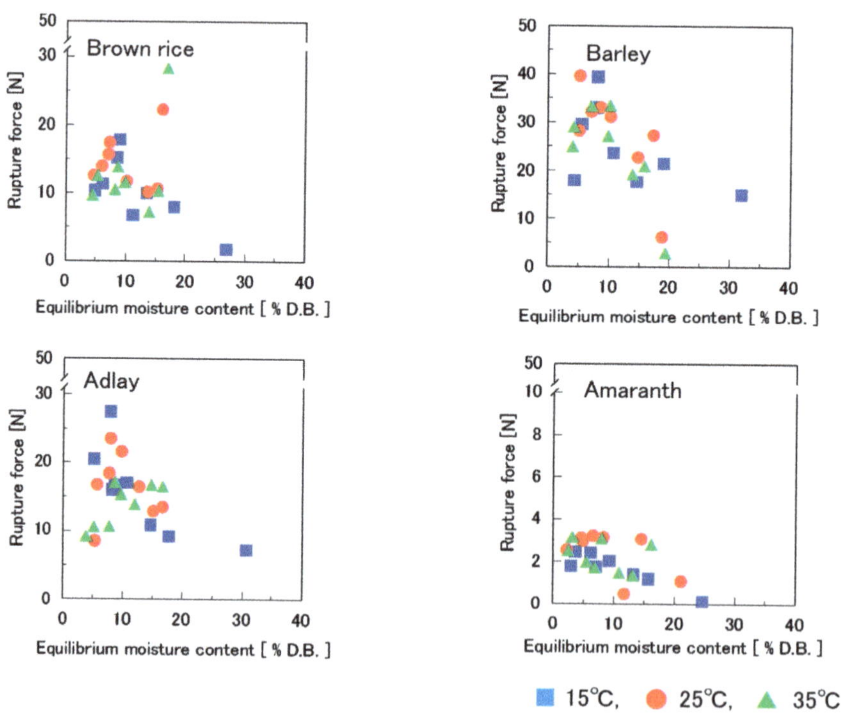

Figure 4. Effect of equilibrium moisture content on the rupture force of four types of puffed cereals at 15 °C, 25 °C, and 35 °C.

The significantly increased rupture force of puffed brown rice at 15% equilibrium moisture content was attributed to the shrinkage and tightening of the samples in the rubbery state with water.

The relationship between the amount of monomolecular adsorption and the rupture force was also examined. The amount of monomolecular adsorption is obtained in the 0.05–0.35 water activity range, which is approximately 4–8% at equilibrium moisture content. In Figure 4, the values of rupture force at 4–8% equilibrium moisture content are approximately 8–20 N for brown rice and adlay, 18–40 N for barley, and 2–4 N for amaranth, with amaranth having the lowest value. The values of monomolecular adsorption at 0.9 MPa in Table 2 show that amaranth has the lowest values at all temperatures (15 °C, 25 °C, and 35 °C).

Figure 5 shows the force-deformation curves of brown rice at 25 °C. The ductile fracture was observed from RH 93.6, 57.6, 57.6, and 32.8% RH at 0.1, 0.7, 0.9, and 1.1 MPa swelling, respectively. The dotted line represents the borderline between brittle and ductile fractures, and the filled area represents the ductile fracture region. The higher the swelling pressure, the lower the RH in the boundary region between the brittle and ductile fractures. Similar trends were observed for the three types of puffed cereals, excluding brown rice; therefore, the results for brown rice were considered representative.

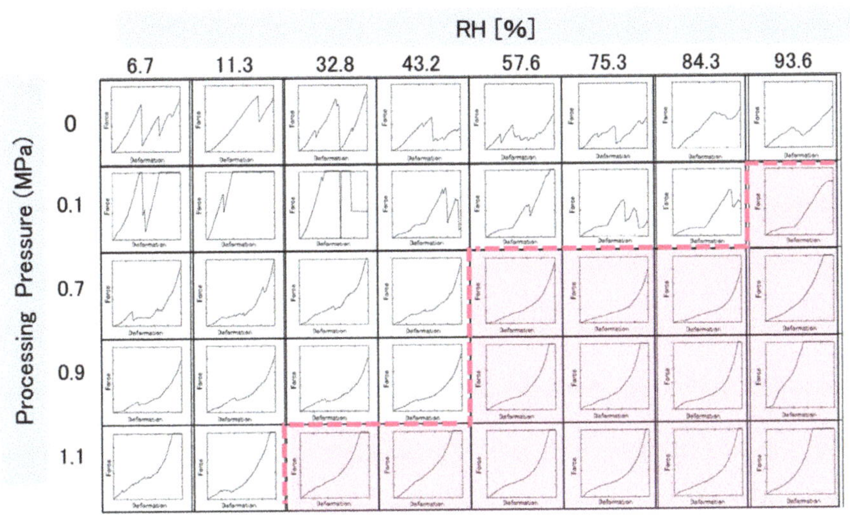

Figure 5. Force-deformation curves of puffed brown rice at various relative humidity (RH).

Figure 6 shows the moisture sorption isotherms of the four types of puffed cereals. The transition from brittle to ductile fracture is indicated by the moisture sorption isotherm, with the circled region indicating the boundary. All the samples with swelling pressures between 0.7 and 1.1 MPa exhibit ductile fracture when the moisture content was approximately >8%. This demonstrates that the crispiness of puffed cereals can be effectively assessed by identifying the brittle to ductile transition in the fracture curve, which occurs at a moisture content of 8%. In a previous study [34], at water contents higher than 8.5% for CCF (commercial corn flakes), the compression curves showed fewer fluctuations, reflecting a decrease in brittle fracture. Similarly, in the present study, the fracture curve changed from brittle to ductile fracture at >8% moisture content.

The products produced by baking or extruder exhibit a water activity value of 0.5, or an equilibrium moisture content of approximately 8–10%. Additionally, the critical moisture content for crispiness was in the water activity value range of 0.45–0.55 [35]. In the present study, samples other than amaranth with swelling pressures of 0.7–1.1 MPa exhibited brittle fracture up to 43.2% RH and ductile fracture from 57.6%RH, similar to previous studies. Amaranth was the only exception, with 0.7 MPa showing a brittle fracture up to

57.6%RH and a ductile fracture from 75.3%RH. According to a previous study [36], the 0.7 MPa amaranth sample had a low swelling ratio of 1.13, which was close to that of the raw sample. In other words, the 0.7 MPa amaranth sample with insufficient swelling was considered to be in a glassy state, while the other samples with advanced swelling absorbed water and became rubbery, and therefore, the brittle fracture zone of amaranth was under 75.3%RH.

This indicates that the crispiness of puffed cereals can be assessed effectively by the brittle and ductile fractures in the force-deformation curve, with brittle fractures transitioning to ductile fractures beyond 8% moisture content.

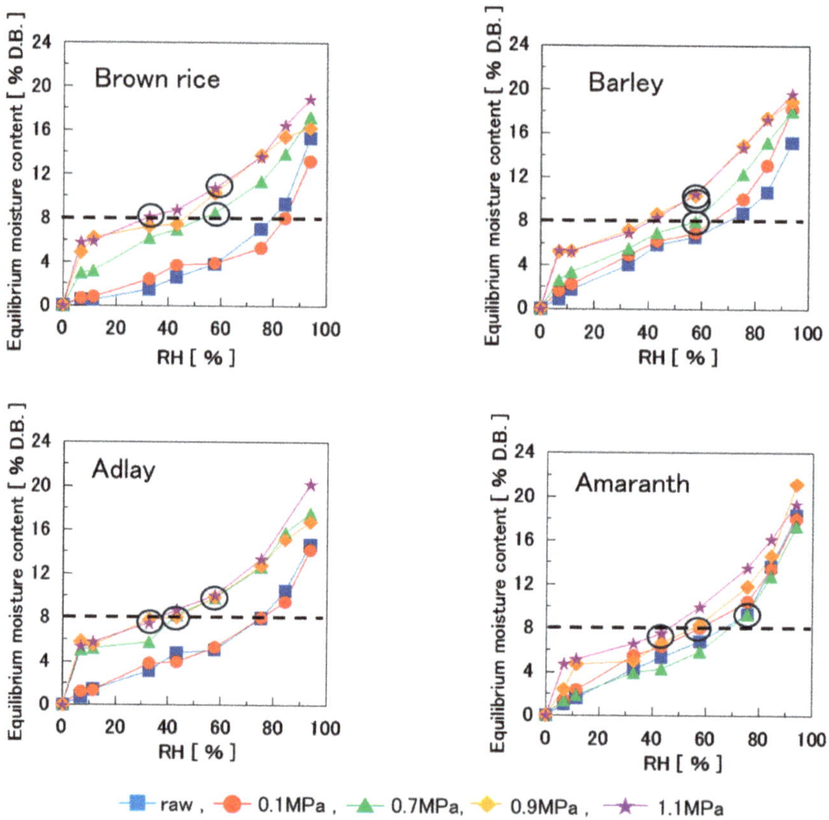

Figure 6. Effect of relative humidity on the equilibrium moisture content of four types of puffed cereals at 25 °C.

3.4. Microstructure of Puffed Cereals

Figure 7 shows cross-sectional photographs of the internal microstructure of various puffed cereals at 90.8–95.4% RH humidification. Numerous large pores were observed in brown rice, barley, and adlay at 25 °C and 35 °C humidification, whereas smaller pores were observed at 15 °C humidification. In amaranth, the pores became smaller with decreasing temperature (35 °C > 25 °C > 15 °C).

Figure 3 shows that the equilibrium moisture content at 25 °C and 35 °C did not increase with an increase in relative humidity, but the moisture content at 15 °C increased significantly. This was thought to be due to the cell walls of the puffed cereals becoming thicker and the pores becoming smaller as a result of water absorption. Conversely, the equilibrium moisture content of the amaranth sample at approximately 90% RH was 35 °C

< 25 °C < 15 °C, and the equilibrium moisture content increased as the temperature decreased, which is consistent with the pore size in the electron micrograph shown in Figure 7.

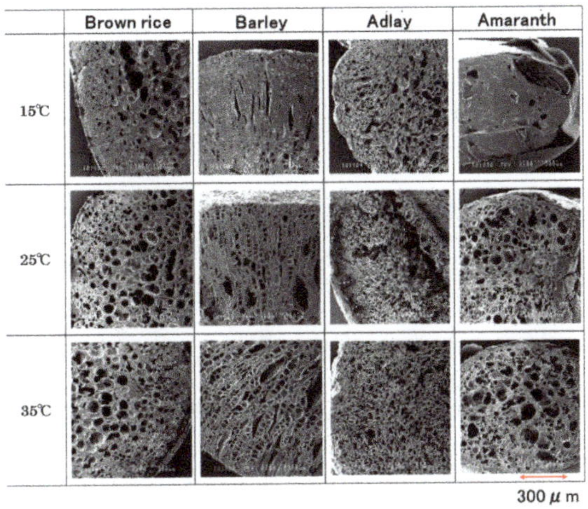

Figure 7. Scanning electron microscope images of the microstructure of four kinds of puffed cereals of 0.9 MPa at high humidity (100×).

Figure 8 shows the relationship between the internal microstructure and force-deformation curve during the compression of four types of puffed cereals swollen at 0.9 MPa and humidified at 25 °C. When the puffs were moistened at 6.7% RH, the pores were relatively large, and the continuous phase was significantly thin. The force-deformation curve exhibited a clear break point, indicating that the puffs were brittle with a low rupture force. The internal microstructure at 43.2 and 75.3% RH demonstrated that the pore size did not change much. However, the continuous phase constituting the pores gradually became thicker as the RH increased.

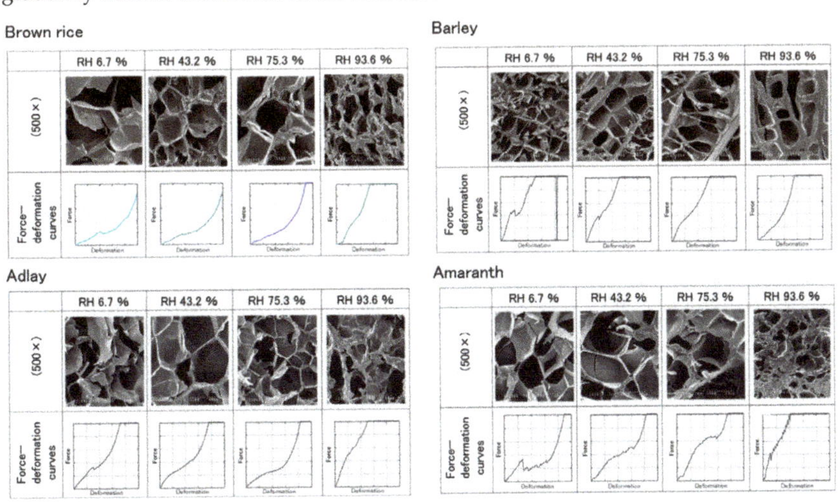

Figure 8. Scanning electron microscope images of microstructure and force-deformation curves of four types of puffed cereals (0.9 MPa/25 °C).

The force-deformation curve at 43.2% RH demonstrated that the reduction in force became smaller after fracture, and at humidity control > 75.3% RH, the force-deformation curve exhibited a ductile fracture. At 93.6% RH, the pores were smaller, and the continuous phase was considerably thicker. The force-deformation curve exhibited a significant increase in force with slight deformation, indicating that the moist specimen shrunk during compression and did not fracture.

The pore size of barley was different from that of brown rice, adlay, and amaranth, which had fibrous pores. Because of this shape, the cell walls were thicker, and the angle of the rising curve at the beginning of compaction was greater.

3.5. Measurement of Tg

Determining Tg is critical for determining the shelf life of food. At temperatures below Tg, Brownian micromotion and molecular migration are reduced, thereby stabilizing food quality and enhancing shelf life. Conversely, above Tg, molecules exhibit enhanced Brownian micromotion and greater molecular migration, leading to deterioration of the food product. Therefore, knowing the Tg is critical for identifying the temperature at which the food product can be stored to maintain quality and extend its shelf life.

The Tg values of various puffed cereals conditioned at 25 °Care illustrated in Figure 9. The Tg ranged from −16.6 °C to 80.2 °C. Tg decreased as the equilibrium moisture content increased, which is consistent with the trends for corn cakes [20], wafers [9], corn flakes [34], and extruded cereal-based products [37].

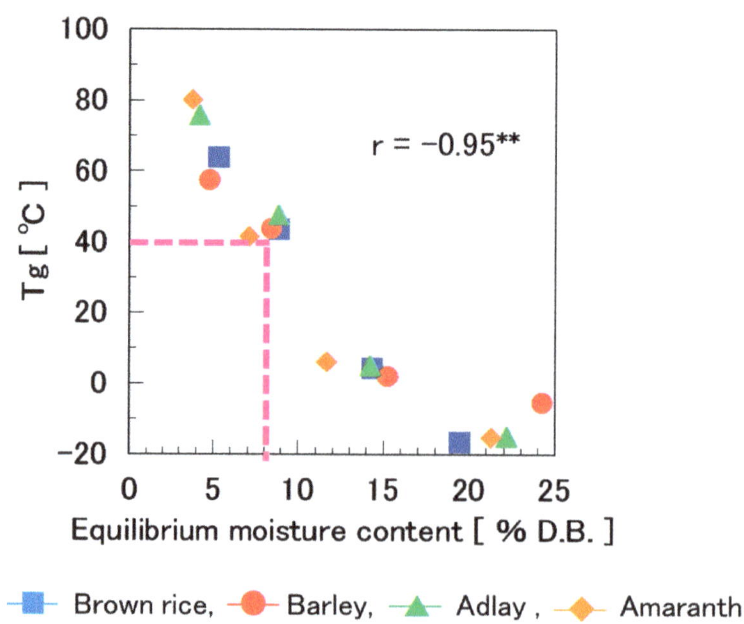

Figure 9. Tg in the second scan among the four types of puffed cereals (25 °C). r = −0.95 **, ** correlation is significant at $p < 0.01$.

In the previous section, it was demonstrated that the boundary region where the rupture properties transition from brittle to ductile fracture occurs at a moisture content of 8%. Additionally, it was identified that the Tg of these puffed cereals with 8% moisture content was approximately 40 °C. These puffed cereals exhibited a crispy texture in a glassy state when stored at 25 °C. However, it was inferred that they shifted to a rubbery state at temperatures above 40 °C, and the crispy texture was lost.

In a previous study [37], Young's modulus results showed that water acts as an anti-plasticizer at low aw, while exhibiting a plasticizing effect at high aw, and a stability map can explain the brittle-ductile transition that occurred when it was below Tg. In the present study, as in the previous studies [34,37], the threshold between brittle and ductile fracture was almost at Tg.

Figure 10 shows the effect of moisture content on the apparent elastic modulus of various puffed cereals. In all samples, the apparent elastic modulus decreased as the moisture content increased.

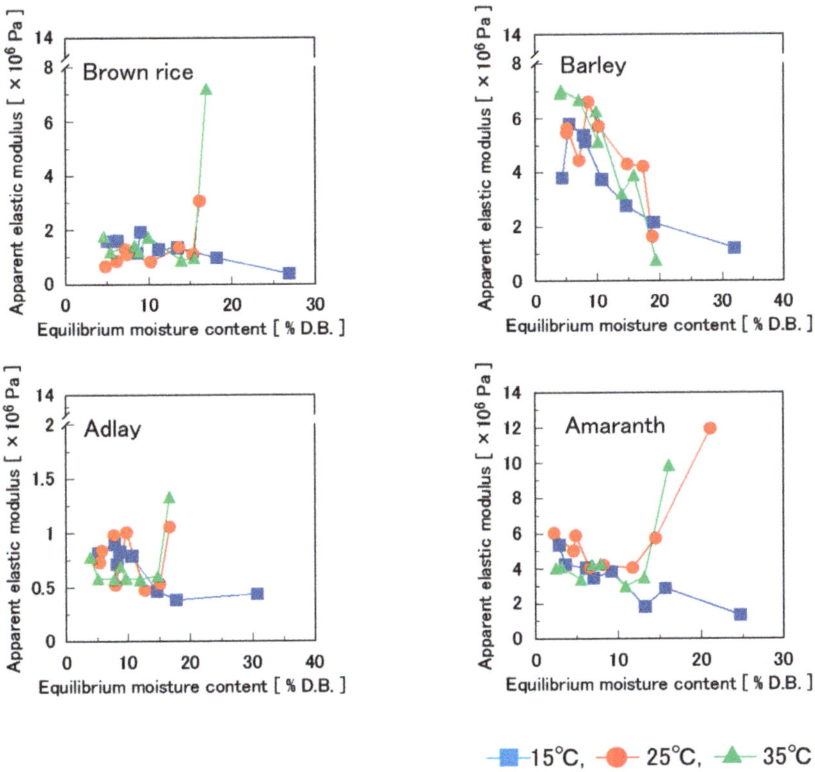

Figure 10. Effect of equilibrium moisture content on the apparent elastic modulus of four kinds of puffed cereals.

Based on a previous study that assessed the effect of moisture adsorption on the rupture properties of commercial confectioneries [32], the apparent elastic modulus decreased with increasing humidity, and the transition was relatively small up to an RH of approximately 43%; however, it decreased significantly at higher RH. Similar results were obtained in this study. This is attributed to the glass transition of various puffed cereals as a result of moisture content, which causes their rupture properties to transition to softness. The moisture content at which the apparent elastic modulus decreased rapidly was approximately 8–10% in all cases. In this study, it was assumed that puffed cereals transitioned from glass to rubber.

There are certain exceptions. For brown rice, adlay, and amaranth (excluding barley), the apparent elastic modulus was significantly higher under the highest moisture content conditions at 25 °C and 35 °C. The significantly higher apparent elastic modulus was attributed to the shrinkage and tightness of the samples in the rubbery state containing moisture.

4. Conclusions

Four types of puffed cereals were stored under different temperatures (15 °C, 25 °C, and 35 °C) and humidity (6–94%) conditions. The samples stored at 25 °C and 35 °C were not affected by relative humidity (RH), whereas those stored at 15 °C had higher moisture content at higher humidity. Electron micrographs of the internal structure of the samples stored at 15 °C showed that the cell walls became thicker, and the pore size of the puffs became smaller due to moisture absorption. The rupture force at this time showed a low value, indicating that the puffs were in a rubbery state following moisture absorption. The rupture force peaked at 8% RH and then decreased with an increase in relative humidity.

The physical properties changed from brittle to ductile fracture around a moisture content of 8%, and the transition point between glass and rubber was considered to be 8%.

The Tg of the puffed cereals at 8% moisture content was approximately 40 °C. It was inferred that the puffed cereals were glassy and crispy when stored at <40 °C, but changed to rubbery and lost their crispy texture when stored at >40 °C.

The results indicated that a crispy texture can be maintained by keeping the moisture content at <8% and the storage temperature at <40 °C.

Author Contributions: Conceptualization, A.T. and K.F.; Data curation, A.T.; Formal analysis, A.T. and K.F.; Funding acquisition, K.F.; Investigation, A.T.; Methodology, A.T. and K.F.; Project administration, K.F.; Resources, K.F.; Software, A.T.; Supervision, K.F.; Validation, K.F.; Visualization, A.T. and K.F.; Writing—original draft, A.T.; Writing—review & editing, K.F. All authors have read and agreed to the published version of the manuscript.

Funding: This work was supported by a JSPS Grant—in-Aid for Scientific Research (C) Grant Number JP17K00827. Any opinions, findings, conclusions, or recommendations expressed in this material are those of the authors and do not necessarily reflect the views of the authors' organization, JSPS.

Institutional Review Board Statement: Not applicable.

Informed Consent Statement: Not applicable.

Data Availability Statement: The data presented in this study are available on request from the corresponding author.

Acknowledgments: We would like to thank the Mitake Foods Corporation for providing barley for this study. We would also like to thank Sekizawa Shoten Ltd. for their cooperation in the expansion of cereals. We would also like to thank Mamiko Sato for her guidance in taking the electron microscope photographs.

Conflicts of Interest: The authors declare no conflicts of interest.

References

1. Japan Society for Food Engineering (Ed.) *Food Engineering*; Asakura Publishing Co., Ltd.: Tokyo, Japan, 2012; pp. 94–111, 115–117, 120–125.
2. Matsuno, R.; Yano, T. *Physical Chemistry of Foods*; Buneido Publishing Co., Ltd.: Tokyo, Japan, 1996; pp. 38–57, 82–98.
3. Kubota, S.; Ishitani, T.; Sano, Y. *Food and Water*; Kourin Publishing Co., Ltd.: Tokyo, Japan, 2008; pp. 205–232, 276–279.
4. Kato, H.; Kurata, T. *Food Preservation Science*; Buneido Publishing Co., Ltd.: Tokyo, Japan, 1999; pp. 135–149.
5. Mohsenin, N.N. *Physical Properties of Foods*; Kourin Publishing Co., Ltd.: Tokyo, Japan, 1982; pp. 31–33.
6. Noguchi, S. *Food and Water Science*; Saiwai Publishing Co., Ltd.: Tokyo, Japan, 1992; pp. 169–193.
7. Japan Society for Food Engineering. *Food Engineering Handbook*; Asakura Publishing Co., Ltd.: Tokyo, Japan, 2006; pp. 455–456, 503–504, 600–603, 621–623.
8. Sampaio, M.R.; Marcos, K.S.; Moraes, I.C.F.; Perez, V.H. Moisture adsorption behavior of biscuits formulated using wheat, oatmeal and passion fruit flour. *J. Food Process Preserv.* **2009**, *33*, 105–113. [CrossRef]
9. Martínez-Navarrete, N.; Moraga, G.; Talens, P.; Chiralt, A. Water sorption and the plasticization effect in wafers. *Int. J. Food Sci. Technol.* **2004**, *39*, 555–562. [CrossRef]

10. Van Hecke, E.; Allaf, K.; Bouvier, J.M. Texture and structure of crispy-puffed food products part II: Mechanical properties in puncture. *J. Texture Stud.* **1998**, *29*, 617–632. [CrossRef]
11. Barrett, A.M.; Normand, M.D.; Peleg, M.; Ross, E. Characterization of the jagged stress-strain relationships of puffed extrudates using the fast Fourier transform and fractal analysis. *J. Food Sci.* **1992**, *57*, 227–232. [CrossRef]
12. Norton, C.R.T.; Mitchell, J.R.; Blanshard, J.M.V. Fractal determination of crisp or crackly textures. *J. Texture Stud.* **1998**, *29*, 239–253. [CrossRef]
13. Roopa, B.S.; Mazumder, P.; Bhattacharya, S. Fracture behavior and mechanism of puffed cereal during compression. *J. Texture Stud.* **2009**, *40*, 157–171. [CrossRef]
14. Barrett, A.H.; Kaletunc, G. Quantitative description of fracturability changes in puffed corn extrudates affected by sorption of low levels of moisture. *Cereal Chem.* **1998**, *75*, 695–698. [CrossRef]
15. Fujii, K. Development of a novel foam food made from wheat flour and soy protein isolate. *Soy Protein Res. Jpn.* **2002**, *5*, 51–57.
16. Sato, Y.; Noguchi, S. Water sorption characteristics of dietary fibers. *J. Home Econ. Jpn.* **1993**, *44*, 625–631.
17. Higo, A.; Wada, Y.; Sato, Y. Relationship between physical properties and water layers in commercial confectionary products during moisture sorption processing. *J. Jpn. Soc. Food Sci. Technol.* **2012**, *59*, 572–582. [CrossRef]
18. Suzuki, T. Glassy states of food and its utilization. *Food Technol.* **2006**, *426*, 1–9.
19. Kawai, K.; Kurosaki, K.; Suzuki, T. Analytical approach for the non-equilibrium state of glassy foods affected by its manufacturing process. *Cryobiol. Cryotechnol.* **2008**, *54*, 71–77.
20. Li, Y.; Kloeppel, K.M.; Hsieh, F. Texture of glassy corn cakes as a function of moisture content. *J. Food Sci.* **1998**, *63*, 869–872. [CrossRef]
21. Kagawa, Y. *VII Food Ingredient List 2016*; Kagawa Nutrition University Press: Tokyo, Japan, 2016; pp. 6, 12, 18.
22. Takahashi, S.; Kuno, M.; Nishizawa, K.; Kainuma, K. New method for evaluating the texture and sensory attributes of cooked rice. *J. Appl. Glycosci.* **2000**, *47*, 343–353. [CrossRef]
23. Ida, R. Adaptability of a moisture analyzer equipped with a halogen heater to the measurement of water content of unhulled rice. *Jpn. J. Crop. Sci.* **2016**, *85*, 173–177. [CrossRef]
24. Goshima, G.; Aoyama, H.; Nishizawa, K.; Tsuge, H. Popping mechanism of popcorn grain. *J. Jpn. Soc. Food Technol.* **1988**, *35*, 147–153.
25. Wada, Y.; Shimoda, Y.; Higo, A. Effect of absorped moisture on the textural properties of low-moisture foods. *J. Home Econ. Jpn.* **1997**, *48*, 597–606.
26. Wada, Y.; Higo, A. Effect of moisture and temperature on the textural properties of commercial confectionery. *J. Jpn. Soc. Food Sci. Technol.* **2007**, *54*, 253–260. [CrossRef]
27. Mazumder, P.; Roopa, B.S.; Bhattacharya, S. Textural attributes of a model snack food at different moisture contents. *J. Food Eng.* **2007**, *79*, 511–516. [CrossRef]
28. Higo, A.; Wada, Y. Effect of modified wheat starches on the textural properties of baked products, such as cookies. *J. Food Sci. Technol.* **2008**, *55*, 224–232.
29. Chang, Y.P.; Cheah, P.B.; Seow, C.C. Plasticizing-antiplasticizing effects of water on physical properties of tapioca starch films in the glassy state. *J. Food Sci.* **2000**, *65*, 445–451. [CrossRef]
30. Katz, E.E.; Labuza, T.P. Effect of water activity on the sensory crispness and mechanical deformation of snack food products. *J. Food Sci.* **1981**, *46*, 403–409. [CrossRef]
31. Chang, Y.P.; Cheah, P.B.; Seow, C.C. Variations in flexural and compressive fracture behavior of a brittle cellular food (dried bread) in response to moisture sorption. *J. Texture Stud.* **2000**, *31*, 525–540. [CrossRef]
32. Wada, Y.; Ogawa, K.; Higo, A. Changes in breaking properties of commercial confectionery with moisture absorption and their factors. *J. Jpn. Soc. Food Sci. Technol.* **2002**, *49*, 771–781. [CrossRef]
33. Higo, A.; Wada, Y.; Sato, Y. Relationship between physical properties and mono-and multilayer water contents of a white wheat baked product at various temperatures and relative humidity. *J. Jpn. Soc. Food Sci. Technol.* **2014**, *61*, 486–496. [CrossRef]
34. Farroni, A.E.; Matiacevich, S.B.; Guerrero, S.; Alzamora, S.; Buera, M.D.P. Multi-level approach for the analysis of water effects in corn flakes. *J. Agric. Food Chem.* **2008**, *56*, 6447–6453. [CrossRef]
35. Heidenreich, S.; Jaros, D.; Rohm, H.; Ziems, A. Relationship between water activity and crispness of extruded rice crisps. *J. Texture Stud.* **2004**, *35*, 621–633. [CrossRef]
36. Takahashi, A.; Fujii, K. Examination of the mechanical properties of puffed cereals. *J. Cookery Sci. Jpn.* **2021**, *54*, 14–23.
37. Masavang, S.; Roudaut, G.; Champion, D. Identification of complex glass transition phenomena by DSC in expanded ce-real-based food extrudates: Impact of plasticization by water and sucrose. *J. Food Eng.* **2019**, *245*, 43–52. [CrossRef]

foods

MDPI

Article

Prediction of Pasta Colour Considering Traits Involved in Colour Expression of Durum Wheat Semolina

Antonio Troccoli [1,†], Donatella Bianca Maria Ficco [1,†], Cristiano Platani [2], Maria Grazia D'Egidio [3] and Grazia Maria Borrelli [1,*]

1 Consiglio per la Ricerca in Agricoltura e l'Analisi dell'Economia Agraria, Centro di Ricerca Cerealicoltura e Colture Industriali, S.S. 673, m 25200, 71122 Foggia, Italy; antonio.troccoli@crea.gov.it (A.T.); donatellabm.ficco@crea.gov.it (D.B.M.F.)

2 Consiglio per la Ricerca in Agricoltura e l'Analisi dell'Economia Agraria, Centro di Ricerca Orticoltura e Florovivaismo, Località Stella, Via Salaria, 1, 63030 Monsampolo del Tronto, Italy; cristiano.platani@crea.gov.it

3 Consiglio per la Ricerca in Agricoltura e l'Analisi dell'Economia Agraria, Centro di Ricerca Ingegneria e Trasformazioni Agroalimentari, Via Manziana, 30, 00189 Roma, Italy; mariagrazia.degidio@gmail.com

* Correspondence: graziamaria.borrelli@crea.gov.it

† These authors contributed equally to this work.

Abstract: Colour plays an important role among the quality traits of durum wheat, attracting consumer attention for the pasta market. The traits involved in colour expression are affected by genotype, environment, and processing. In the present study, based on eighteen durum wheat genotypes grown in eight environments, the effects of different traits related to colour expression were evaluated. Carotenoid pigments, such as lutein and β-carotene content; yellow and brown indices; and lipoxygenase, peroxidase, and polyphenoloxidase activities were analysed in semolina. The effects of processing were evaluated by measuring both the content of carotenoid pigments and colorimetric indices in pasta. The genotype, the environment, and their interaction were significant for all traits, although with a strong prevalence of genotypic effects, except for the brown index. After processing, a decrease in carotenoid content and the yellow index (86.7% and 16.0%, respectively) was observed, while the brown index increased (8.2%). A multiple regression analysis was performed on semolina traits, and the yellow index emerged as the main predictor for pasta colour, strengthening this trait as a fast and reliable criterion of selection. A High-Performance Index tool was also used to identify the genotype and environment that better combine all traits, positively influencing colour expression. All this information can be used in durum wheat breeding programmes for the prediction of pasta colour.

Keywords: durum wheat genotypes; environment; oxidative enzymes; carotenoid pigments; colorimetric indices; pasta; high-performance index

1. Introduction

Durum wheat (*Triticum turgidum* L. ssp. *durum*) represents an important component of the Mediterranean diet. It is the tenth most cultivated cereal in the world, and among wheats, it is the second most produced crop after bread wheat [1]. Durum wheat covers about 13.7 million ha, representing approximately less than 7% of the total global wheat cultivation area, with a global production of 34.3 million tons, considering the average values of 2018–2022 [2]. It is the main crop in many areas of the Mediterranean basin, whose countries account for almost 50% of the cultivated area and global production of durum wheat [3]. Mediterranean countries are also the main consumers of products derived from durum wheat, such as pasta, bread, bulgur, and couscous. Currently, Canada is the leading producer of durum wheat, followed by Italy, Algeria, and Turkey [1,4], and it is the largest durum wheat exporter in the world, with its main destinations being Algeria, Italy, and Morocco [1,5].

Ensuring durum wheat quality is essential to maintaining competitiveness in the market. Consumer choices play an important role in directing breeding towards the improvement of traits linked to the quality of durum wheat end products. Among these end products, pasta represents the most common and popular staple cereal food thanks to its sensory and nutritional properties [6,7], and it is the most widespread cereal food in Italian cuisine. Its quality is the complex result of the influence of different factors which intervene in various stages of the durum wheat supply chain [8]. Quality is mainly influenced by genotype but also by the environment (weather and soil), crop management, and the primary (milling) and secondary (pasta making) processing steps [9,10].

Oxidative properties that are associated with colour are particularly important for durum wheat commercial value, since consumers prefer semolina and pasta with a bright yellow colour. Colour is influenced by the interaction between a yellow component and a brown one, both resulting from the balance between the oxidative and antioxidative components of grain.

The yellow component is mainly affected by (i) the endogenous content of carotenoid pigments in kernels, antioxidant compounds that act as scavengers of free radicals, providing significant health benefits; (ii) their residual content after milling; and (iii) their degradation that occurs during pasta processing due to oxidative enzymes, especially lipoxygenase (LOX) and, to a minor extent, peroxidase (POD) [11–14]. Among the carotenoid pigments, lutein and, to a minor extent, β-carotene, which is an important precursor of vitamin A in mammalians, are mainly present in durum wheat grain [15].

The brown component is the result of the oxidation of compounds such as polyphenols, which occurs in the endosperm, during kernel maturation, and in semolina and dough during processing, due to enzymes such as peroxidase (POD) and polyphenoloxidase (PPO), and it is also a result of the ash content, an indicator of residual bran fraction [16–20].

Genotype, environment, and their interaction affect all the cited traits in grain and semolina to a different extent [11,21–25], together with physical and chemical variations that occur during milling and pasta processing [11,12].

Two approaches were used in the present study to evaluate the colour traits of durum wheat products. Firstly, the components of the variance for genotype, environment, and their interaction were estimated in terms of the colour-related traits of semolina and pasta in a set of eighteen durum wheat genotypes grown throughout the Italian territory in eight different climate areas. Secondly, the semolina trait that could be considered as the best predictor of the final pasta colour was identified by multiple regression analysis.

Finally, to identify the genotype and environment that best combine the colour traits studied, a High-Performance Index (HPI) tool was used. This is a mathematical tool employed to simultaneously evaluate, in a simple and objective way, multiple traits that are different in their units of measurement and also have opposite effects [26]. The use of this tool allows us to simplify the interpretation of data, and its application can be easily extended to other studies by adapting the procedure to the specific objective.

2. Materials and Methods

2.1. Plant Materials

Eighteen varieties of durum wheat, randomly selected from those mainly released in the past 90 years, were grown in the 1999–2000 growing season throughout the Italian territory in eight environments with different agro-climatic characteristics (Table S1), included in the Italian National Cereal Test Network. The experimental design was a lattice square with three completely randomized replicates. The plots were 10 m^2 with eight rows, spaced 17 cm apart. The cultivation of durum wheat varieties was carried out in different locations within each agro-climatic environment using standardized cultural practices. For each variety, equal amounts of grain samples collected from the different locations of the same environment were mixed to obtain a composite sample [27]. The appropriateness of this standardized sampling procedure, usually applied in the Italian National Cereal Test Network, was assured by the small genotype × location interaction for the main quality traits analysed [28]. The detailed thermo-pluviometric trend for all climate areas was described in Desiderio et al. [29].

A flow chart showing the entire experimental process is displayed in Figure 1.

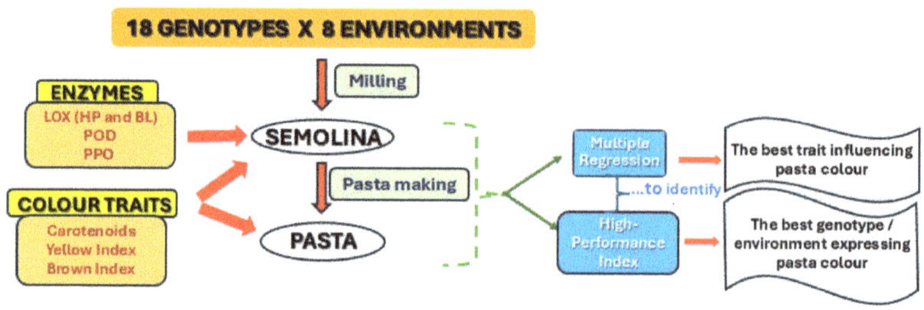

Figure 1. Experimental process.

2.2. Semolina Production

Cleaned seeds were conditioned overnight to 16.5% moisture and processed by an MLU 202 experimental mill (Bühler Brothers, Uzwill, Switzerland). The extraction rate was on average 65%. Semolina was stored at 4 °C up to analysis.

2.3. Pasta Processing

Semolina was processed into spaghetti (1.7 mm diameter) using a 2 kg capacity laboratory plant (NAMAD, Rome, Italy). A low-temperature drying procedure (50 °C for 20 hr) was applied in the pilot plant (AFREM, Lyon, France). Dried pasta, finely ground by a Tecator Cyclotec 1093 laboratory mill (International PBI, Milano, Italy) and sieved (<500 µm), was stored at 4 °C up to analysis.

2.4. Enzyme Assays in Semolina

Semolina's hydroperoxidation and bleaching activities of LOX, POD, and PPO activities were determined in triplicate and expressed as µmoles of changed substrate $min^{-1}g^{-1}$. Crude extracts for LOX and POD analysis were obtained as described by Borrelli et al. [12], whereas those for PPO analysis were obtained according to González et al. [30]. The protein content of extracts was evaluated by the Lowry et al. [31] method, using crystalline bovine serum albumin as a standard.

2.4.1. Hydroperoxidation and Bleaching Activities of LOX

Both LOX activities were evaluated using a λ18 UV/vis Spectrometer (Perkin Elmer, Norwalk, CT, USA), equipped with a water-jacketed cell holder. The optimal pH for both activities was chosen according to Barone et al. [32].

LOX hydroperoxidation activity (Hp) was determined at 25 °C and at pH 6.6 by measuring conjugate diene absorption at 234 nm, according to Borrelli et al. [12]. One unit of enzymatic activity corresponded to the production of 1 µmol of conjugate hydroperoxydienoic min^{-1}, using a molar extinction coefficient of 28 mM cm^{-1}.

The β-carotene bleaching activity (BL) of LOX was evaluated at 25 °C and at pH 5.2 by measuring the decrease in absorbance at 460 nm, as described by Borrelli et al. [12]. One unit of enzyme activity corresponded to the destruction of 1 μM of β-carotene min^{-1}, using a molar extinction coefficient of 123.5 mM cm^{-1}.

2.4.2. Peroxidase Activity

Peroxidase (POD) activity was determined at 30 °C by measuring the slope from the linear increase in absorbance at 470 nm at pH 4.2, due to the oxidation of guaiacol to tatraidroguaiacol, in the presence of hydroxide peroxide, using a λ20 UV/vis Spectrometer (Perkin Elmer), as described by Borrelli et al. [12]. One unit of enzyme activity was defined as the change in one absorbance unit per min with an ε value of 26.6 $mM^{-1}.cm^{-1}$.

2.4.3. Polyphenoloxidase Activity

Polyphenoloxidase (PPO) activity was evaluated at 25 °C by measuring the increase in absorbance at 480 nm, as a result of DOPA conversion into dopachrome, according to Okot-Kotber et al. [33], with little modifications. The reaction mixture (3 mL) contained 10 mM L-DOPA in 0.1 M phosphate buffer, pH 6.5, and a suitable amount of extract to give a linear initial velocity of the reaction. One unit of PPO activity was defined as the amount of enzyme resulting in a change in absorbance of 0.001 min^{-1}. PPO activity was discriminated from that of POD using tropolone as a selective inhibitor for PPO.

2.5. Laboratory Analyses of Semolina and Pasta

2.5.1. Determination of β-Carotene and Lutein Content

β-carotene and lutein contents were measured by HPLC with an isocratic solvent programme, using an Agilent HPLC system 1100 (Agilent Technologies, Waldbronn, Germany), and expressed as μg/g, dry matter (DM), according to Borrelli et al. [12]. Analyses were carried out in duplicate, and quantification was performed with reference to a calibration curve obtained using relative standards.

2.5.2. Yellow and Brown Indices

A chromameter (CR200, Minolta, Osaka, Japan) was used to determine semolina and pasta yellow (b*) (YI) and brown (100-L*) (BI) indices, according to CIELAB colour space system [34] which measures lightness (L*: 100 white, 0 black), redness (a*: +60 red, −60 green), and yellowness (b*: +60 yellow, −60 blue) values. Each index was the average of three measurements.

2.6. Procedure for Calculating High-Performance Index (HPI)

To identify superior genotypes which better express traits influencing the colour of durum wheat products, the HPI was used. This descriptive tool is a dimensionless index derived from a mathematical algorithm developed in Microsoft Excel by Troccoli et al. [26]. The algorithm requests some constraints on analytical variables to be specified: in particular, the optimal condition consists of low LOX, POD, and PPO activities in semolina and a high content of pigments and the YI, as well as a low BI, in semolina and pasta. Starting from all the analytical variables detected in semolina and pasta, the constraints imposed on each of them are used by the mathematical algorithm to firstly calculate the standardized deviations from the overall mean of each variable and then to assign a positive or negative score that increases (up to a maximum of 2) as the size of the deviation increases, obtaining the final value of the HPI of each genotype by adding all the scores, according to the detailed procedure described in Troccoli et al. [26]. Considering that each variable comes from eight environments, the absolute maximum HPI score that each genotype can achieve is "2 × n variables × m environments" (2 × 12 × 8= 192). The same approach was applied considering the environment as a categorical variable. The HPI is graphically represented with a spider chart, sorting the data in descending order. The chart is composed of axes that extend from a central point, each of which represents the HPI value for every genotype.

2.7. Statistical Analysis

ANOVA-Type II and variance component analysis were performed considering both environment and genotype as random factors.

The mean values of genotype x environment interaction traits were used to obtain a Pearson correlation matrix and the p-values of the correlation coefficient.

A multiple regression analysis was performed on all semolina traits used as the predictors (independent variables) of the yellow index of pasta (dependent variable). Mallow's Cp statistics were used as a criterion to find the best regression model involving a subset of predictor variables. The significance of the regression model was tested with an ANOVA. A collinearity test and other statistical parameter calculations were performed on the predictor variables included in the regression model. The Durbin–Watson statistic test was conducted on the residuals to exclude the autocorrelation phenomenon.

Based on the simple correlation matrix between all traits analysed in the eighteen genotypes, a Principal Component Analysis (PCA) was performed using the mean genotypic values.

Statistical analyses were performed using the STATISTICA program (StatSoft Italia srl, vers. 8.0, 2007). PCA was performed using PAST software 4.17c (http://palaeo-electronica.org/2001_1/past/issue1_01.htm, accessed on 08 January 2025).

Chemicals: Linoleic acid, β-carotene, carotenoid standards, Tween 20, DOPA, tropolone, and LOX soybean (type IV) were purchased from Sigma Aldrich Chemical Co. (St. Louis, MO, USA). Reagents for HPLC and gas chromatography were purchased from J.T. Baker (Deventer, Holland). Tween 80 was purchased from Merck (Darmstadt, Germany).

3. Results and Discussion

3.1. Contribution of Genotype, Environment, and Their Interaction to Colour-Related Traits in Semolina and Pasta

Among quality traits, the colour of the end products, semolina and pasta, plays an important role both for aesthetic aspects, widely appreciated by consumers, and for nutritional implications, mainly linked to the presence of carotenoid pigments.

The importance of genotype, environment, and their interaction in terms of the traits directly and indirectly related to semolina colour was evaluated by ANOVA-Type II (Table 1).

The genotype, the environment, and their interaction were highly significant for all traits (Table 1). To estimate the percentage of total variance due to environment, genotype, and their interaction on the analysed traits, a variance component analysis was performed. Excluding BL, BI-S, Lut-P, and BI-P, most traits were strongly influenced by genotype, ranging from 50% (β-Car-P) to 90% (YI-S) of the total variance (Figure 2; Table S2). According to Schulthess et al. [35], the environment played an important role (above 65%) only for the BI-S and BI-P traits, while a moderate effect ranging from 4% (HP) to 21% (β-Car-S) on the remaining traits was found (Figure 1; Table S2). This effect could be explained with the correlation between the BI and ash content, which is significantly influenced by the environment [8,36,37]. The minor effects of the environment on traits related to the yellow colour component are in agreement with Sisson et al. [38].

Table 1. Mean values of traits related to semolina and pasta colour of eighteen genotypes grown in eight Italian environments.

	HP (EU g⁻¹)	BL (EU g⁻¹)	POD (EU g⁻¹)	PPO (EU g⁻¹)	Lut-S (µg g⁻¹, DM)	ß-Car-S (µg g⁻¹, DM)	YI-S	BI-S	Lut-P (µg g⁻¹, DM)	ß-Car-P (µg g⁻¹, DM)	YI-P	BI-P
Genotype (G; *n* = 24)												
Arcobaleno	1.284 ± 0.427	0.006 ± 0.003	154.492 ± 45.203	273.034 ± 113.304	1.718 ± 0.675	0.485 ± 0.164	23.438 ± 0.748	14.113 ± 1.201	0.165 ± 0.054	0.047 ± 0.018	19.425 ± 1.054	15.438 ± 1.337
Claudio	1.442 ± 0.471	0.010 ± 0.007	118.625 ± 36.392	207.330 ± 51.962	2.148 ± 0.607	0.560 ± 0.16	22.759 ± 0.993	13.500 ± 0.994	0.181 ± 0.06	0.054 ± 0.026	18.450 ± 0.978	14.638 ± 1.116
Colosseo	0.937 ± 0.419	0.005 ± 0.002	138.693 ± 39.417	140.928 ± 46.412	1.031 ± 0.247	0.210 ± 0.211	19.783 ± 0.928	13.213 ± 0.888	0.259 ± 0.085	0.039 ± 0.017	17.775 ± 0.517	14.125 ± 0.747
Creso	0.671 ± 0.428	0.005 ± 0.005	131.700 ± 40.375	256.388 ± 63.158	1.117 ± 0.229	0.177 ± 0.037	18.688 ± 1.174	13.875 ± 1.184	0.124 ± 0.043	0.047 ± 0.018	16.525 ± 1.323	14.838 ± 1.163
Duilio	4.640 ± 1.610	0.004 ± 0.003	15.030 ± 10.055	60.240 ± 31.166	1.092 ± 0.309	0.339 ± 0.094	21.753 ± 0.924	13.413 ± 1.313	0.148 ± 0.024	0.044 ± 0.015	18.450 ± 1.006	14.613 ± 1.237
Duprì	0.921 ± 0.277	0.007 ± 0.004	8.291 ± 3.496	64.113 ± 22.515	2.946 ± 0.716	0.745 ± 0.227	26.324 ± 1.034	13.775 ± 0.905	0.204 ± 0.041	0.068 ± 0.038	21.400 ± 1.143	14.788 ± 0.926
Flaminio	0.970 ± 0.252	0.001 ± 0.001	17.086 ± 5.515	84.943 ± 38.013	2.336 ± 0.495	0.652 ± 0.184	25.441 ± 0.672	13.450 ± 1.250	0.219 ± 0.060	0.046 ± 0.016	21.200 ± 1.089	14.471 ± 1.212
Gianni	0.870 ± 0.151	0.004 ± 0.002	9.318 ± 4.214	61.018 ± 28.986	1.084 ± 0.202	0.170 ± 0.043	18.709 ± 0.893	12.925 ± 0.994	0.135 ± 0.025	0.024 ± 0.014	16.338 ± 1.081	14.000 ± 1.068
Iride	1.052 ± 0.267	0.007 ± 0.003	106.818 ± 30.023	137.060 ± 54.426	1.844 ± 0.351	0.361 ± 0.074	22.889 ± 1.018	13.838 ± 1.034	0.181 ± 0.036	0.027 ± 0.017	19.063 ± 1.068	14.825 ± 1.193
Meridiano	1.312 ± 0.281	0.001 ± 0.003	110.448 ± 37.772	227.706 ± 69.657	2.621 ± 0.658	0.776 ± 0.215	25.670 ± 1.290	14.225 ± 1.004	0.416 ± 0.250	0.157 ± 0.092	21.088 ± 1.339	15.288 ± 1.244
Nefer	0.967 ± 0.331	0.020 ± 0.008	142.755 ± 33.476	268.312 ± 88.313	1.376 ± 0.481	0.344 ± 0.095	23.761 ± 0.739	14.188 ± 0.793	0.236 ± 0.086	0.045 ± 0.023	20.263 ± 0.826	15.150 ± 0.956
Parsifal	0.509 ± 0.217	0.004 ± 0.002	26.010 ± 23.127	96.948 ± 46.591	1.695 ± 0.297	0.379 ± 0.025	20.867 ± 0.772	12.788 ± 0.967	0.192 ± 0.078	0.036 ± 0.016	17.850 ± 0.621	14.050 ± 1.087
Preco	5.056 ± 0.861	0.002 ± 0.002	34.454 ± 10.969	78.097 ± 35.682	2.952 ± 1.002	0.923 ± 0.339	31.482 ± 0.708	14.300 ± 1.181	0.464 ± 0.247	0.145 ± 0.077	25.925 ± 2.177	15.425 ± 1.257
Saadi	0.411 ± 0.231	0.005 ± 0.003	137.448 ± 48.702	171.027 ± 41.246	1.811 ± 0.411	0.473 ± 0.120	22.777 ± 0.794	14.000 ± 0.945	0.160 ± 0.071	0.055 ± 0.031	19.213 ± 1.084	15.275 ± 1.056
San Carlo	1.655 ± 0.665	0.006 ± 0.006	107.216 ± 42.485	285.593 ± 126.047	1.809 ± 0.452	0.705 ± 0.243	24.445 ± 0.783	13.900 ± 0.892	0.204 ± 0.048	0.049 ± 0.02	20.350 ± 1.471	15.238 ± 0.994
Simeto	4.062 ± 1.547	0.007 ± 0.004	167.517 ± 30.868	494.108 ± 146.923	1.711 ± 0.359	0.319 ± 0.071	23.013 ± 1.054	14.575 ± 1.057	0.222 ± 0.075	0.038 ± 0.02	19.388 ± 1.411	15.438 ± 1.056
Torrebianca	4.898 ± 1.315	0.005 ± 0.003	97.698 ± 34.32	246.375 ± 57.289	1.567 ± 0.396	0.339 ± 0.095	22.989 ± 0.706	14.200 ± 0.919	0.221 ± 0.070	0.042 ± 0.014	18.625 ± 0.506	14.863 ± 1.308
Verdi	1.524 ± 1.230	0.005 ± 0.002	51.122 ± 17.611	192.983 ± 46.37	1.356 ± 0.388	0.408 ± 0.132	25.033 ± 1.343	13.713 ± 1.088	0.273 ± 0.076	0.110 ± 0.030	20.563 ± 1.438	15.788 ± 1.183
$F_{(17,119)}$	39.014	10.066	58.806	25.122	29.661	34.272	204.441	8.630	8.322	12.666	39.059	7.625
p-value	0.000	0.000	0.000	0.000	0.000	0.000	0.000	0.000	0.000	0.000	0.000	0.000

Table 1. *Cont.*

	HP (EU g^{-1})	BL (EU g^{-1})	POD (EU g^{-1})	PPO (EU g^{-1})	Lut-S (μg g^{-1}, DM)	ß-Car-S (μg g^{-1}, DM)	YI-S	BI-S	Lut-P (μg g^{-1}, DM)	ß-Car-P (μg g^{-1}, DM)	YI-P	BI-P
						Environment (E; n = 54)						
ENV1	2.207 ± 1.814	0.007 ± 0.004	88.085 ± 53.558	150.317 ± 91.262	1.939 ± 0.627	0.524 ± 0.239	23.687 ± 2.818	14.894 ± 0.767	0.257 ± 0.108	0.091 ± 0.065	19.711 ± 1.39	16.061 ± 0.854
ENV2	2.536 ± 2.342	0.004 ± 0.003	63.798 ± 44.962	143.255 ± 95.172	1.584 ± 0.537	0.501 ± 0.235	22.418 ± 3.144	13.928 ± 0.541	0.186 ± 0.079	0.071 ± 0.039	18.333 ± 2.457	15.067 ± 0.635
ENV3	2.021 ± 2.121	0.009 ± 0.009	76.794 ± 51.64	230.835 ± 166.08	1.578 ± 0.632	0.504 ± 0.265	22.737 ± 3.093	14.078 ± 0.705	0.159 ± 0.068	0.059 ± 0.030	18.944 ± 2.372	14.989 ± 0.810
ENV4	1.780 ± 1.622	0.003 ± 0.004	74.848 ± 50.823	175.371 ± 127.671	2.258 ± 0.784	0.533 ± 0.286	23.518 ± 2.993	12.744 ± 0.630	0.265 ± 0.079	0.060 ± 0.031	20.228 ± 2.512	13.806 ± 0.569
ENV5	1.722 ± 1.604	0.003 ± 0.003	60.804 ± 41.421	132.449 ± 68.127	2.223 ± 0.759	0.497 ± 0.252	23.304 ± 3.334	13.072 ± 0.604	0.344 ± 0.259	0.082 ± 0.088	20.017 ± 2.651	14.106 ± 0.644
ENV6	1.589 ± 1.312	0.006 ± 0.004	90.043 ± 62.742	208.868 ± 104.753	1.902 ± 0.700	0.520 ± 0.282	22.942 ± 3.145	13.217 ± 0.637	0.218 ± 0.081	0.052 ± 0.035	19.389 ± 2.692	14.361 ± 0.730
ENV7	1.329 ± 1.171	0.007 ± 0.006	106.996 ± 66.394	217.845 ± 142.554	1.892 ± 0.692	0.491 ± 0.207	24.833 ± 2.831	15.406 ± 0.665	0.169 ± 0.083	0.034 ± 0.024	20.511 ± 2.305	16.761 ± 0.787
ENV8	1.564 ± 1.255	0.005 ± 0.005	138.508 ± 84.528	228.261 ± 158.994	0.941 ± 0.517	0.148 ± 0.072	23.146 ± 2.790	12.878 ± 0.701	0.180 ± 0.059	0.029 ± 0.018	19.261 ± 2.055	14.070 ± 0.739
$F_{(7,119)}$	5.273	6.661	27.058	7.434	32.645	26.788	27.362	79.108	9.413	9.267	9.500	69.328
p-value	0.000	0.000	0.000	0.000	0.000	0.000	0.000	0.000	0.000	0.000	0.000	0.000
						G × E Interaction (n = 3)						
$F_{(119,288)}$	249.274	143.747	185.877	144.206	142.449	13.295	8.766	17.325	1593.217	1949.488	21.520	20.032
p-value	0.000	0.000	0.000	0.000	0.000	0.000	0.000	0.000	0.000	0.000	0.000	0.000
Overall mean	1.843 ± 1.726	0.006 ± 0.006	87.485 ± 62.699	185.9 ± 128.421	1.79 ± 0.768	0.465 ± 0.266	23.323 ± 3.077	13.777 ± 1.126	0.222 ± 0.132	0.06 ± 0.05	19.549 ± 2.414	14.903 ± 1.216

HP = hydroperoxidation activity of LOX; BL = bleaching activity of LOX; POD = peroxidase activity; PPO = polyphenoloxidase activity; Lut = lutein; β-Car = β-carotene; YI = yellow index; BI = brown index; S = semolina; P = pasta; ENV1 = Po Valley; ENV2 = North-Central Adriatic Coast; ENV3 = North-Central Tyrrhenian Coast; ENV4 = North-Central Mountainous Apennines; ENV5 = South Mountainous Apennines; ENV6 = Adriatic–Ionic; ENV7 = Sicily; ENV8 = Sardinia. Values are reported as mean ± Standard deviations (n = 3). For G, E, and G × E factors, number of data (n) used for mean value is shown in brackets: for G: n = 24 corresponds to 3 replicates (rep) × 8 environments; for E: n = 54 corresponds to 3 rep × 18 genotypes; for G × E interaction, n = 3 corresponds to rep considered for analysis.

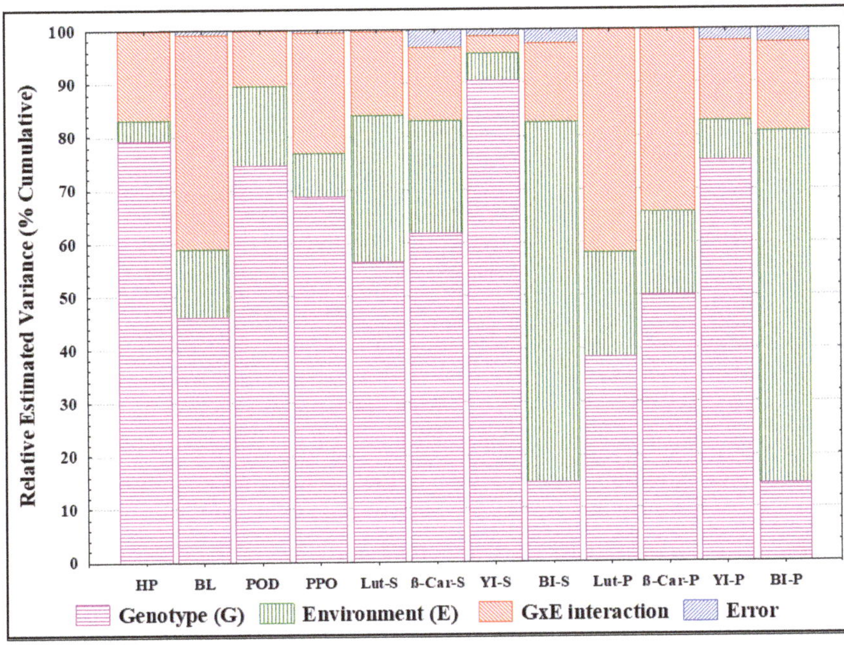

Figure 2. Relative variance component, expressed in percentage of total variance, for all traits evaluated in semolina and pasta of eighteen genotypes grown in eight different Italian environments. Columns indicate cumulative sums of components. HP = hydroperoxidation activity of LOX; BL = bleaching activity of LOX; POD = peroxidase activity; PPO = polyphenoloxidase activity; Lut = lutein; β-Car = β-carotene; YI = yellow index; BI = brown index; S = semolina; P = pasta.

The effects of the G × E interaction were quite evident on Lut-P (42%), the BL activity of LOX (40%), and β-Car-P (34%) traits but less impactful on the remaining traits. The error component of total variance on all traits can be considered completely marginal.

Carotenoid pigment content is a quantitative character regulated by various genes with additive effects and a minor effect of the environment [39]. The results of our study confirmed the main effects of genotype on yellow pigment and enzyme activities in semolina. These results agree with previous studies [14,21–25,40,41], although some of them evidenced a marginal effect of environmental conditions, including year, location, agronomic practices, and their interaction [40–43].

POD and PPO showed great variability among genotypes with respect to HP and BL (Table 1). The lowest HP values were found in Saadi, Parsifal, and Creso, while the highest ones were in Preco, Torrebianca, Duilio, and Simeto. Semolina in Simeto, in addition to having the highest POD and PPO values, had a very negative outcome in terms of its final colour (Table 1). In general, in semolina, the lutein and β-carotene content and the YI were high in Preco and Duprì and low in Gianni, whereas the BI was the highest in Simeto and the lowest in Parsifal and Gianni. This trend was also generally found in pasta. The increase in carotenoid content in newer varieties with respect to older ones (1.194 vs. 0.793 μg g^{-1}, db, on average) confirmed the positive results of breeding programmes aimed at improving the yellow colour of durum wheat due to the high value of the heritability of carotenoid pigment content [21,24,25,42].

3.2. Effects of Processing on Pasta Colour

3.2.1. Variation in Colour-Related Traits in Pasta

The effects of processing on parameters directly related to pasta colour, such as both carotenoid pigments and yellow and brown indices, are shown in Figure 3.

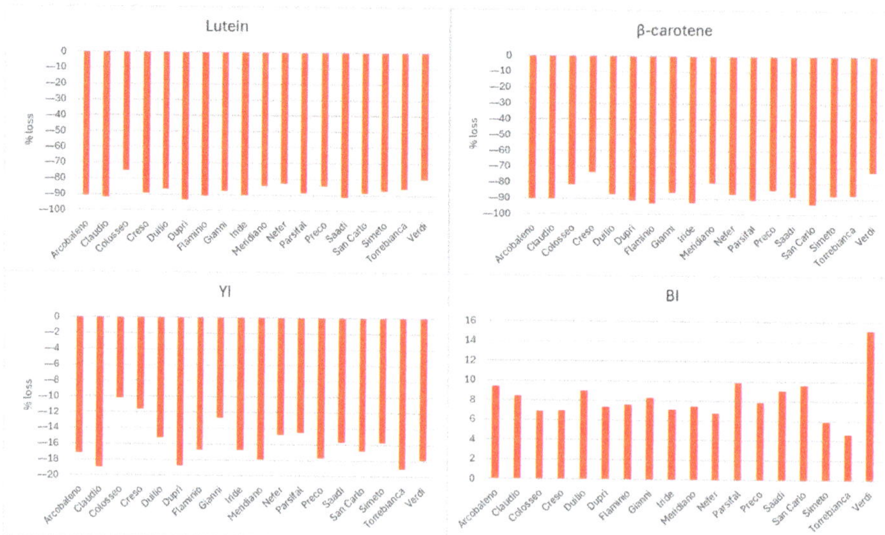

Figure 3. Percentage variation in colour traits in eighteen genotypes across eight environments.

Considering the mean genotypic values, decreases of 86.7% of carotenoid pigment content, on average, and 16.0% of the YI were found after processing, while the BI showed an increase of 8.2%. These variations are of the same order of magnitude as those found by Borrelli et al. [12]. The differences in the extent of pigment content and YI variation after pasta making could be due to the different analysis methods used (chemical vs. physical/colorimetric). It should be noted, however, that since the YI is an overall measure of yellowness, its value includes the presence of other yellow carotenoid pigments, even if they are present in small amounts [25].

3.2.2. Multiple Regression Analysis for Pasta Colour

To define the relationship between all traits involved in this study, a simple Pearson correlation analysis was performed. The results are shown in Table 2.

Correlation analysis indicated that YI-S had the highest and positive significant association with YI-P (r = 0.93, p-value = 0.000), followed by, in decreasing order, β-Car-S (r = 0.66, p-value = 0.000), Lut-S (r = 0.64, p-value = 0.000), HP (r = 0.26, p-value = 0.002), and BI-S (r = 0.21, p-value = 0.012).

Regarding enzymatic activities, HP did not show any correlation with BL and the POD and PPO enzymes, which instead were correlated with each other.

LOX's HP and BL activities significantly influenced carotenoid pigment content in pasta in the opposite way. HP showed a positive correlation, while BL showed a negative one with carotenoid pigment, confirming their involvement in the degradation of carotenoids during processing (Table 2). Despite HP and BL being the result of the activity of the same enzyme, different roles for them have been described [44]. LOX catalyses, in the first instance, the hydroperoxidation of polyunsaturated fatty acids (HP), producing free and peroxy radicals. This reaction mainly occurs during mixing, due to the concurrence of a greater availability of oxygen provided with water and of mechanical action that promotes substrate–enzyme–O_2 contact, other than the modification of surface hydrophobicity, which might result in the release and activation of enzymes and lipids. In the subsequent phases of pasta processing, carotenoid pigment degradation (BL) prevails, and it is due to a coupled oxidation mechanism, mediated by reactive oxygen species originating during fatty acid oxidation [12]. The same was highlighted during bread making, when the greatest decreases in carotenoids in flour were observed during the kneading stage [45]. These different reactions are not closely related. In fact, other than the fact that they seem to act in different phases of the pasta making process [12], LOX isoenzymes could intervene to different extents in pigment degradation [46]. This could also explain the lack of a significant correlation between the two activities observed in semolina (Table 2) and the negative significant correlation observed only between BL activity and both pigments in semolina.

Table 2. Pearson's correlation matrix and *p*-values for all traits involved in colour expression.

	HP	BL	POD	PPO	Lut-S	ß-Car-S	YI-S	BI-S	Lut-P	ß-Car-P	BI-P
BL	−0.10 *p* = 0.239										
POD	−0.12 *p* = 0.165	**0.34 *p* = 0.000**									
PPO	0.09 *p* = 0.265	**0.37 *p* = 0.000**	**0.71 *p* = 0.000**								
Lut-S	0.05 *p* = 0.585	**−0.23 *p* = 0.006**	**−0.27 *p* = 0.001**	**−0.20 *p* = 0.019**							
ß-Car-S	0.13 *p* = 0.118	**−0.18 *p* = 0.033**	**−0.30 *p* = 0.000**	**−0.17 *p* = 0.036**	**0.86 *p* = 0.00**						
YI-S	**0.30 *p* = 0.000**	−0.07 *p* = 0.389	−0.14 *p* = 0.085	−0.06 *p* = 0.513	**0.66 *p* = 0.00**	**0.71 *p* = 0.00**					
BI-S	**0.19 *p* = 0.022**	**0.32 *p* = 0.000**	**0.24 *p* = 0.003**	**0.28 *p* = 0.001**	0.09 *p* = 0.261	**0.23 *p* = 0.007**	**0.33 *p* = 0.000**				
Lut-P	**0.19 *p* = 0.019**	**−0.23 *p* = 0.006**	−0.12 *p* = 0.147	−0.10 *p* = 0.246	**0.50 *p* = 0.000**	**0.43 *p* = 0.000**	**0.50 *p* = 0.000**	−0.03 *p* = 0.681			
ß-Car-P	**0.19 *p* = 0.026**	**−0.21 *p* = 0.013**	**−0.19 *p* = 0.022**	−0.14 *p* = 0.091	**0.50 *p* = 0.000**	**0.57 *p* = 0.000**	**0.50 *p* = 0.000**	0.16 *p* = 0.052	**0.80 *p* = 0.00**		
BI-P	0.10 *p* = 0.229	**0.26 *p* = 0.002**	**0.21 *p* = 0.013**	**0.21 *p* = 0.010**	0.06 *p* = 0.440	**0.22 *p* = 0.009**	**0.35 *p* = 0.000**	**0.93 *p* = 0.00**	−0.04 *p* = 0.635	**0.19 *p* = 0.026**	
YI-P	**0.26 *p* = 0.002**	−0.08 *p* = 0.354	−0.14 *p* = 0.095	−0.09 *p* = 0.286	**0.64 *p* = 0.000**	**0.66 *p* = 0.00**	**0.93 *p* = 0.00**	**0.21 *p* = 0.012**	**0.53 *p* = 0.000**	**0.45 *p* = 0.000**	**0.23 *p* = 0.005**

HP = hydroperoxidation activity of LOX; BL = bleaching activity of LOX; POD = peroxidase activity; PPO = polyphenoloxidase activity; Lut = lutein; β-Car = β-carotene; YI = yellow index; BI = brown index; S = semolina; P = pasta. The means of the G × E interaction of the traits (n = 144) were used for analysis. The values of significant correlation coefficients are in bold.

A highly significant correlation was found between PPO and POD activities. Both enzymes were shown to be involved in brownness although with different reaction mechanisms [17–20,47]. In fact, these enzymes have some common phenolic substrates and other specific ones for each of them, whose oxidation leads to the browning of products [47]. This explains the significant correlation between their activity and BI of semolina and pasta. The different levels of significance observed for correlations with the BI in semolina and pasta could also be due to the nonenzymatic browning caused by the Maillard reaction that takes place during the drying phase of the process (Table 2). This reaction is highly dependent on the amount and type of substrate available (sugars and proteins) and moisture content, as well as the time–temperature combinations of the drying treatment [48,49], although we can assume that the degree of involvement in our study is low because of the low-temperature drying cycle applied.

The high correlation between the BI in semolina and the BI in pasta indicates that pasta brownness would also be dependent on the inherent brownness of semolina as well as pasta processing [17].

Both in semolina and pasta, lutein and β-carotene correlated with each other and with the YI, confirming that they are mainly responsible for the bright yellow colour in semolina and contribute to the commercial value of pasta.

The strength of the relationship between the YI and pigment content or their biosynthetic enzymes was also evidenced by the identification of quantitative trait loci (QTLs) that often co-mapped in the same chromosome region, increasing the effectiveness of marker-assisted selection (MAS) programmes for these traits [39,50,51]. This was also confirmed in other cereal species [52–54].

Even if it is reasonable that the YI of semolina should predict the YI of pasta, due to the high correlation ($r = 0.93$) between these traits, previous studies [11,55] demonstrated that pigment losses can occur during pasta making due to LOX activities. Moreover, POD and PPO, influencing BI-P, could negatively affect yellowness [19]. For this reason, the mean values of the G × E interaction were used in a multiple regression model, considering YI-P as the dependent variable and all the parameters analysed in semolina as predictors or regressors or independent variables to select variables with greater weight in predicting YI-P.

For the YI-P regression model, an ANOVA showed a highly significant F value (Table 3) with an estimated model error of 0.878. Regarding independent variables, the selection method based on Mallow's Cp statistic extracted only two variables, which were always present in all proposed subset models, with the first subset having the lowest Cp value (0.790) corresponding to the BI-S (β-coefficient = −0.107) and YI-S (β-coefficient = 0.962) variables (Table 4). The ANOVA showed, for these two variables, as well as for the intercept, highly significant univariate tests in addition to high values of the coefficient of multiple regression ($R = 0.932$), of determination ($R^2 = 0.868$), and of correct determination ($R^2_{adj} = 0.866$) (Table 3).

Table 3. ANOVA for regression model (summary).

Dependent Variable	Source of Variation	df	Sum of Square	Mean Square	F	p-Value	Statistics	Value
YI-P	**Regression**	2	715.281	357.641	464.089	0.000	R	0.932
	Error	141	108.659	0.771			R^2	0.868
	Total	143	823.940				R^2_{adj}	0.866
							Error of Model	0.878
	A univariate significance test for the YI-P variable of the subset of the best predictors chosen according to the Mallow Cp procedure							
	Intercept	1	23.156	23.156	30.048	0.000		
	BI-S	1	8.481	8.481	11.005	0.001		
	YI-S	1	679.191	679.191	881.346	0.000		
	Error	141	108.659	0.771				

Table 4. Collinearity statistics for predictors included in regression model for YI-P dependent variable.

Predictors IN	Tolerance	VIF (Variance Inflation Factor)	R^2	B-Coefficient	β-Coefficient	Partial Correlation	Semi-Partial Correlation	t	p-Value
BI-S	0.891	1.122	0.109	−0.231	−0.107	−0.269	−0.101	−3.317	0.001
YI-S	0.891	1.122	0.109	0.752	0.962	0.928	0.908	29.687	0.000
Durbin–Watson statistic (d) test for residual autocorrelation in the regression model (n = 144)									
	d lower tabulated	**d value**	**d upper tabulated**		**Serial correlation**				
	1.468	1.723	1.767		0.137				

YI = yellow index; BI = brown index; S = semolina; P = pasta.

The collinearity statistics for the BI-S and YI-S predictors included in the multiple regression model for the YI-P dependent variable showed very favourable values of tolerance, VIF (Variance Inflation Factor), and residual R^2, demonstrating that the two predictors fit the model of regression very well for estimating YI-P (Table 4). This was also supported by the value of the Durbin–Watson statistic (d) for the estimation of residual autocorrelation (Table 4). The β-coefficients of the BI-S and YI-S variables included in the multiple regression model contribute 10.7% and 96.2%, respectively, to the YI-P variable, but their behaviour is different since BI-S contributes minimally and negatively to the estimate of YI-P, as compared to YI-S.

These results underline the usefulness of evaluating the YI of semolina, commercially identified as being yellow in colour, as a fast and reliable method for performing a screening for pigment content and, ultimately, for large-scale selection in durum wheat breeding programmes for pasta colour [15,24,25,38,56]. Due to the very high correlation of the YI (Minolta b* index) with the quantity of yellow pigment in semolina measured by chemical extraction and quantified by spectrophotometric and/or by HPLC methods, the use of light reflectance via a Minolta chromameter, requiring no chemicals, could represent a rapid alternative to time-consuming quantification. Previously, it had already been demonstrated that the colour measurement of a small amount of whole meal samples using a scanning NIR instrument correlated well (r = from 0.74 to 0.90) with semolina's b* values measured with a Minolta chromameter, offering a speedy prediction of semolina colour especially in early generation selection of breeding [57,58]. Moreover, since semolina is the raw material used for pasta production, from the point of view of the supply chain, knowing the level of pigments in semolina in the early stages of selection, when its quantity is not enough for pasta pilot plants, is very important for making a prediction of the final pasta colour.

Our results do not agree with previous studies. In fact, Borrelli et al. [11] found that LOX activities were prevalent in influencing the loss of pigments during pasta processing. Furthermore, the identification of the deletion of the *Lpx-B1.1* allele in durum wheat resulting in a 4.5-fold reduction in LOX activity was determined to lead to a significant improvement in pasta colour [55]. This was also confirmed by Fu et al. [59]. This discrepancy could be due to the greater variability in genotypes and environments (each including different locations) investigated, which gives greater strength to the results. The involvement of the BI, although minimal, would seem to indirectly indicate the role of POD and PPO in the final pasta colour.

3.3. Principal Component Analysis

To establish a stronger correlation among the traits analysed in the eighteen genotypes, a PCA was performed, and the results are shown in Figure 4.

About 83% of the total variability is explained by the components PC1 (55.1%) and PC2 (27.6%). Three groupings are evident. The first group is significantly and positively associated with PC1 and includes all traits directly related to yellowness, analysed in semolina and pasta, which are highly correlated to each other (Table 2), and it is mainly linked to the genotype Verdi. In the second group, positively associated with PC2, are the two enzymes PPO and POD, mainly responsible for brownness [10–13], which are strictly related to Arcobaleno, Nefer, Saadi, and Torrebianca. Finally, in the third group, there are BI-S and BI-P that resulted from POD and PPO activities as well as from the level of the grey component due to the accumulation of minerals in grain (ash content) or that is produced during the processing of pasta [10]. The two vectors of BI-S and BI-P are about at 45° with respect to the point of insertion of two axes and, therefore, are not significantly associated with the components PC1 and PC2. Among genotypes, only the San Carlo genotype is aligned on these components. HP and BL were not significantly correlated with any other variable and were not selected in PCA.

For the other genotypes, PC2 clearly distinguishes Simeto and, to a lower extent, Arcobaleno, Nefer, and Saadi, from Flaminio, while Meridiano, as opposed to Colosseo, is well represented by the PC1 component with Iride and Claudio which reflect the same trend with respect to Verdi. The remaining genotypes, almost all in the lower quadrants, are a bit scattered and can be considered isolated cases, with Preco being the most isolated by far.

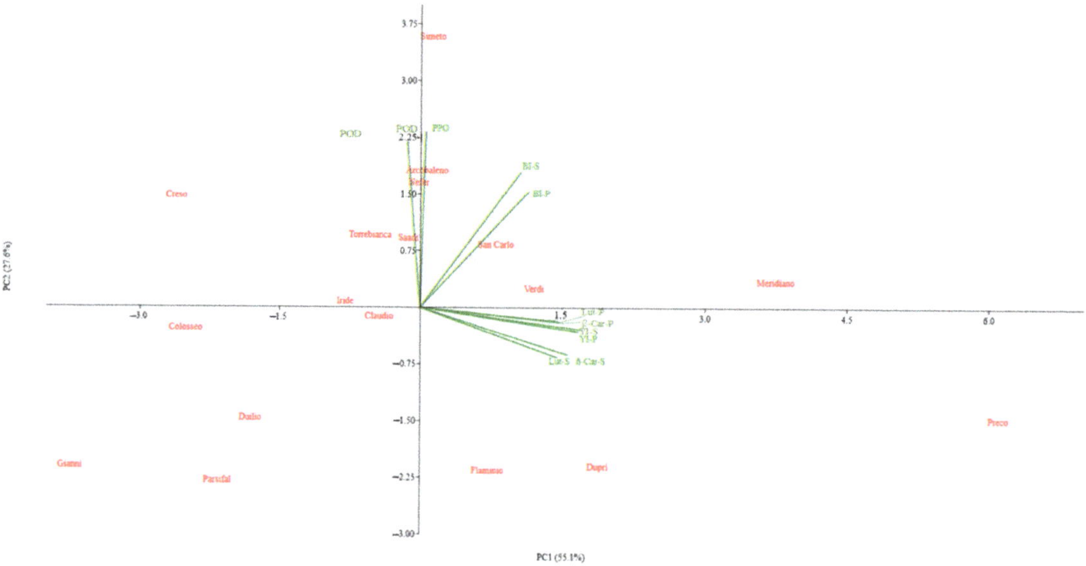

Figure 4. A PCA biplot showing the distribution of the traits analysed in semolina and pasta and of genotypes along the two principal factors. POD = peroxidase activity; PPO = polyphenoloxidase activity; Lut = lutein; β-Car = β-carotene; YI = yellow index; BI = brown index; S = semolina; P = pasta.

3.4. High-Performance Index (HPI)—Tool for Descriptive Identification of Genotype with Best Combination of Traits Involved in Colour of Pasta

Most of the traits considered in this study showed a strong genotypic component, and each trait played its own role, positive or negative, in determining the pasta colour of each genotype. A descriptive method for determining the relative contribution of each trait and identifying the genotype that reaches the best compromise among all the traits that affect semolina and pasta colour was previously developed by Troccoli et al. [26], using a mathematical algorithm named the High-Performance Index (HPI). In HPI calculations, the enzymatic activities of LOX, POD, and PPO in semolina and the BI in semolina and pasta were considered negative variables for the final colour (lower levels were desirable), while carotenoid content and the YI in semolina and pasta were considered positive variables (higher levels were welcomed). After standardizing the deviation values of each variable with respect to its overall mean value, the algorithm checked whether the variable was specified as negative or positive and, for each deviation, calculated a positive or negative score that varies from 0 to a maximum of 2 as the deviation value increases. The score assigned to each analytical variable represents the specific contribution of that trait and therefore its performance. The partial score of the row-by-column crossing of each cell represents, for each genotype, the Specific Performance Index (SPI) for each variable. The joint use of the HPI and SPI helps to immediately visualize how much each variable (SPI) contributes to the final HPI.

In Table S3, for each genotype, the partial score attributed to each variable is reported, and their sum was used to calculate the HPI. Considering the mean values of the G $\times$ E interaction, each of the twelve variables has eight replicate values, one for each environment. Therefore, for each genotype, the maximum expected HPI value is, in absolute value, 192 (2 $\times$ 8 Env $\times$ 12 Var), if all 12 variables obtain a score of 2 in each environment. The HPI was represented with a radar chart (Figure 5), showing that the best performance was obtained by the Preco genotype which achieved the highest HPI value (71.6), since most of the variables were favourable, presenting a positive SPI (Table S3).

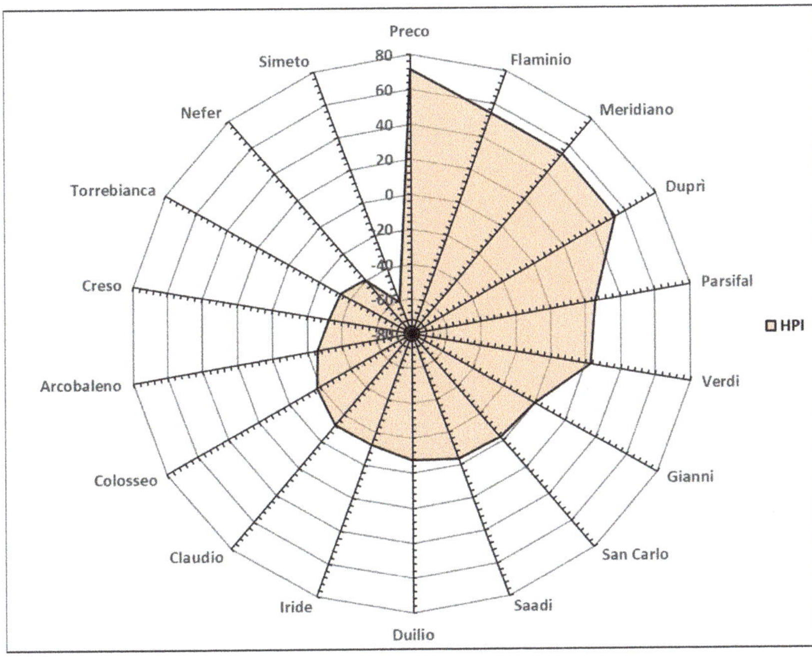

Figure 5. Spider chart for High-Performance Index (HPI) for semolina and pasta of eighteen genotypes grown in eight different Italian environments.

In detail, the Preco genotype reached the maximum negative value (2 × 8 Env) of the SPI for HP, since this trait had the highest value in semolina (Table 1), but this negative characteristic was largely compensated for by the positive SPI values of the other traits, excluding the low values of the BI of semolina and pasta. The Preco genotype was followed by a group of genotypes comprising Flaminio, Meridiano, and Duprì that showed HPI values close to each other (54.3 to 53.3) due to the positive score for most traits. On the contrary, the most negative HPI value was attributed to Simeto, since all the variables had negative SPI scores, due to high enzymatic activities and low pigment contents (Table 1 and Table S3).

The same tool was used to identify the most promising environment for colour expression (Table 5).

A narrower range for the environmental HPI was evidenced as compared with genotype one (Table S3), further confirming the influence of genotype on colour traits. It is interesting to observe that the environment negatively affected both enzyme activities, producing positive SPI scores, and colour-related traits, producing negative scores (Table 4). The environments with the best HPI values were the North-Central Tyrrhenian Coast (ENV3), under sub-littoral conditions, and the North-Central Mountainous Apennines (ENV4), generally characterized by abundant autumn rainfall, a dry winter with low temperatures, and limited rainfall and a high maximum temperature during the grain filling stage until ripening. Insular environments (ENV7 and ENV8), together with the North-Central Adriatic Coast (ENV2), showed intermediate performance. Finally, the South-Central Mountainous Apennines (ENV5), characterized by different rainfall distributions and quantities across locations and lower temperatures during winter and higher ones in the final period of the growth cycle with respect to multi-year seasonal averages [29], had a more negative HPI value and showed a highly negative SPI value for β-Car-P.

Table 5. For each environment, the cumulative score (Specific Performance Index) of the eighteen genotypes, attributed to each analytical variable, was calculated using the HPI tool, according to the procedure described in Troccoli et al. [19]. For each environment, the HPI is the sum of the score of the 12 variables.

	Specific Performance Index (SPI; *n* = 18) for Negative (NV *) or Positive (PV *) Variable												
Specific Attributes	**NV**	**NV**	**NV**	**NV**	**PV**	**PV**	**PV**	**NV**	**PV**	**PV**	**PV**	**NV**	**HPI**
Environment	**HP**	**BL**	**POD**	**PPO**	**Lut-S**	**β-Car-S**	**YI-S**	**BI-S**	**Lut-P**	**β-Car-P**	**YI-P**	**BI-P**	
ENV3	7.03	4.93	−0.03	2.00	−0.48	−0.35	−0.23	−0.20	−0.90	−0.58	−0.20	−0.2	10.80
ENV4	4.83	6.13	0.12	1.73	−0.20	−0.63	−0.23	−0.23	−0.23	−0.78	−0.18	−0.25	10.10
ENV7	4.38	3.78	−0.08	1.75	−0.20	−0.20	−0.18	−0.23	−0.70	−1.68	−0.23	−0.25	6.17
ENV2	4.75	2.63	0.60	1.88	−0.30	−0.28	−0.20	−0.18	−1.08	−1.50	−0.20	−0.2	5.92
ENV8	3.38	5.00	−0.23	1.30	−1.15	−1.10	−0.20	−0.28	−0.25	−1.13	−0.30	−0.175	4.88
ENV6	3.30	1.95	−0.08	0.02	−0.23	−1.05	−0.18	−0.23	−0.25	−2.35	−0.18	−0.275	0.47
ENV1	2.98	0.33	−0.20	1.38	−0.18	−0.23	−0.23	−0.23	−0.55	−2.88	−0.20	−0.25	−0.25
ENV5	4.55	3.78	0.18	−0.15	−0.25	−0.30	−0.23	−0.25	−4.23	−7.53	−0.23	−0.225	−4.88

HP = hydroperoxidation activity of LOX; BL = bleaching activity of LOX; POD = peroxidase activity; PPO = polyphenoloxidase activity; Lut = lutein; β-Car = β-carotene; YI = yellow index; BI = brown index; S = semolina; P= pasta; ENV1: Po Valley; ENV2: North-Central Adriatic Coast; ENV3: North-Central Tyrrhenian Coast; ENV4: North-Central Apennines; ENV5: Central Mountainous Apennines (South); ENV6: Adriatic–Ionic; ENV7: Sicily; ENV8: Sardinia. * A variable classified as "negative" (NV) or "positive" (PV) is considered favourable if the initial content of specific variable results respectively lower or higher than the overall mean value, obtaining a positive score.

However, from a broader perspective, it is necessary to consider the effects of some abiotic stresses (mainly drought, salinity, and high temperatures) on the pigment content of cereal grains [36,60–62]. This is even more important when considering the scenario of climate change in which stress conditions are increasingly present and involve larger geographical areas, orienting the choice towards more tolerant genotypes, which are capable of facing emerging challenges and assuring at the same time a good yield and quality performance.

4. Conclusions

Knowledge of the relative contribution of genotype, environment, and their interaction to colour expression is very important for setting up appropriate breeding programmes in agreement with the quality requirements of the food market. The results reported in this study confirm the importance of the genotypic component in terms of traits related to colour expression.

For the multiple regression model, we selected semolina's YI as the key trait to be used in predicting the bright yellow colour of pasta so appreciated by consumers. This model can be considered very robust because it derives from a multi-environmental dataset. Although the data presented come from experimental trials carried out in the past, this feature makes them widely applicable.

The YI is closely linked to the content of carotenoid pigments. During pasta processing, other traits, such as enzymes, intervene to modify the content of carotenoid pigments, even if they seem to contribute marginally to the final pasta colour.

Furthermore, since the analysed traits intervene in different ways in determining colour expression, the HPI could represent an innovative tool to use to simplify data interpretation, as it provides a unique value to express the variability in the effects of each trait. The consistency of the dataset makes this index able to provide reliable indications for making the best choice of genotypes/environments for producing more coloured pasta.

In a future study, it would be interesting to verify these results in more recent genotypes, mostly cultivated along the Italian territory, improved for production and other qualitative characteristics.

Supplementary Materials: The following supporting information can be downloaded at https://www.mdpi.com/article/10.3390/foods14030392/s1, Table S1: Year of release, country, and pedigree information of eighteen durum wheat genotypes grown in 1999–2000 crop years in eight climate areas; Table S2: Variance components for all traits measured in semolina and pasta of eighteen durum wheat genotypes grown in eight environments; Table S3: For each genotype, the cumulative score (Specific Performance Index) of the eight environments attributed to each analytical variable was calculated using the HPI tool, according to the procedure described in Troccoli et al. [19]. For each genotype, the HPI is the sum of the score of the 12 variables.

Author Contributions: Conceptualization, G.M.B. and M.G.D.; methodology, G.M.B., M.G.D. and C.P.; validation, G.M.B.; formal analysis, G.M.B., C.P. and D.B.M.F.; data curation, A.T., D.B.M.F. and G.M.B.; writing—original draft preparation, A.T., D.B.M.F. and G.M.B.; writing—review and editing, G.M.B. and M.G.D.; supervision, G.M.B. All authors have read and agreed to the published version of the manuscript.

Funding: This research received no external funding.

Institutional Review Board Statement: Not applicable.

Informed Consent Statement: Not applicable.

Data Availability Statement: The original contributions presented in this study are included in the article/Supplementary Materials. Further inquiries can be directed to the corresponding author.

Acknowledgments: The authors are grateful to Anna Maria Mastrangelo for the critical revision of the manuscript.

Conflicts of Interest: The authors declare no conflicts of interest.

References

1. Martínez-Moreno, F.; Ammar, K.; Solís, I. Global changes in cultivated area and breeding activities of durum wheat from 1800 to date: A historical review. *Agronomy* **2022**, *12*, 1135. [CrossRef]
2. Blanco, A. Structure and trends of worldwide research on durum wheat by bibliographic mapping. *Int. J. Plant Biol.* **2024**, *15*, 132–160. [CrossRef]
3. Grosse-Heilmann, M.; Cristiano, E.; Deidda, R.; Viola, F. Durum wheat productivity today and tomorrow: A review of influencing factors and climate change effects. *Resour. Environ. Sustain.* **2024**, *17*, 100170. [CrossRef]
4. Xynias, I.N.; Mylonas, I.; Korpetis, E.G.; Ninou, E.; Tsaballa, A.; Avdikos, I.D.; Mavromatis, A.G. Durum wheat breeding in the mediterranean region: Current status and future prospects. *Agronomy* **2020**, *10*, 432. [CrossRef]
5. WITS (World Integrated Trade Solution). Cereals; Durum Wheat Imports by Country. 2023. Available online: https://wits.worldbank.org/trade/comtrade/en/country/ALL/year/2021/tradeflow/Imports/partner/WLD/product/100110 (accessed on 3 January 2025).
6. Dello Russo, M.; Spagnuolo, C.; Moccia, S.; Angelino, D.; Pellegrini, N.; Martini, D. Nutritional quality of pasta sold on the italian market: The food labelling of italian products (FLIP) study. *Nutrients* **2021**, *13*, 171. [CrossRef]
7. Bresciani, A.; Pagani, M.A.; Marti, A. Pasta-making process: A narrative review on the relation between process variables and pasta quality. *Foods* **2022**, *11*, 256. [CrossRef] [PubMed]
8. Troccoli, A.; Borrelli, G.M.; De Vita, P.; Fares, C.; Di Fonzo, N. Durum wheat quality: A multidisciplinary concept. A review. *J. Cereal Sci.* **2000**, *32*, 99–113. [CrossRef]
9. Hare, R. Durum Wheat: Grain-Quality Characteristics and Management of Quality Requirements. In *Cereal Grains: Assessing and Managing Quality*, 2nd ed.; Wrigley, C., Batey, I., Miskelly, D., Eds.; Woodhead Publishing: Cambridge, UK, 2017; Chapter 6; pp. 135–151. [CrossRef]
10. Martínez-Peña, R.; Rezzouk, F.Z.; Díez-Fraile, M.d.C.; Nieto-Taladriz, M.T.; Araus, J.L.; Aparicio, N.; Vicente, R. Genotype-by-environment interaction for grain yield and quality traits in durum wheat: Identification of ideotypes adapted to the Spanish region of Castile and León. *Eur. J. Agron.* **2023**, *151*, 126951. [CrossRef]

11. Borrelli, G.M.; Troccoli, A.; Di Fonzo, N.; Fares, C. Durum wheat lipoxygenase activity and other quality parameters that affect pasta color. *Cereal Chem.* **1999**, *76*, 335–340. [CrossRef]
12. Borrelli, G.M.; De Leonardis, A.M.; Fares, C.; Platani, C.; Di Fonzo, N. Effects of modified processing conditions on oxidative properties of semolina dough and pasta. *Cereal Chem.* **2003**, *80*, 225–231. [CrossRef]
13. Borrelli, G.M.; De Leonardis, A.M.; Platani, C.; Troccoli, A. Distribution along durum wheat kernel of the components involved in semolina colour. *J. Cereal Sci.* **2008**, *48*, 494–502. [CrossRef]
14. Fraignier, M.-P.; Michaux-Ferriere, N.; Kobrehel, K. Distribution of peroxidase in durum wheat (*Triticum durum*). *Cereal Chem.* **2000**, *77*, 11–17. [CrossRef]
15. Ficco, D.B.M.; Mastrangelo, A.M.; Trono, D.; Borrelli, G.M.; De Vita, P.; Fares, C.; Beleggia, R.; Platani, C.; Papa, R. The colours of durum wheat: A review. *Crop Pasture Sci.* **2014**, *65*, 1–15. [CrossRef]
16. Taha, S.A.; Sagi, F. Relationships between chemical composition of durum wheat semolina and quality. II. Ash, carotenoid pigments, and oxidative enzymes. *Cereal Res. Commun.* **1987**, *15*, 123–129.
17. Feillet, P.; Autran, J.-C.; Icard-Vernière, C. Pasta brownness: An Assessment. *J. Cereal Sci.* **2000**, *32*, 215–233. [CrossRef]
18. Taranto, F.; Delvecchio, L.N.; Mangini, G.; Del Faro, L.; Blanco, A.; Pasqualone, A. Molecular and physico-chemical evaluation of enzymatic browning of whole meal and dough in a collection of tetraploid wheats. *J. Cereal Sci.* **2012**, *55*, 405–414. [CrossRef]
19. Cabas-Lühmann, P.A.; Manthey, F.A.; Elias, E.M. Variations of colour, polyphenol oxidase and peroxidase activities during the production of low temperature dried pasta in various durum wheat genotypes. *Int. J. Food Sci. Technol.* **2021**, *56*, 4700–4709. [CrossRef]
20. Harisha, R.; Singh, S.K.; Ahlawat, A.K.; Narwal, S.; Jaiswal, J.P.; Singh, J.B.; Kumar, R.R.; Singhal, S.; Balakrishnan, A.P.; Shukla, P.; et al. Elucidating the effects on polyphenol oxidase activity and allelic variation of polyphenol oxidase genes on dough and whole wheat-derived product color parameters. *Int. J. Food Prop.* **2023**, *26*, 2716–2731. [CrossRef]
21. Elouafi, I.; Nachit, M.M.; Martin, L.M. Identification of a microsatellite on chromosome 7B showing a strong linkage with yellow pigment in durum wheat (*Triticum turgidum* L. var. *durum*). *Hereditas* **2001**, *135*, 255–261. [CrossRef]
22. Clarke, F.R.; Clarke, J.M.; McCaig, T.N.; Knox, R.E.; DePauw, R.M. Inheritance of yellow pigment concentration in four durum wheat crosses. *Can. J. Plant Sci.* **2006**, *86*, 133–141. [CrossRef]
23. Leenhardt, F.; Lyan, B.; Rock, E.; Boussard, A.; Potus, J.; Chanliaud, E.; Remesy, C. Genetic variability of carotenoid concentration, and lipoxygenase and peroxidase activities among cultivated wheat species and bread wheat varieties. *Eur. J. Agron.* **2006**, *25*, 170–176. [CrossRef]
24. Digesù, A.M.; Platani, C.; Cattivelli, L.; Mangini, G.; Blanco, A. Genetic variability in yellow pigment components in cultivated and wild tetraploid wheats. *J. Cereal Sci.* **2009**, *50*, 210–218. [CrossRef]
25. Blanco, A.; Colasuonno, P.; Gadaleta, A.; Mangini, G.; Schiavulli, A.; Simeone, R.; Digesù, A.M.; De Vita, P.; Mastrangelo, A.M.; Cattivelli, L. Quantitative trait loci for yellow pigment concentration and individual carotenoid compounds in durum wheat. *J. Cereal Sci.* **2011**, *54*, 255–264. [CrossRef]
26. Troccoli, A.; De Leonardis, A.M.; Platani, C.; Borrelli, G.M. High Performance Index as a tool to identify the best combination of pearled fractions and durum wheat genotypes for semolina and pasta colour improvement. *Int. J. Food Sci. Technol.* **2021**, *56*, 4799–4806. [CrossRef]
27. D'Egidio, M.G.; Cecchini, C.; Cantone, M.T.; Dottori, A.; Gosparini, E. Caratteristiche qualitative delle varietà in prova nel '99-2000. *Molini d'Italia* **2001**, *1*, 35–49.
28. Mariani, B.M.; D'Egidio, M.G.; Novaro, P. Durum wheat quality evaluation: Influence of genotype and environment. *Cereal Chem.* **1995**, *72*, 194–197.
29. Desiderio, E.; Fornara, M.; Cecchi, V.; Frongia, L. Risultati della sperimentazione condotta nel 1999–2000. *L'Informatore Agrario* **2000**, *35*, 30–66.
30. González, E.M.; de Ancos, B.; Cano, M.P. Partial characterization of peroxidase and polyphenol oxidase activities in blackberry fruits. *J. Agric. Food Chem.* **2000**, *48*, 5459–5464. [CrossRef]
31. Lowry, O.; Rosebrough, N.J.; Farr, A.L.; Randall, R.J. Protein measurement with Folin phenol reagent. *J. Biol. Chem.* **1951**, *193*, 265–275. [CrossRef]
32. Barone, R.; Briante, R.; D'Auria, S.; Febbraio, F.; Vaccaro, C.; Del Giudice, L.; Borrelli, G.M.; Di Fonzo, N.; Nucci, R. Purification and characterization of lipoxygenase activity from durum wheat flour. *J. Agric. Food Chem.* **1999**, *47*, 1924–1931. [CrossRef]
33. Okot-Kotber, M.; Liavoga, A.; Yong, K.-J.; Bagorogoza, K. Activity and inhibition of polyphenol oxidase in extracts of bran and other milling fractions from a variety of wheat cultivars. *Cereal Chem.* **2001**, *78*, 514–520. [CrossRef]
34. Commission Internationale de l'Eclairage. *Publication 15.2. Colorimetry*, 2nd ed.; Commission Internationale de l'Eclairage Central Bureau: Vienna, Austria, 1986.
35. Schulthess, A.; Matus, I.; Schwember, A.R. Genotypic and environmental factors and their interactions determine semolina color of elite genotypes of durum wheat (*Triticum turgidum* L. var. *durum*) grown in different environments of Chile. *Field Crops Res.* **2013**, *149*, 234–244. [CrossRef]

36. Rharrabti, Y.; Villegas, D.; Royo, C.; Martos-Núñez, V.; García del Moral, L.F. Durum wheat quality in Mediterranean environments: II. Influence of climatic variables and relationships between quality parameters. *Field Crops Res.* **2003**, *80*, 133–140. [CrossRef]
37. Ficco, D.B.M.; Beleggia, R.; Pecorella, I.; Giovanniello, V.; Frenda, A.S.; Vita, P.D. Relationship between seed morphological traits and ash and mineral distribution along the kernel using debranning in durum wheats from different geographic sites. *Foods* **2020**, *9*, 1523. [CrossRef] [PubMed]
38. Sissons, M.; Kadkol, G.; Taylor, J. Genotype by environment effects on durum wheat quality and yield-implications for breeding. *Crop Breed. Genet. Genom.* **2020**, *2*, e200018. [CrossRef]
39. Colasuonno, P.; Marcotuli, I.; Blanco, A.; Maccaferri, M.; Condorelli, G.E.; Tuberosa, R.; Parada, R.; de Camargo, A.C.; Schwember, A.R.; Gadaleta, A. Carotenoid pigment content in durum wheat (*Triticum turgidum* L. var *durum*): An overview of quantitative trait loci and candidate genes. *Front. Plant Sci.* **2019**, *10*, 1347. [CrossRef]
40. Ghaed-Rahimi, L.; Heidari, B.; Dadkhodaie, A. Genotype × environment interactions for wheat grain yield and antioxidant changes in association with drought stress. *Arch. Agron. Soil Sci.* **2014**, *61*, 153–171. [CrossRef]
41. Al-Ashkar, I.; Sallam, M.; Al-Suhaibani, N.; Ibrahim, A.; Alsadon, A.; Al-Doss, A. Multiple stresses of wheat in the detection of traits and genotypes of high-performance and stability for a complex interplay of environment and genotypes. *Agronomy* **2022**, *12*, 2252. [CrossRef]
42. Ziegler, J.U.; Wahl, S.; Würschum, T.; Longin, C.F.H.; Carle, R.; Schweiggert, R.M. Lutein and lutein esters in whole grain flours made from 75 genotypes of 5 *Triticum* species grown at multiple sites. *J. Agric. Food Chem.* **2015**, *63*, 5061–5071. [CrossRef] [PubMed]
43. Groth, S.; Wittmann, R.; Longin, C.F.; Böhm, V. Influence of variety and growing location on carotenoid and vitamin E contents of 184 different durum wheat varieties (*Triticum turgidum* ssp. *durum*) in Germany. *Eur. Food Res. Technol.* **2020**, *246*, 2079–2092. [CrossRef]
44. Siedow, J.N. Plant lipoxygenase, structure and function. *Ann. Rev. Plant Physiol. Plant Mol. Biol.* **1991**, *42*, 145–188. [CrossRef]
45. Leenhardt, F.; Lyan, B.; Rock, E.; Boussard, A.; Potus, J.; Chanliaud, E.; Remesy, C. Wheat lipoxygenase activity induces greater loss of carotenoids than vitamin E during breadmaking. *J. Agric. Food Chem.* **2006**, *54*, 1710–1715. [CrossRef] [PubMed]
46. Verlotta, A.; De Simone, V.; Mastrangelo, A.M.; Cattivelli, L.; Papa, R.; Trono, D. Insight into durum wheat Lpx-B1: A small gene family coding for the lipoxygenase responsible for carotenoid bleaching in mature grains. *BMC Plant Biol.* **2010**, *10*, 263. [CrossRef] [PubMed]
47. Nokthai, P.; Lee, V.S.; Shank, L. Molecular modeling of peroxidase and polyphenol oxidase: Substrate specificity and active site comparison. *Int. J. Mol. Sci.* **2010**, *11*, 3266–3276. [CrossRef]
48. Nooshkam, M.; Varidi, M.; Bashash, M. The Maillard reaction products as food-born antioxidant and antibrowning agents in model and real food systems. *Food Chem.* **2019**, *275*, 644–660. [CrossRef] [PubMed]
49. Schutte, M.; Hayward, S.; Manley, M. Nonenzymatic browning and antioxidant properties of thermally treated cereal grains and end products. *J. Food Biochem.* **2024**, *2024*, 3865849. [CrossRef]
50. Pozniak, C.J.; Knox, R.E.; Clarke, F.R.; Clarke, J.M. Identification of QTL and association of a phytoene synthase gene with endosperm colour in durum wheat. *Theor. Appl. Genet.* **2007**, *114*, 525–537. [CrossRef]
51. Singh, A.; Reimer, S.; Pozniak, C.J.; Clarke, F.R.; Clarke, J.M.; Knox, R.E.; Singh, A.K. Allelic variation at *Psy1-A1* and association with yellow pigment in durum wheat grain. *Theor. Appl. Genet.* **2009**, *118*, 1539–1548. [CrossRef] [PubMed]
52. Crawford, A.C.; Francki, M.G. Lycopene-ε-cyclase (e-LCY3A) is functionally associated with quantitative trait loci for flour b * colour on chromosome 3A in wheat (*Triticum aestivum* L.). *Mol. Breed.* **2013**, *31*, 737–741. [CrossRef]
53. Owens, B.F.; Mathew, D.; Diepenbrock, C.H.; Tiede, T.; Wu, D.; Mateos-Hernandez, M.; Gore, M.A.; Rocheford, T. Genome-Wide Association Study and Pathway-Level Analysis of Kernel Color in Maize. *G3 Genes | Genomes | Genet.* **2019**, *9*, 1945–1955. [CrossRef]
54. McDowell, R.; Banda, L.; Bean, S.R.; Morris, G.P.; Rhodes, D.H. Grain yellowness is an effective predictor of carotenoid content in global sorghum populations. *Sci. Rep.* **2024**, *14*, 25132. [CrossRef]
55. Carrera, A.; Echenique, V.; Zhang, W.; Helguera, M.; Manthey, F.; Schrager, A.; Picca, A.; Cervigni, G.; Dubcovsky, J. A deletion at the *Lpx-B1* locus is associated with low lipoxygenase activity and improved pasta color in durum wheat (*Triticum turgidum* ssp. *durum*). *J. Cereal Sci.* **2007**, *45*, 67–77. [CrossRef]
56. Fratianni, A.; Irano, M.; Panfili, G.; Acquistucci, R. Estimation of color of durum wheat. comparison of WSB, HPLC, and reflectance colorimeter measurements. *J. Agric. Food Chem.* **2005**, *53*, 2373–2378. [CrossRef] [PubMed]
57. Sgrulletta, D.; De Stefanis, E. Simultaneous evaluation of quality parameters of durum wheat (*Triticum durum*) by near infrared spectroscopy. *Ital. J. Food Sci.* **1997**, *9*, 295–301.
58. Clarke, J.M.; Clarke, F.R.; Cames, N.P.; McCaig, T.N.; Knox, R.E. Evaluation of predictors of quality for use in early generation selection. In *Options Méditerranéennes, Proceedings of the Séminaries Méditerranéennes Serie A/N.40 "Durum Wheat Improvement in the Mediterranean Region: New challenges", Zaragoza, Spain, 12–14 April 2000*; Royo, C., Nachit, M., Di Fonzo, N., Araus, J.L., Eds.; CIHEAM, CYMMIT, ICARDA: Zaragoza, Spain, 2000; pp. 439–446.

59. Fu, B.X.; Schlichting, L.; Pozniak, C.J.; Singh, A.K. Pigment loss from semolina to dough: Rapid measurement and relationship with pasta colour. *J. Cereal Sci.* **2013**, *57*, 560–566. [CrossRef]
60. Fratianni, A.; Giuzio, L.; Di Criscio, T.; Flagella, Z.; Panfili, G. Response of carotenoids and tocols of durum wheat in relation to water stress and sulfur fertilization. *J. Agric. Food Chem.* **2013**, *61*, 2583–2590. [CrossRef] [PubMed]
61. Beleggia, R.; Ficco, D.B.M.; Nigro, F.M.; Giovanniello, V.; Colecchia, S.A.; Pecorella, I.; De Vita, P. Effect of sowing date on bioactive compounds and grain morphology of three pigmented cereal species. *Agronomy* **2021**, *11*, 591. [CrossRef]
62. Requena-Ramírez, M.D.; Rodríguez-Suárez, C.; Hornero-Méndez, D.; Atienza, S.G. Durum wheat at risk in a climate change scenario: The carotenoid content is affected by short heat waves. *J. Agric. Food Chem.* **2024**, *72*, 20354–20361. [CrossRef]

foods

Article

Designing of an Oat-Mango Molded Snack with Feasible Nutritional and Nutraceutical Properties

Yudit Aimee Aviles-Rivera [1], José Benigno Valdez-Torres [1], Juan Pedro Campos-Sauceda [2], José Basilio Heredia [1], Jeny Hinojosa-Gómez [3] and María Dolores Muy-Rangel [1,*]

[1] Centro de Investigación en Alimentación y Desarrollo, A. C., Subsede Culiacán, Carretera Eldorado Km 5.5, Campo el Diez, Culiacán CP 80110, Sinaloa, Mexico; yaviles221@estudiantes.ciad.mx (Y.A.A.-R.); jvaldez@ciad.mx (J.B.V.-T.); jbheredia@ciad.mx (J.B.H.)

[2] Tecnológico Nacional de México: Instituto Tecnológico de Culiacán, Juan de Dios Bátiz No. 310 pte. Col. Guadalupe, Culiacán CP 80220, Sinaloa, Mexico; juan.cs@culiacan.tecnm.mx

[3] CONAHCYT-Centro de Investigación en Alimentación y Desarrollo, A. C., Subsede Culiacán, Carretera Eldorado Km 5.5, Campo el Diez, Culiacán CP 80110, Sinaloa, Mexico; jeny.hinojosa@ciad.mx

* Correspondence: mdmuy@ciad.mx

Abstract: In recent years, the market has seen a growing demand for healthy and convenient food options, such as fruit and cereal bars, driven by shifts in eating habits. These changes are primarily attributed to time constraints in meal preparation and the need for ready-to-eat foods. Consequently, this has promoted interest in creating a nutritious, high-quality snack combining oats and mango. This study employed a response surface analysis of extreme vertex mixtures, incorporating constraints and three components: oats, mango peel, and dehydrated mango pulp. This resulted in ten different mixtures, each with unique combinations and proportions of the three components. It evaluated the microbiological quality, proximal composition, total phenolic content, tannins, Aw, color, texture, and chemical properties during storage at room temperature. The optimal blend, which demonstrated the best quality characteristics, consisted of 44.38% oats, 5.36% mango peel, and 29.24% mango pulp. This formulation yielded a protein content of 7.1 g, dietary fiber of 20.3 g per 100 g, total phenols of 3.4 mg gallic acid per g, and no pathogenic microorganisms. According to the obtained data, Aw > 0.3, the estimated shelf life could be 12 months at room temperature. Developing a stable oat-mango snack with excellent nutritional, nutraceutical, chemical quality, and microbiological properties is technologically feasible.

Keywords: food design; functional foods; *Mangifera indica*; properties; quality

1. Introduction

Mango (*Mangifera indica* L.) is a delicious tropical fruit and a nutritional powerhouse, making it one of the most sought-after fruits in the global market. Its sensory, nutritional, and functional properties have economically secured its position as the third most important fruit after bananas and oranges [1,2]. In 2022, global production of mango, guava, and mangosteen fruits reached 59.15 million tons, with Mexico ranking fifth with 1.8 mt, following India, China, Indonesia, and Thailand [3].

The "Ataulfo" variety is native to Mexico, and its widespread commercialization worldwide is attributed to its distinctive flavor, firm texture, sweet flesh, low acidity, and intense aroma [1,4]. Due to these appealing characteristics, a portion of the production is allocated for processed foods, including juices, nectars, jams, and canned and dehydrated products [5]. However, this industrial processing generates up to 60% of waste, primarily peel, seed, and fiber, leading to economic losses, environmental challenges, and possible uses [6].

Over the past two decades, research has focused on characterizing and processing mango peels for their potential reuse in food development. This not only offers a new source of income for the mango processing industry [5,6] but also emphasizes the nutritional value of this by-product. Mango peels are rich in dietary fiber, polyphenols, proteins, and carotenoids, making them a promising source of functional ingredients [6–9].

Mango by-products have been utilized in producing various processed foods, including fermented beverages, jellies, chips, pasta, cookies, bread, and snacks [10–13]. The results demonstrate that these products are rich in phenolic compounds, exhibit antioxidant activity, possess probiotic potential, and are high in dietary fiber. Consuming products containing these compounds is associated with several health benefits. For instance, a high dietary fiber intake is crucial in addressing obesity, coronary heart disease, cancer, and diabetes problems [14–16]. Additionally, the bioactive compounds in these products play a significant role by neutralizing free radicals, thus preventing cellular damage [17]. This information is essential for understanding the health benefits of these products.

The competitive landscape of the food industry continually drives the search for innovative product designs [18]. In recent years, there has been a growing demand for healthy foods that are convenient and sensorially appealing. As a result, fruit and protein bars, along with a variety of other healthy snacks, are gaining popularity. This research focuses on developing and optimizing a snack made from oats, mango pulp, and mango peel flour, offering substantial benefits in terms of nutritional value, nutraceutical properties, and quality stability at room temperature. This presents a promising opportunity for the food industry.

2. Materials and Methods

2.1. Materials

In this study, ripe mangos of the Ataulfo variety, cultivated in the South Pacific region of Chiapas, México, were used. Mango fruits were washed with water and sanitized with 20 ppm sodium hypochlorite for 15 min. Once sanitized, the fruits were peeled, cut into pieces, and crushed in a T12L stainless steel industrial blender (Rbanda, Mexico) to a homogeneous paste and stored at −17 °C. The fruit peels were placed in a stainless-steel strainer and blanched in water at 90 °C for 120 s. To inactivate the enzymes responsible for deterioration reactions such as browning, they were cut and dried for 20 h at 50 °C ± 2 °C in a hot air dehydrator (Excalibur Food Dehydrator, Burleigh Heads, QLD, Australia) until reaching a constant weight. The peel was pulverized in a mill (Pulvex 200) and stored at 2 °C. Commercial oat flakes Granvita (Grupo Industrial Vida S.A. de C.V.) were used, which contained protein = 10%, fat = 10%, total carbohydrates = 70%, and dietary fiber = 13.3% on dry matter.

2.2. Snack Design, Preparation, and Characterization

All mango-oat combinations described in Table 1 were added to 21% of the complement (15% sucrose, 5.3% butter, 0.5% salt, and 0.2% baking powder) to achieve 100%. All components were mixed to make a homogeneous mass, which was molded into an 8 cm long and 7 mm thick indented cylindrical shape. Each sample was left to rest for 15 min at room temperature and baked at 180 °C for 6 min, then allowed to rest for 8 min at room temperature (to allow the water in the product to migrate from the inside to the outside to reach equilibrium), followed by a second cooking stage of 10 min at 180 °C. The samples were removed from the oven, cooled to room temperature, and meticulously packaged in Ziploc® plastic bags for further analysis, ensuring the product's quality.

2.3. Proximal Analysis

Moisture, ash, protein (N × 5.83), fat, and carbohydrates (by difference) contents were determined according to AOAC methods [19]. Total dietary fiber was evaluated by the Megazyme total dietary fiber assay procedure, which is based on the AOAC 991.43 method (AOAC, 1998).

Table 1. Experimental runs, primary components (proportions), and experimental region of the extreme vertices design.

Treatments	Oat Flakes	Mango Peel Flour	Mango Pulp	Experimental Region
1	0.30	0.00	0.49	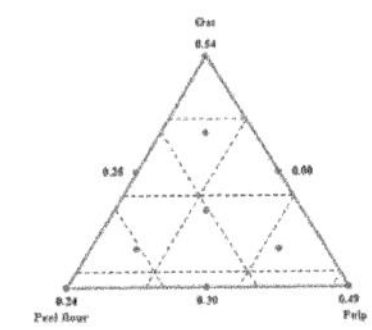
2	0.54	0.00	0.25	
3	0.30	0.24	0.25	
4	0.30	0.12	0.37	
5	0.42	0.00	0.37	
6	0.42	0.12	0.25	
7	0.38	0.08	0.33	
8	0.34	0.04	0.41	
9	0.46	0.04	0.29	
10	0.34	0.16	0.29	

2.4. Hardness

The hardness of the samples was determined using a TA-XT2i texturometer (AMETEX, Cassatt Road Berwyn, FL, USA) measured at 2 points on the surface of the product with a double blade attachment at a speed of 20 mm/s and a contact force of 1 Newton until total breakage of the sample structure [20]. The results were reported in Newton (N).

2.5. Color Analysis

The color attributes of the samples were obtained in the CIELCh color space using a spectrophotometer (CM-700d; Konica Minolta, Ramsey, NJ, USA). The luminosity (L), hue angle (Hue = arc tan b/a), and chromaticity (Chroma = $\sqrt{}$ [a^2 + b^2]) were obtained with OnColor QC version 5 (CyberChrome, Stone Ridge, NY, USA) [21,22].

2.6. Chemical Analysis

The samples' pH, acidity, and total soluble solids (TSS) were carried out according to [19]. To 10 g of sample, 50 mL of distilled H_2O pH 7.0 was added, liquefied, and filtered. A total of 50 mL of filtered sample was taken and placed in a plastic cup. The pH and percentage of acidity were determined with a Mettler Toledo T50 automatic titrator using the LabX program, where the dilution factor and the milliequivalent of citric acid were indicated. One milliliter of the sample was placed in the Metter Toledo S470 refractometer cavity for total soluble solids. The result was multiplied by the dilution factor and reported in degrees Brix (°Brix).

2.7. Antioxidant Capacity

Extract preparation for antioxidant capacity involved 2 g of dry sample being added to 10 mL methanol, shaken at 200 rpm for 2 h, and centrifuged at 27,670× *g* for 20 min at 4 °C [23], using a centrifuge mod Z36 HK (Hermle Labortechnik, Wehigen, Germany).

The Folin Ciocalteu method was used to quantify the total phenolic compounds. A total of 10 μL of the extract, 230 μL of distilled water, and 10 μL of the 2 N Folin Ciocalteu reagent were placed in a clear flat-bottom 96-well transparent plate, incubated for 3 min, 25 μL of Na_2CO_3 4 N was added, and incubated for 2 h in the dark. The absorbance was measured at 725 nm using a Synergy HT Microplate Reader (BioTek, Instruments, Inc., Winooski, VT, USA). The results were expressed as mg of gallic acid equivalents (GAE) per gram [23].

The antioxidant capacity was determined by the DPPH (2,2-diphenyl-1-picrylhydrazyl) method. A 20 μL sample extract was prepared and added to the microplate. A 20 μL sample extract, using Trolox (0, 0.05, 0.1, 0.2, 0.25, 0.3, 0.35, 0.4, 0.5, 0.6, 0.8, and 1.0 mg·mL^{-1}); and 200 μL of DPPH (2,2-definil-1-picrylhydrazyl radical) reagent were added and incubated at 20 °C for 30 min, under dark conditions to determine the calibration curve. Absorbance was measured at 540 nm using a Synergy microplate reader HT (BioTek, Winooski, VT, USA). The results were expressed as mg Trolox equivalents/100 g sample dry weight [24].

2.8. Total Tannins

Total tannins were obtained as the difference between the content of phenolic compounds and the non-tannin phenolics compounds. First, it was necessary to determine the content of phenolic compounds using the Folin Ciocalteu assay, and then the content of non-tannin phenolics was subtracted from that value. For this, tannin precipitation was performed with PVPP. A total of 100 mg of polyvinyl polypyrrolidone (PVPP) and 1 mL of extract were added to Eppendorf tubes, shaken in a Vortex Maximix LP (Thermo Scientific, Waltham, MA, USA), and incubated for 15 min at 4 °C. Then, they were shaken rapidly and centrifuged for 10 min at 4427× g. The supernatant was recovered, and the content of phenolic compounds (non-tannin compounds) was determined using the Folin Ciocalteu assay. The results were expressed as mg catechin equivalents (CE)/g [25,26]. Total tannins = (the content of phenolic compounds minus the non-tannin phenolics compounds).

2.9. Water Activity

A 2 g mango bar was placed in a cell in the equipment cavity of Aqua LabCX-2 (Cromer, Australia) dew point hygrometer at 25 °C, previously calibrated with a NaCl standard of 0.75 water activity (Aw). Finally, Aw was measured and recorded in the sample.

2.10. Microbiological Analysis

Escherichia coli (*E. coli*), fecal coliforms, and total coliforms were determined following the Mexican Official Standard NOM-210-SSA1-2014 [27], appendix H, to *Salmonella* spp. appendix A and appendix B, the procedure for food. A sample of 25 g was placed in a sterile bag, 225 mL of phosphate butter was added, and the sample was homogenized manually for 2 min (stock sample). Coliforms from this homogenate were analyzed with decimal dilutions made using 9 mL of phosphate buffer. Biochemical tests were performed to evaluate *E. coli*, and the results were reported as the most probable number/mL (MPN/g) for *Salmonella* expressed as presence-absence in 25 g of sample. The fungi and yeasts were analyzed according to the Mexican Official Standard NOM-111-SSA1-1994 [28] for products intended for human consumption by plate count.

2.11. Data Analysis

A mixture design of degree 2, with extreme vertices, was implemented with three components (oats, mango peel flour, and mango pulp) with the following constraints: oats 30–55%, mango peel flour 0–30%, and mango pulp 25–65%. Oats and mango (peeled flour and pulp) constituted a balanced 79% of the total mixture, and 21% was the complement in the formulation (15% sucrose, 5.3% butter, 0.5% salt, and 0.2% baking powder). The mixture design consisted of 10 mixtures [29] and each mixture was replicated thrice (Table 1). Statistical significance of coefficients for the terms in the models were taken as $p < 0.05$.

3. Results and Discussion

3.1. Nutritional Characterization of Snack Ingredients

Table 2 shows the nutritional and nutraceutical content of the components used to prepare the oat-mango snack. The moisture content of the mango pulp, at 79%, served as the primary water source in the mixture, facilitating the kneading process with the other ingredients. The nutritional and nutraceutical characteristics of the snack were meticulously estimated, considering the protein contribution from oats (10%), dietary fiber (21.5%), and primarily tannins from mango peel. These estimations provide a solid basis for confidence in the snack's health benefits. The raw materials' analyzed content was consistent with values reported by various authors for oat and mango fruits and peels [5,7,8].

Table 2. Characteristics of primary components for oat-mango snacks.

Components	Mango Pulp (FM)	Mango Peel Flour (DM)	Oat Flakes (DM)
Moisture, %	79.3 ± 0.1 *	8.3 ± 0.28	4.5
Protein, %	1.2 ± 0.1	3.8 ± 0.13	10.0

Table 2. *Cont.*

Components	Mango Pulp (FM)	Mango Peel Flour (DM)	Oat Flakes (DM)
Dietary fiber, %	3.12 ± 0.8	21.50 ± 1.6	13.3
Total phenols, mg GAE·g^{-1}	1.6 ± 0.05	127.3 ± 0.25	NR
Total tannins, mg CAT·g^{-1}	0.05 ± 0.001	50.58 ± 1.2	NR

* Mean and standard deviation of three replicates. GAE: gallic acid equivalents. CAT: Catechin. NR. Not reported. FM: Fresh matter. DM: Dry matter.

3.2. Optimization Process

Table 3 presents the protein, dietary fiber, and bioactive compound contents across the ten treatments derived from the experimental design, which were influenced by the proportions of oats, mango peel, and pulp. Higher ratios of oats and mango peel flour in the combinations enhanced the protein and dietary fiber content. Conversely, including mango pulp and mango peel flour improved the total phenol values, along with the incorporation of tannins. It was reported that mango peel is rich in total phenols, which help eliminate free radicals and function as natural antioxidants [8]. These findings offer a significative promise for future food development, opening opportunities for the creation of healthier and nutritious snack.

Table 3. Means ± Standard deviation of the response variables of 10 mixtures of oats-mango snacks.

Mixtures			Response Variables			
Oat	Peel	Pulp	Protein (%)	Dietary Fiber (%)	Total Phenols (**)	Total Tannins (***)
0.30	0.00	0.49	5.7 ± 0.4 *	29.1 ± 0.4	1.0 ± 0.1	0.2 ± 0.05
0.54	0.00	0.25	6.4 ± 0.1	31.7 ± 0.1	0.7 ± 0.01	0.1 ± 0.01
0.30	0.24	0.25	6.0 ± 0.9	23.1 ± 0.6	14.2 ± 0.4	10.6 ± 0.1
0.30	0.12	0.37	4.6 ± 0.2	21.1 ± 0.2	6.6 ± 0.2	5.3 ± 0.1
0.42	0.00	0.37	6.3 ± 0.1	19.4 ± 0.1	0.2 ± 0.1	0.1 ± 0.01
0.42	0.12	0.25	6.5 ± 0.2	23.6 ± 0.2	6.0 ± 0.2	5.0 ± 0.2
0.38	0.08	0.33	5.5 ± 0.5	22.3 ± 0.5	6.2 ± 0.1	4.0 ± 0.3
0.34	0.04	0.41	6.7 ± 0.1	22.3 ± 0.5	2.3 ± 0.2	2.0 ± 0.3
0.46	0.04	0.29	6.2 ± 0.6	20.7 ± 0.6	2.0 ± 0.1	1.5 ± 0.2
0.34	0.16	0.29	5.1 ± 0.1	20.6 ± 0.1	8.4 ± 0.4	7.5 ± 0.2

* Mean and standard deviation of three replicates, ** (mg GAE·g^{-1} snack), *** (mg CAT·g^{-1} snack). GAE: gallic acid equivalents. CAT: Catechin.

The dietary fiber functionality of oat-mango snacks, determined by the content and ratio of soluble and insoluble fractions [30], significantly influences glucose levels. The soluble fiber in these snacks helps stabilize glucose levels, providing health benefits. Furthermore, soluble fiber serves as a prebiotic, nourishing beneficial gut bacteria, which promotes a balanced microbiota and can enhance overall consumer health [16,31]. Furthermore, consuming foods that contain tannins could be therapeutic; they prevent cancer and the formation of carcinogens. Similarly, tannins enhance body's natural detoxification defenses [25].

3.3. Regression Models

The ANOVA for the response variable is shown in Table 4. Protein and dietary fiber were adjusted by second-order models, while total phenols and total tannins were adjusted by linear models. The significant regression coefficients ($p < 0.05$) for each response are shown in Table 5. The corresponding polynomial models and their goodness of fit (R^2) are given in Table 6.

Table 4. Analysis of variance for protein, dietary fiber, total phenols, and total tannins.

Protein (%)						
Source	DF	Seq SS	Adj SS	Adj MS	F-Value	*p*-Value
Regression	3	2.354	2.354	0.7846	2.70	0.013
Linear	2	1.191	1.352	0.6762	2.41	0.017

Table 4. *Cont.*

Source	DF	Seq SS	Adj SS	Adj MS	F-Value	*p*-Value
			Protein (%)			
Quadratic	1	1.163	1.163	1.1627	4.14	0.08
Peel * Pulp	1	1.163	1.163	1.1627	4.14	0.08
Residual Error	6	1.686	1.986	0.2010		
Total	9	4.040				
			Dietary fiber (%)			
Regression	3	106.72	106.72	35.573	6.35	0.027
Linear	2	20.61	101.87	50.936	9.10	0.015
Quadratic	1	86.11	86.11	86.109	15.38	0.008
Oat * Pulp	1	86.11	86.11	86.109	15.38	0.008
Residual Error	6	33.59	33.59	5.598		
Total	9	140.31				
			Total phenols (mg gallic acid equivalents·g^{-1} snack)			
Regression	2	172.498	172.498	86.2492	125.15	0.000
Linear	2	172.498	172.498	86.2492	125.15	0.000
Residual Error	7	4.824	4.824	0.6892		
Total	9	177.323				
			Total tannins (mg catechin·g^{-1} snack)			
Regression	2	106.550	106.550	53.2749	689.54	0.000
Linear	2	106.550	106.550	53.2749	689.54	0.000
Residual Error	7	0.541	0.541	0.0773		
Total	9	107.091				

DF = Degree of Freedom; Seq SS = Sequential sums of squares; Adj SS = Adjusted sums of squares; Adj MS = Adjusted mean squares. * = Interaction effect.

Table 5. Estimated regression coefficients.

Terms	Response Variables			
	Protein (%)	Dietary Fiber (%)	Total Phenols *	Total Tannins **
Oat	8.82	151.561	−0.5102	−0.4845
Mango peel flour	26.84	−68.235	55.1952	43.0926
Mango pulp	7.15	178.793	1.5076	0.6740
Mango peel flour * Mango pulp	−83.21	--	--	--
Oat * Mango pulp	--	−716.065	--	--

* (mg GAE·g^{-1} snack), ** (mg CAT·g^{-1} snack). GAE: gallic acid equivalents. CAT: Catechin. Mango peel flour.

Table 6. Regression models and coefficient of determination.

Response Variable	Models	R^2 (%)
Protein (%)	8.82 O + 26.84 Pe + 7.15 Pu − 83.21 Pe * Pu	58.3
Dietary fiber (%)	151.56 O − 68.24 Pe + 178.79 Pu − 716.07 O * Pu	76.1
Total phenols *	− 0.51 O + 55.20 Pe + 1.51 Pu	97.3
Total tannins **	− 0.48 O + 43.09 Pe + 0.67 Pu	99.5

* (mg GAE·g^{-1} snack), ** (mg CAT·g^{-1} snack). GAE: Gallic acid equivalents. CAT: Catechin. O = Oat, Pe = Mango peel flour, Pu = Mango pulp.

3.4. Optimization

Using Minitab 19, an overlap of the contour plots of all response models over the restricted experimental region showed a small area where all responses intersect (Figure 1). A feasible solution in this region was oat = 0.441 (44.1%), mango peel flour = 0.0648 (6.5%), and mango pulp = 0.284 (28.4%). The response values at this point were protein (%) = 6.12, dietary fiber = 23.45 (%), total phenols (mg GAE·g^{-1} snack) = 3.67, and total tannins mg Catechin·g^{-1} snack = 2.89. This result was corroborated by the optimization procedure based on the optimal desirability functions for the four desired variables of 0.805.

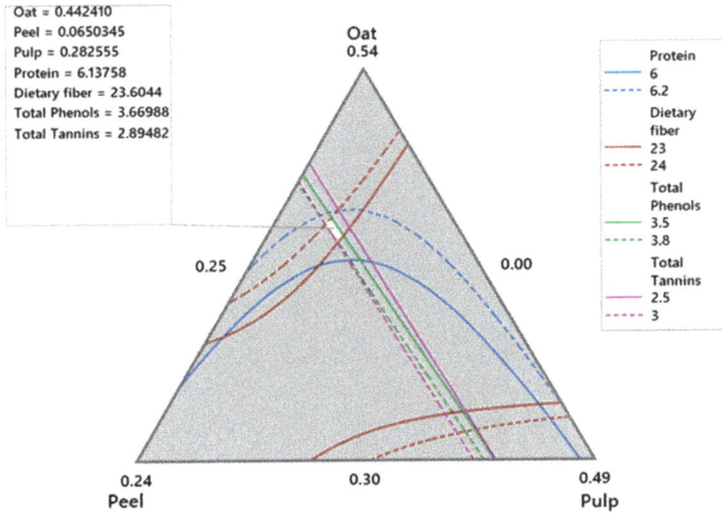

Figure 1. Overlaid contours of response variables of the oat-mango snack.

3.5. Physicochemical Properties of the Optimized Mango-Oat Snack

The snack made with the feasible oat-mango blend demonstrated important protein, dietary fiber, total phenols, and tannin contributions (Table 7), which were consistent with the results predicted by the model (Figure 1), indicating excellent reproducibility of the experiment. Overall, the optimized snack contained 3.8% moisture (Table 8), which falls within the acceptable range for baked goods according to NOM-247-SSA1-2008 [32] (moisture < 8%). The protein content is primarily derived from oats, followed by mango peel. With a high dietary fiber content of 20.3% (with a ratio of 3:1 insoluble to soluble fiber), this snack has three times the fiber found in other snacks enhanced with mango peel [4,33,34], potentially offering significant health benefits.

Table 7. Results for predicted and experimental values in the feasible mixture of dehydrated mango and walnuts for conditions of 60 °C and 3 min.

Values	Response Variables			
	Protein (%)	Total Dietary Fiber (TDF, %)	Total Phenols (**)	Total Tannins (***)
Predicted by model	6.1252	23.4483	3.6687	2.891
Experimental data	6.6 ± 0.6 *	20.3 ± 0.4	3.4 ± 0.1	2.2 ± 0.1

* Mean and standard deviation of three replicates, ** (mg GAE·g^{-1} snack), *** (mg CAT·g^{-1} snack). GAE: gallic acid equivalents. CAT: Catechin.

Table 8. Quality characteristics of the optimized oat-mango snack expressed on dry matter.

Nutritional/Nutraceutical Quality		Physicochemical Characteristics	
Moisture	* 3.82 ± 0.2	Aw	* 0.31 ± 0.002
Fat	7.4 ± 0.6	Texture (N)	60.6 ± 2.6
Ash	2.3 ± 0.2	pH	4.8 ± 0.02
Total carbohydrates	79.8 ± 0.3	TSS (°Brix)	22.0 ± 1.0
Sugars	18.9 ± 0.7	Titratable acidity (% citric acid)	0.08 ± 0.003
Insoluble dietary fiber (IDF), %	15.2 ± 0.6		
Soluble dietary fiber (SDF), %	5.1 ± 0.2		
SDF/TDF ratio	0.22 ± 0.3		
DPPH antioxidant capacity, μmol Trolox·g^{-1}	41.4 ± 0.5		

* Mean and standard deviation of three repetitions. TDF: Total dietary fiber (%, Table 5). DPPH = 2,2-diphenyl-1-picrylhydrazyl. Aw: water activity. TSS: total soluble solids. N: Newtons.

A serving of baked cereal products with fruit is 40 g by regulations (Grain-based bars with or without filling or coating, e.g., breakfast bars, granola bars, rice cereal bars, piece or g [35] from USA) and NOM-086-SSA1-1994 [36] from México. In this context, a serving of the oat-mango snack (40 g) provides 8.1 g of dietary fiber, which accounts for 29% of the recommended daily intake (DI), where 20% DI or more of a nutrient per serving is considered high from USA [35], qualifying it as "high in dietary fiber". The insoluble dietary fiber in the snack primarily comes from the mango peel, while the soluble dietary fiber is derived mainly from the oat content [37]. The nutritional properties and quality of fiber are influenced by the composition and ratio of soluble and insoluble fractions; oats contain over 50% soluble fiber [38], whereas mango peel and pulp are richer in insoluble fiber [39]. However, a 40 g serving of the oat-mango snack contains 6.5 g of sugars, representing 26% of the daily value recommended by the World Health Organization (maximum 25 g/days). This suggests that sensory studies should be conducted to reduce the 15% added sugar content and assess its impact on acceptability.

Fiber-rich by-products such as oatmeal and mango can be incorporated into food products as substitutes for flour, fat, or sugar, enhancing water and oil retention and improving emulsion and oxidative stability [38]. Dietary fiber plays a crucial role in the body, it helps reduce glucose absorption in the bloodstream, leading to better insulin control and maintaining a healthy intestinal ecosystem, among other benefits [16]. The consumption of foods with soluble dietary fiber (SDF) to a total dietary fiber (TDF) ratio of 0.5 is associated with these positive characteristics [40]. In the optimized oat-mango snack, this ratio was 0.22, indicating a higher contribution of insoluble dietary fiber (Table 8).

In recent years, the food industry has increasingly focused on developing products that contain bioactive ingredients, utilizing by-products from fruit processing and marketing them as functional foods. These foods offer health benefits beyond basic nutrition, often due to the inclusion of bioactive compounds. Our study found that the oat-mango snack has a phenolic content of 3.4 mg of GAE per gram (Table 7), comparable to the 5.3 mg of GAE per gram reported by [4] for bread made with mango peel flour.

Mango peel, a significant contributor to our oat and mango snacks, is rich in antioxidant phenolics and tannin compounds. Tannin incorporation was meticulously considered in our mixture combinations, resulting in a snack with beneficial properties that boasted 2.2 mg of catechin per gram (Table 7). Although specific values for tannin consumption are unavailable, it is generally recommended not to exceed 2.5 g per day [41].

The antioxidant capacity of the optimal oat and mango snack was 41.4 μmol ET$\cdot$g^{-1} (Table 8), this value is in a range of results found in various foods [33,34,42]. It is important to note that the antioxidant capacity of a product is influenced not only by the content of phenolic compounds but also by the presence and levels of other bioactive compounds [40].

The optimized oat-mango snack offers benefits with its excellent physical, chemical, and physicochemical qualities (Table 8). Products with water activity (Aw) values less than 0.3, such as our snack, exhibit superior stability due to the low availability of water, which inhibits microbial growth and reduces biochemical changes during storage and distribution [43]. The low Aw value in the oat-mango snack was achieved using mango peel, oats, and dehydrated ingredients, with only the moisture from the mango pulp being added. Additionally, the snack achieved 22°Brix and 60.6 N, values comparable to those reported for cookies made with 20% mango peel flour [44].

The color of food plays a crucial role in its marketing and consumption. An appealing color can enhance a product's acceptability, often independent of its flavor and nutritional content [45]. The oat-mango snacks exhibit orange hues (Figure 2), characteristic of both the pulp and peel of the mango [46,47]. These snacks display luminosity values of 57.2, chromaticity of 41.6, and a hue angle of 79.1, indicating that the optimized snack appears dark orange due to the heat treatment during baking at 180 °C. In contrast, Ataulfo mango pulp, fresh ripe, has luminosity values of 60.6, chromaticity of 63.4, and a hue angle of 62.9 [44]. Ref. [47] reported a luminosity of 51.5 and a hue angle of 55.7 for Ataulfo mango pulp powder. The color values for the oat-mango snack align closely with those of the Ataulfo mango, falling within similar orange ranges. The combination of the ingredients and the baking process results in lower color saturation and a slight product darkening, as reflected in the reduced luminosity values.

oat-mango molded snack	Portion a*, b*, chromaticity, Hue diagram (MINOLTA, 2007)	Data color
	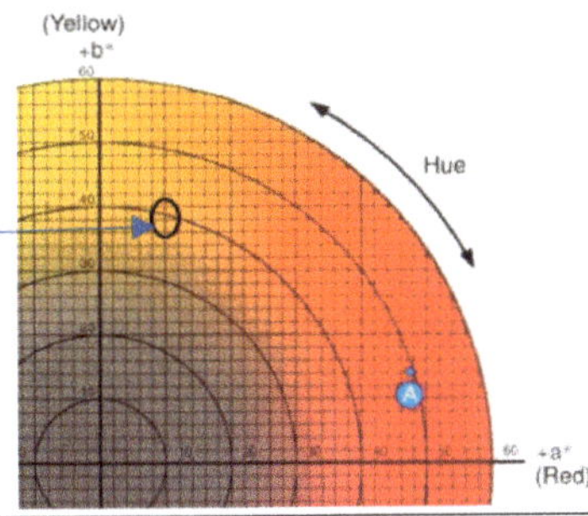	Luminosity = 57.2±2.6* Color a = 8.1±1.9 Color b = 40.7±1.5 Chromaticity = 41.6±1.5 Hue angle = 79.1±2.7 *Mean and SD. n=3

Figure 2. Picture and the color variables in the oat-mango snack.

3.6. Microbiological

The feasible oat-mango snack demonstrated acceptable microbiological quality, attributed to the high quality of the ingredients and the baking process conducted at 180 °C. No presence of *Escherichia coli*, total coliforms, fecal coliforms, fungi, or yeasts was detected. Additionally, *Salmonella* spp. was absent in 25 g of the sample, and only 50 colony-forming units of aerobic mesophiles were found per gram, significantly below the limit of 30,000 CFU·g^{-1} established for cookies according to NMX-F-006-1983 [48]. Furthermore, the aerobic mesophile counts are acceptable for cereal-based foods, edible seeds, flours, semolina, mixtures, and bakery products NOM-247-SSA1-2008 [32].

4. Conclusions

From a technological standpoint, the oat and mango snacks offer acceptable nutritional and nutraceutical benefits and low water activity (Aw) values that enhance stability. The optimal formulation identified through the optimization process for the specified study region consists of 44.1% oats, 6.5% mango peel, and 28.5% mango pulp. The mixture yields a snack rich in protein, high in dietary fiber, and contains significant phenolic compounds; however, adjustment of the added sugars is necessary. Ultimately, it successfully elaborates a snack that is microbiologically safe, nutritious, easy to handle, visually appealing, and capable of maintaining its shape during distribution. The stability of this product ensures a longer shelf life, making it an attractive option for food product developers, including food design and nutrition innovation. Moreover, this approach opens new avenues for utilizing mango peel as a valuable ingredient in fortified foods for human consumption. However, it is crucial to continue research on suitable packaging and shelf life for the mango oat snack to ensure it retains its nutritional and nutraceutical quality as much as possible.

Author Contributions: Conceptualization, M.D.M.-R. and J.B.H.; methodology, Y.A.A.-R., J.B.V.-T., J.P.C.-S. and J.H.-G.; software, Y.A.A.-R. and J.B.V.-T.; validation, Y.A.A.-R., J.B.V.-T., J.P.C.-S. and M.D.M.-R.; formal analysis, Y.A.A.-R., J.P.C.-S., J.B.H. and J.H.-G.; investigation, Y.A.A.-R.; resources, Y.A.A.-R. and M.D.M.-R.; data curation, Y.A.A.-R., J.B.V.-T. and J.P.C.-S.; writing—original draft preparation, Y.A.A.-R., J.P.C.-S., J.B.H. and M.D.M.-R.; writing—review and editing, J.B.V.-T., J.B.H. and M.D.M.-R.; visualization, M.D.M.-R. and J.B.V.-T.; supervision, M.D.M.-R.; project administration, M.D.M.-R.; funding acquisition, M.D.M.-R. and J.B.V.-T. All authors have read and agreed to the published version of the manuscript.

Funding: Consejo Nacional de Humanidades, Ciencia y Tecnologías (CONAHCYT) for the financial support (FORDECYT # 292474).

Institutional Review Board Statement: Not applicable.

Informed Consent Statement: Not applicable.

Data Availability Statement: All data generated or analyzed during this study are included in this published article.

Acknowledgments: The authors thank the Centro de Investigación en Alimentación y Desarrollo, A. C. (CIAD, A. C.) Culiacán, Sinaloa, México for its facilities. and to the for conducting this research, as well as the support of the Laura Aracely Contreras Angulo, and Verónica Pérez Rubio, researchers at the CIAD-Culiacán, Sinaloa, Mexico.

Conflicts of Interest: The authors declare no conflict of interest.

References

1. Yahia, E.; Ornelas-Paz, J.; Brecht, J.; García-Solís, P.; Maldonado, M. The contribution of mango fruit (*Mangifera indica* L.) to human nutrition and health. *Arab. J. Chem.* **2023**, *16*, 104860. [CrossRef]
2. Afifa, A.; Farooq, U.; Akram, K.; Hayat, Z.; Shafi, A.; Sarfraz, F.; Sidhu, M.; Rehman, H.; Aftab, S. Therapeutic potentials of bioactive compounds from mango fruit wastes. *Trends Food Sci. Technol.* **2016**, *53*, 102–112. [CrossRef]
3. FAO Statistics Division. Major Tropical Fruits Market Review-Preliminary Results 2022. Food and Agriculture Organization of the United Nations Rome. 2023. Available online: https://openknowledge.fao.org/server/api/core/bitstreams/852265a4-9006-4d54-a792-51f1d9c44673/content (accessed on 18 April 2024).
4. Ramírez-Maganda, J.; Blancas-Benítez, F.J.; Zamora-Gasga, V.M.; García-Magaña, M.; Bello-Pérez, L.A.; Tovar, J.; Sáyago-Ayerdi, S.G. Nutritional properties and phenolic content of a bakery product substituted with a mango (*Mangifera indica*) "Ataulfo" processing by-product. *Food Res. Int.* **2015**, *73*, 117–123. [CrossRef]
5. Marçal, S.; Pintado, M. Mango peels as food ingredient/additive: Nutritional value, processing, safety and applications. *Trends Food Sci. Technol.* **2021**, *114*, 472–489. [CrossRef]
6. García-Mahecha, M.; Soto-Valdez, H.; Carvajal-Millan, E.; Madera-Santana, T.; Lomelí-Ramírez, M.; Colín-Chávez, C. Bioactive compounds in extracts from the agro-industrial waste of mango. *Molecules* **2023**, *28*, 458. [CrossRef]
7. Pacheco-Jiménez, A.A.; Basilio Heredia, J.; Gutiérrez-Grijalva, E.P.; Quintana-Obregón, E.A.; Muy-Rangel, M.D. Potencial industrial de la cáscara de mango (*Mangifera indica* L.) para la obtención de pectina en México. *TIP Rev. Espec. Cienc. Químico-Biológicas* **2022**, *25*, 1–13. [CrossRef]
8. Quintana-Obregón, E.; San-Martín-Hernández, C.; Muy-Rangel, M.; Vargas-Ortiz, M. Valoración de polvos de cáscara de mango (*Mangifera indica* L.) como una alternativa para la generación de alimentos funcionales. *TIP Rev. Espec. Cienc. Químico-Biológicas* **2019**, *22*, 1–5. [CrossRef]
9. Brecht, J.K.; Sidhu, J.S. *Handbook of Mango Fruit: Production, Postharvest Science, Processing Technology and Nutrition*; John Wiley & Sons Ltd.: Chichester, UK, 2017; pp. 1–6. [CrossRef]
10. Kumar, A.G.; Singh, P.G.; Beer, K.; Pongener, P.; Ravi, C.S.; Singh, S.; Verma, A.; Singh, A.; Thakur, M.; Tripathy, S.; et al. A review on valorization of different byproducts of mango (*Mangifera indica* L.) for functional food and human health. *Food Biosci.* **2022**, *48*, 101783. [CrossRef]
11. Mayo-Mayo, G.; Navarrete-García, A.; Maldonado-Astudillo, Y.I.; Jiménez-Hernández, J.; Santiago-Ramos, D.; Arámbula-Villa, G.; Álvarez-Fitz, P.; Ramirez, M.; Salazar, R. Addition of roselle and mango peel powder in tortilla chips: A strategy for increasing their functionality. *J. Food Meas. Charact.* **2020**, *14*, 1511–1519. [CrossRef]
12. Azura, N.; Radhiah, S.; Zunairah, W.; Shazini, N.; Hanani, N.; Ismail-Fitry, M.R. Physicochemical, cooking quality and sensory characterization of yellow alkaline noodle: Impact of mango peel powder level. *Food Res.* **2020**, *4*, 70–76. [CrossRef]
13. Vicenssuto, G.M.; de Castro, R.J. Development of a novel probiotic milk product with enhanced antioxidant properties using mango peel as a fermentation substrate. *Biocatal. Agric. Biotechnol.* **2020**, *24*, 101564. [CrossRef]
14. Waddell, I.S.; Orfila, C. Dietary fiber in the prevention of obesity and obesity-related chronic diseases: From epidemiological evidence to potential molecular mechanisms. *Crit. Rev. Food Sci. Nutr.* **2023**, *63*, 8752–8767. [CrossRef] [PubMed]
15. Niño-Medina, G.; Muy-Rangel, D.; De La Garza, A.L.; Rubio-Carrasco, W.; Pérez-Meza, B.; Araujo-Chapa, A.P.; Gutiérrez-Álvarez, K.A.; Urías-Orona, V. Dietary fiber from chickpea (*Cicer arietinum*) and soybean (*Glycine max*) husk byproducts as baking additives: Functional and nutritional properties. *Molecules* **2019**, *24*, 991. [CrossRef] [PubMed]
16. Sáyago-Ayerdi, S.G.; Velázquez-López, C.; Montalvo-González, E.; Goñi, I. By-product from decoction process of *Hibiscus sabdariffa* L. calyces as a source of polyphenols and dietary fiber. *J. Sci. Food Agric.* **2014**, *94*, 898–904. [CrossRef]
17. Geiner, S.T.; Ari, A.; Dohi, H.; Dikdik, K. A Comprehensive review of free radicals, oxidative stress, and antioxidants: Overview, clinical applications, global perspectives, future directions, and mechanisms of antioxidant activity of flavonoid compounds. *J. Chem.* **2024**, 1–21. [CrossRef]
18. Parasecoli, F. Food design, nutrition, and innovation. *Front. Public Health* **2022**, *10*, 1039795. [CrossRef]

19. AOAC. *Official Methods of Análisis*, 16th ed.; Association of Official Analytical Chemists: Arlington, VA, USA, 1995; Volume 1.
20. Márquez-Villacorta, L.F.; Pretell Vásquez, C.C. Evaluación de características de calidad en barras de cereales con alto contenido de fibra y proteína. *Biotecnol. Sect. Agropecu. Agroind.* **2018**, *16*, 69–78. [CrossRef]
21. Konica Minolta. *Precise Color Communication: Color Control from Perception to Instrumentation*; Konica Minolta Sensing Inc.: Osaka, Japan, 2007.
22. Báez-Sañudo, M.A.; Muy-Rangel, M.D.; Heredia, J.B.; Adriana, J. Reducing postharvest softening of papaya (*Carica papaya* cv. Maradol) by using an aqueous 1-methylcyclopropene application. *Int. J. Environ. Agric. Res.* **2017**, *3*, 144–151.
23. Swain, T.; Hillis, W. The phenolics constituents of *Prunus domestica* I.—The quantitative analysis of phenolics constituents. *J. Sci. Food Agric.* **1959**, *10*, 63–68. [CrossRef]
24. Singh, S.; Singh, R.P. In vitro methods of assay of antioxidants: An overview. *Food Rev. Int.* **2008**, *24*, 392–415. [CrossRef]
25. Barman, K. Biodegradation of Tanniniferous Feeds and Their Influence on Nutrient Utilization and Productivity of the Dairy Animals. Ph.D. Thesis, NDRI, Karnal, India, 2004.
26. Heil, M.; Baumann, B.; Andary, C.; Linsenmair, E.K.; McKey, D. Extraction and quantification of "condensed tannins" as a measure of plant anti-herbivore defence? Revisiting an old problem. *Naturwissenschaften* **2002**, *89*, 519–524. [CrossRef] [PubMed]
27. *NOM-210-SSA1-2014-DOF: 26-06-2015*; Productos y Servicios, Métodos de Prueba Microbiológicos, Determinación de Microorganismos Indicadores, Determinación de Microorganismos Patógenos. Diario Oficial de la Federación: Ciudad de Mexico, Mexico, 2015. Available online: https://dof.gob.mx/nota_detalle.php?codigo=5398468&fecha=26/06/2015#gsc.tab=0 (accessed on 13 May 2024).
28. *NOM-111-SSA1-1994-DOF: 13-09-1995*; Bienes y Servicios. Método para la Cuenta de Mohos y Levaduras en Alimentos. Diario Oficial de la Federación: Ciudad de Mexico, Mexico, 1995. Available online: https://dof.gob.mx/nota_detalle.php?codigo=4881226&fecha=13/09/1995#gsc.tab=0 (accessed on 15 May 2024).
29. Minitab. Minitab Statistical Software. Simple Correspondence Analysis, Version 20.4. 2021. Available online: www.minitab.com (accessed on 23 August 2024).
30. García-Gabarra, A. Ingesta de nutrientes: Conceptos y recomendaciones internacionales (1st part). *Nutr. Hosp.* **2006**, *21*, 291–299. [PubMed]
31. Castañeda-Guillo, C. Actualización en prebióticos. *Rev. Cuba. Pediatr.* **2018**, *90*, e648.
32. *NOM-247-SSA1-2008-DOF: 27-07-2009*; Productos y Servicios, Cereales y Sus Productos, Cereales, Harinas de Cereales, Sémolas o Semolinas, Alimentos a Base de: Cereales, Semillas Comestibles, de Harinas, Sémolas o Semolinas o Sus Mezclas, Productos de Panificación—Disposiciones y Especificaciones Sanitarias y Nutrimentales. Diario Oficial de la Federación: Ciudad de Mexico, Mexico, 2009. Available online: https://dof.gob.mx/nota_detalle.php?codigo=5100356&fecha=27/07/2009#gsc.tab=0 (accessed on 7 June 2024).
33. Chen, Y.; Zhao, L.; He, T.; Ou, Z.; Hu, Z.; Wang, K. Effects of mango peel powder on starch digestion and quality characteristics of bread. *Int. J. Biol. Macromol.* **2019**, *140*, 647–652. [CrossRef]
34. Sairam, S.; Krishna, G.; Urooj, A. Physico-chemical characteristics of defatted rice bran and its utilization in a bakery product. *J. Food Sci. Technol.* **2011**, *48*, 478–483. [CrossRef]
35. FDA. U.S. Food and Drug Administration. Health Daily Value and Percent Daily Value on the Nutrition and Supplement Facts Labels. 2023. And Food Labeling: Serving Sizes of Foods that Can Reasonably Be Consumed at One Eating Occasion. 2018. Available online: https://www.fda.gov/Foodguidances (accessed on 11 October 2024).
36. *NOM-086-SSA1-1994-DOF: 26-06-1996*; Bienes y Servicios. Alimentos y Bebidas no Alcohólicas con Modificaciones en Su Composición—Especificaciones Nutrimentales. Diario Oficial de la Federación: Ciudad de Mexico, Mexico, 1996. Available online: https://dof.gob.mx/nota_detalle.php?codigo=4890075&fecha=26/06/1996#gsc.tab=0 (accessed on 10 June 2024).
37. Pathak, D.; Majumdar, J.; Raychaudhuri, U.; Chakraborty, R. Characterization of physicochemical properties in whole wheat bread after incorporation of ripe mango peel. *J. Food Meas. Charact.* **2016**, *10*, 554–561. [CrossRef]
38. Elleuch, M.; Bedigian, D.; Roiseux, O.; Besbes, S.; Blecker, C.; Attia, H. Dietary fibre and fibre-rich by-products of food processing: Characterisation, technological functionality and commercial applications: A review. *Food Chem.* **2011**, *124*, 411–421. [CrossRef]
39. Pacheco-Jiménez, A.; Lizardi-Mendoza, J.; Heredia, J.B.; Gutiérrez-Grijalva, E.; Quintana-Obregón, E.; Muy-Rangel, M.D. Physicochemical characterization of pectin and mango peel (*Mangifera indica* L.) from Mexican cultivars. *Heliyon* **2024**, *10*, e35184. [CrossRef]
40. Blancas-Benitez, F.J.; Avena-Bustillos, R.J.; Montalvo-González, E.; Sáyago-Ayerdi, S.G.; McHugh, H.T. Addition of dried 'Ataulfo' mango (*Mangifera indica* L.) by-products as a source of dietary fiber and polyphenols in starch molded mango snacks. *J. Food Sci. Technol.* **2015**, *52*, 7393–7400. [CrossRef]
41. Sharma, K.; Kumar, V.; Kaur, J.; Tanwar, B.; Goyal, A.; Sharma, R.; Gat, Y.; Kumar, A. Health effects, sources, utilization and safety of tannins: A critical review. *Toxin Rev.* **2019**, *40*, 432–444. [CrossRef]
42. Hernández-Sánchez, C.; Moreno, I.; Cameán, A.; González, A.G.; González-Weller, D.; Castilla, A.; Gutiérrez, A.; Rubio, C.; Hardisson, A. Differentiation of mangoes (*Magnifera indica* L.) conventional and organically cultivated according to their mineral content by using support vector machines. *Talanta* **2012**, *97*, 325–330. [CrossRef] [PubMed]
43. Rockland, L.B.; Nishi, S.K. Influence of water activity on food product quality and stability. *Food Technol.* **1980**, *34*, 42–46.
44. Ajila, C.M.; Naidu, K.A.; Bhat, S.G.; Rao, U.J. Bioactive compounds and antioxidant potential of mango peel extract. *Food Chem.* **2007**, *105*, 982–988. [CrossRef]

45. Krokida, M.K.; Maroulis, Z.B. Kinetics on color changes during drying of some fruits and vegetables. *Dry. Technol.* **1998**, *16*, 667–685. [CrossRef]
46. Jiménez-Maldonado, M.I.; Tovar-Pedraza, J.M.; León Félix, J.; Muy Rangel, M.D.; Islas-Osuna, M.A. Respuesta fisiológica y calidad de mango cv. Ataulfo infectado por *Colletotrichum* spp. *Rev. Mex. Cienc. Agríc.* **2022**, *13*, 1103–1115. [CrossRef]
47. San-Martín-Hernández, C.; Pérez-Rubio, V.; Muy-Rangel, M.D.; Vargas-Ortiz, M.A.; Quintana-Obregón, E.A. Caracterización del polvo y pectina del pericarpio del mango "Ataulfo" maduro y análisis FODA para su procesamiento. *TIP Rev. Espec. Cienc. Químico-Biológicas* **2020**, *23*, 1–10. [CrossRef]
48. *NMX-F-006-1983*; Alimentos-Galletas-Food-Cookie-Normas Mexicanas. Dirección General de Normas, Secretaría de Comercio y Fomento Industrial. Available online: https://sitios1.dif.gob.mx/alimentacion/docs/NMX-F-006-1983_GALLETAS.pdf (accessed on 3 June 2024).

foods

MDPI

Review

Research Progress on the Quality, Extraction Technology, Food Application, and Physiological Function of Rice Bran Oil

Wengong Huang [1,2], Baohai Liu [1,2], Dongmei Shi [1,2], Aihua Cheng [1,2], Guofeng Chen [1,2], Feng Liu [1,2], Jiannan Dong [1,2], Jing Lan [1,2], Bin Hong [3], Shan Zhang [3] and Chuanying Ren [3,*]

1 Safety and Quality Institute of Agricultural Products, Heilongjiang Academy of Agricultural Sciences, Harbin 150086, China; huangwengong1736@163.com (W.H.); shslbh@163.com (B.L.); shidongmei@126.com (D.S.); aihcheng@163.com (A.C.); hljjiance@163.com (G.C.); 13766881790@126.com (F.L.); jiannan_dong@163.com (J.D.); lanjing1968@163.com (J.L.)

2 Key Laboratory of Quality and Safety of Cereals and Their Products, State Administration for Market Regulation, Harbin 150086, China

3 Food Processing Research Institute, Heilongjiang Academy of Agricultural Sciences, Harbin 150086, China; gru.hb@163.com (B.H.); zhangshanfood@163.com (S.Z.)

* Correspondence: chuanying1023@163.com; Tel.: +86-0451-86610253

Abstract: Rice bran oil is recommended by the World Health Organization as one of the three major healthy edible oils (along with corn and sesame oils), owing to its unique fatty acid composition and functional components. This study screened, organized, and analyzed a large number of studies retrieved through keyword searches, and investigated the nutritional value and safety of rice bran oil. It reviews the stability of raw rice bran materials and the extraction and refining process of rice bran oil and discusses food applications and sub-health regulations. Research has found that a delayed stabilization treatment of rice bran seriously affects the overall quality of rice bran oil. Compared with traditional solvent extraction, the new extraction technologies have improved the yield and nutritional value of rice bran oil, but most of them are still in the research stage. Owing to the lack of economical and applicable supporting production equipment, extraction is difficult to industrialize, which is a challenging research area for the future. Rice bran oil has stronger antioxidant stability than other edible oils and is more beneficial to human health; however, its application scope and consumption are limited owing to the product price and lack of understanding. Rice bran oil has significant antioxidant, anti-inflammatory, anti-cancer, hypoglycemic, lipid-lowering, and neuroprotective effects. Further exploratory research on other unknown functions is required to lay a scientific basis for the application and development of rice bran oil.

Keywords: rice bran oil; nutrition; new oil extraction; improve food characteristics; biological activity

1. Introduction

Rice is the world's second-largest food crop after wheat. The crop is grown in 114 countries, and more than 650 million tons are produced annually. Although China's rice planting area ranks second globally, its total output ranks first, and its yield per unit area is approximately 50% higher than the world average [1]. By 2023, China's total rice production is expected to reach 210 million tons, accounting for approximately one-third of the world's output. Many types of rice are cultivated in China, and the proportion of irrigated areas is much higher than the world average [2]. These varieties are famous for their high yields and high quality. Rice is the world's leading cereal crop and serves as the staple food for approximately half of the global population [3].

Rice bran is an underutilized byproduct of brown rice milling, accounting for 5% to 10% of the total rice mass fraction. The crop is enriched in lipids (12–23%), protein (14–16%), and crude fiber (8–10%) [4]. It also contains abundant non-starch polysaccharides, vitamins, minerals, phenolic acids, flavonoids, tocopherols, tocotrienols, and γ-glutamine [5]. One of the most common utilizations of rice bran is the extraction of rice bran oil. Among all vegetable oils, rice bran oil is rich in antioxidant compounds that provide several beneficial properties [6]. The lipids are extracted into the pale yellow and translucent rice bran oil, which has a slightly nutty taste. Rice bran oil is typically composed predominantly of triglycerides (81–84%), with small quantities of monoacylglycerol (1–2%), diacylglycerol (2–3%), free fatty acids (FFAs, 2–6%), waxes (3–4%), glycolipids (0.8%), phospholipids (1–2%), and unsaponifiable matter (4%) [7]. The World Health Organization (WHO), China Grain and Oil Association, American Heart Association, Indian Council of Medical Research, and National Institute of Nutrition of India have all described rice bran oil as a "healthy" edible oil [8].

Owing to the lack of systematic and comprehensive discussions on the characteristics and applications of rice bran oil in the literature, in this study, the nutritional and safety qualities of rice bran oil were analyzed, and the stabilization of rice bran raw materials and the extraction and refining process of rice bran oil were reviewed. Finally, its application in food and sub-health regulations is discussed.

2. Methodology

In this review, the nutritional characteristics, raw material safety, extraction and refining technology, application in food, and physiological function of rice bran oil were comprehensively and systematically studied and discussed. Key databases, such as Web of Science, PubMed, X-MOL, MDPI, and Springer, were selected to ensure a comprehensive and in-depth search of relevant studies. The keywords of the literature search were rice bran oil, nutrition, stabilized rice bran, extraction, application, and biological activity. First, duplicate and non-English studies were mostly removed, and the title and abstract were carefully checked to confirm whether they met the criteria. The studies were then classified according to the predetermined titles, and studies from 2020 to 2024 were selected. Finally, the selected studies were summarized and analyzed. A total of 98 out of 268 articles were selected, forming a solid basis for this review.

3. Nutritional Quality of Rice Bran Oil

Rice bran oil is a nutrient-rich vegetable oil. The fatty acid composition, γ-glutamate, vitamin E, plant sterols, and phenolic compounds are conducive to human absorption, with a rate exceeding 90%. The primary nutritional components are listed in Table 1.

Table 1. Main nutrients of rice bran oil.

Nutrient Composition	Content	Reference
Triglyceride	81~84%	[9]
Diglycerol	2~3%	
Monoglycerin	1~2%	
Free fatty acid	2~6%	
Wax	3~4%	
Glycolipid	0.8%	
Phospholipid	1~2%	
γ-Glutamin	0.9~2.9%	[10]
Tocopherol	0.10~0.14%	
Phytosterol	2~3%	
Sterol ester	2.8~3.1%	
Saturated fatty acid	23.63~24.7%	[11]
Palmitic acid	13.9~27.4%	
Stearic acid	1.5~4.7%	
Monounsaturated fatty acids	43.71~44.3%	
Oleic acid	35.9~49.2%	
Polyunsaturated fatty acids	31~33.6%	
Linoleic acid	27.3~41.0%	

The contents of unsaturated fatty acids in rice bran oil extracted from different varieties or strains of rice vary. In one study, the three most abundant and health-related unsaturated fatty acids (*oleic acid*, *linoleic acid*, and *alpha-linolenic acid*) were freely present in rice bran oil from the nine parent lines (two sterile male lines and seven male lines) and seven hybrid rice lines. The hybrid lines had the highest oil contents, whereas the male lines had the highest contents of two of the three free unsaturated fatty acids (*linoleic* and *oleic acids*) [12]. After fermentation of rice bran with lactic acid bacteria (*Lactobacillus acidophilus*, *L. bulgaricus*, and *Bifidobacterium bifidum*), the nutritional content of rice bran oil changed significantly, with increases in total protein (29.52%), fat (5.38%), ash (48.47%), crude fiber (38.96%), and water (61.04%). The carbohydrate content was reduced (36.61%). Increased contents of free amino acids, total phenols, tannins, flavonoids, γ-oryzanol, vitamin E, and antioxidant activities have been described [13].

3.1. Free Fatty Acids (FFA)

The unsaturated fatty acid contents of edible vegetable oils from different raw materials differ significantly (Figure 1). The saturated, monounsaturated, and polyunsaturated fatty acids ratio in rice bran oil is 0.6:1.1:1, which is close to the golden share ratio advocated by the WHO and is a hallmark of "healthy" edible oils [14]. The triglycerides in rice bran oil are mainly composed of the following three fatty acids: palmitic acid (C16:1), oleic acid (C18:1), and linoleic acid (C18:2). These fatty acids accounted for more than 90% of the total fatty acids in rice bran oil. Oleic acid is the most abundant FFA in rice bran oil, accounting for 42% of the total triglycerides, followed by linoleic acid (32%); however, myristic and stearic acids are relatively limited [9]. The fat content of Indian rice bran is as high as 24% and is mainly composed of oleic acid (48.48%), linoleic acid (35.26%), palmitic acid (14.54%), and FFAs (8.15%), with abundant antioxidants, including glutamic acid, tocopherol, and tocopherol [15].

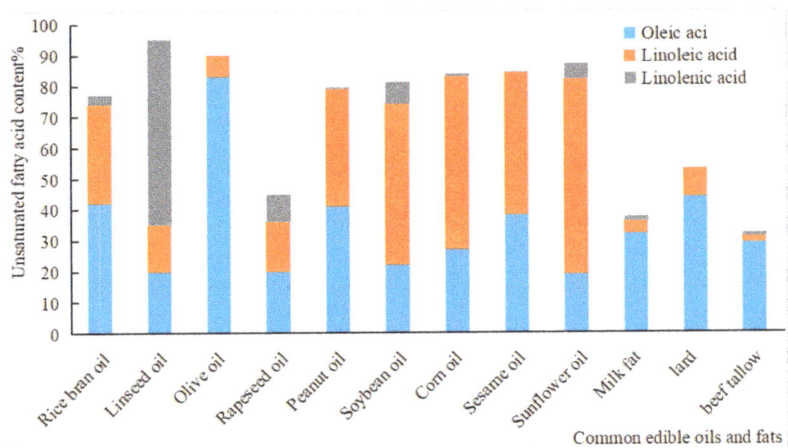

Figure 1. Contents of unsaturated fatty acids in common edible fats (average value).

3.2. γ-Oryzanol

In addition to fatty acids, rice bran oil contains non-saponifiable lipids with a high nutritional value. The most important of these lipids is γ-oryzanol. Oryzanol lipid-binding protein is composed of ferulic acid and phytosterols. The γ-oryzanol constituent 2,4-methylenecycloartanyl ferulate is abundant in purified rice bran oil. We have also described two additional novel γ-oryzanol compounds in rice bran oil, as follows: cyclobranyl ferulate and cyclosadyl ferulate [16]. Crude rice bran oil contains methyl ferulate (0.3%), cycloxybromelane ferulate (0.14%), 24-methylene cycloxybromelane ferulate (0.56%), ferulatecanolasterol (0.49%), and β-sitosterol ferulate (0.24%) [17]. The content of γ-oryzanol in rice bran oil ranges from 0.20% to 2.72%, owing to different rice varieties, weather, and planting areas [18]. γ-Oryzanol is unstable under high-temperature conditions (above 100 °C) and can also be lost under the influence of chemical solvents [19].

3.3. Vitamin E

Plant sterols comprise 4-methyl-free sterols, 4-methyl-sterols, and 4,4′-dimethyl-sterols. The main non-methyl-sterols are B-sitosterol, stigmosterol, and rapeseed sterol [20]. Plant sterols, which mainly exist in plant seeds, are structurally similar to animal sterols. The only differences were the number of methyl groups associated with c-4 and the number of c-11 side chains [21]. Slight differences in these side chains result in differences in physiological functions [22]. Rice bran oil contains a large number of phytosterols, including β-sitosterol (0.90–1.70%), canola sterol (0.50–0.66%), and stigasterol (0.27–0.25%). These sterols have antioxidant properties and constitute a central portion of the non-juice component of nutritional supplements [23].

3.4. Phytosterol

Plant sterols are divided into 4-methyl-free sterols, 4-methyl-sterols, and 4,4′-dimethyl-sterols. The main non-methyl-sterols are B-sitosterol, stigmosterol, rapeseed sterol, and so on [24]. The structures of plant sterols, which mainly exist in plant seeds, are similar to those of animal sterols. The only difference was the number of methyl groups associated with c-4 and the difference in the c-11 side chain. Slight differences in these side chains result in different physiological functions [25]. Rice bran oil contains a large number of phytosterols, including β-sitosterol (0.90–1.70%), canola sterol (0.50–0.66%), and stigasterol (0.27–0.25%). These sterols have antioxidant properties and constitute the main portion of the non-juice component of nutritional supplements [26].

3.5. Phenolic Compounds

Rice bran oil is rich in phytochemicals, including vitamins, cereals, fatty acids, and phenolic compounds [27]. Phenolic compounds in rice bran oil may reduce oxidative stress and cardiovascular disease risk factors by lowering low-density lipoprotein cholesterol levels [28]. In one study, the content of free phenols in bran oil increased gradually with increasing storage time, the contents of combined phenols and total phenols first increased and then decreased, and antioxidant activity was also significantly affected. The total phenolic content of the rice bran oil did not exceed 2 mg/100 g [29].

4. Safety Quality of Rice Bran Oil

Rice bran has good prospects for processing into rice bran oil because it is rich in unsaturated fatty acids (*oleic acid, linoleic acid*, etc.) and healthy bioactive substances. However, the hydrolysis and oxidation rancidity of these lipids during storage are significant problems that directly affect the quality of rice bran oil.

4.1. Oil Oxidation Mechanism

There are three mechanisms for oil oxidation (Figure 2) [30]. The first is autooxidation, which is the reaction of free radicals between activated alkene-containing substrates and ground-state oxygen (Figure 2a). Autooxidation includes the following three stages: chain initiation, chain transfer, and chain termination. The second mechanism is photosensitive oxidation, which is the direct oxidation reaction between unsaturated double bonds and singlet oxygen under the action of a photosensitizer (Figure 2b). The reaction forms a six-membered ring transition state, forming a trans-configuration of hydroperoxide. The third mechanism is reactions that occur in fats, as catalyzed by two types of enzymes (Figure 2c). One is the specific action of fat oxidase on polyunsaturated fatty acids with 1,4 cis and cis-pentadiene structures. The other is the keto-type spoilage caused by the action of enzymes produced by some microorganisms during reproduction that mainly occurs between the α- and β-carbon positions of saturated fatty acids and is also known as the β-oxidation reaction. Oleic acid and linoleic acid account for more than 70% of the FFAs in rice bran oil and are the main causes of oxidation of rice bran oil (Figure 3). In the oleic acid molecule, the hydrogens of the 8- and 11-position carbon atoms have the same activity; therefore, it can generate two different free radicals and four kinds of hydroperoxides. In the linoleic acid molecule, because only the 11-position hydrogen is particularly active, only one free radical and two hydroperoxides are generated.

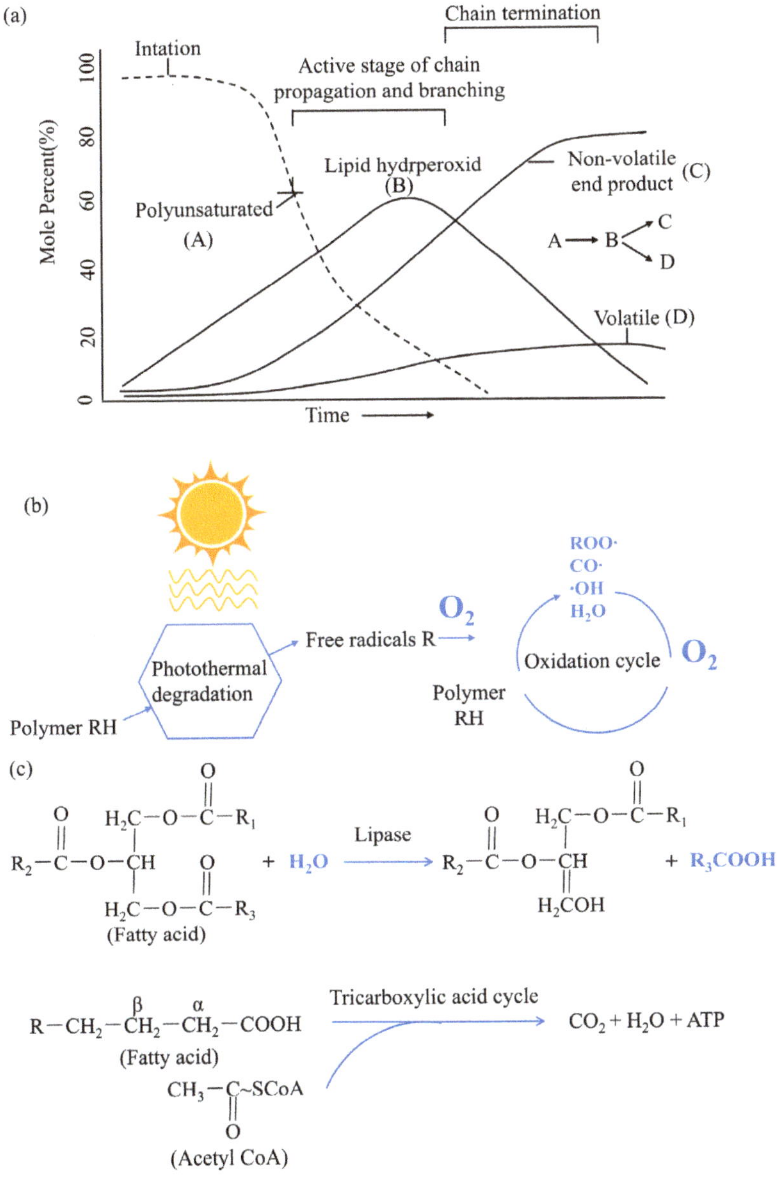

Figure 2. Three mechanisms for oil oxidation: (**a**) automatic oxidation; (**b**) photothermal oxidation; and (**c**) enzymatic hydrolysis and β-oxidation.

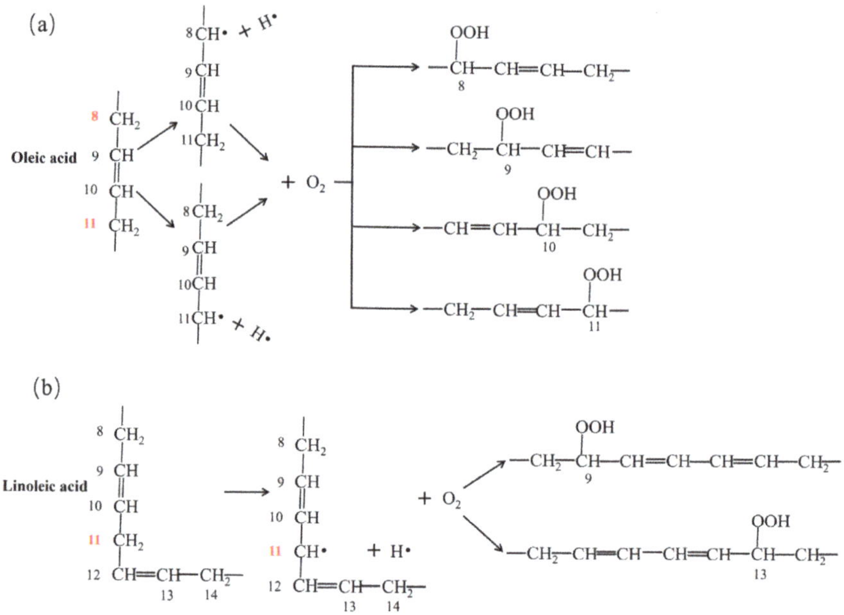

Figure 3. Oxidation mechanism for oil: (**a**) oleic acid; (**b**) linoleic acid.

FFA content and peroxidation value (PV) are essential indices for evaluating the degree of rancidity of rice bran [31]. In the oil oxidation mechanism, the oxidation rate of rice bran oil is directly proportional to the number of double bonds in unsaturated fatty acids, oxygen partial pressure, temperature, surface area, water activity, light, lipase activity, and some metal elements. Therefore, the stabilization treatment of rice bran mainly reduces the hydrolytic rancidity reaction and inhibits the growth of FFAs by passivating lipase activity and reducing water activity (Table 2). These actions prolong the shelf life of rice bran [32]. The latter authors reported that after 14 days of storage at room temperature, the levels of bioactive compounds and functional lipids (25 acylglycerols and 53 phospholipids) were significantly reduced. These significant reductions indicated a rancid process. Phospholipid and glycerolipid metabolic pathways are key for lipid metabolism [33]. Therefore, stabilizing rice bran is key to improving the safety of rice bran oil.

Table 2. Techniques and characteristics of rice bran stabilization.

Stabilization Technique	Advantages and Disadvantages
Vapor treatment	The stabilization effect is general but increases the moisture content and requires further drying.
Hot air drying	It reduces the moisture content of rice bran, and the stabilization effect is better, but the treatment time is longer.
Microwave treatment	The stabilization effect is better, and the content of active ingredients can be increased, but the uniformity of treatment needs to be further improved.
Infrared radiation	The lipase inhibition rate is very high, and the stability effect is very good, but the uniformity needs to be improved.
Coupled processing	The stabilization effect is the best, but the processing cost is higher.

4.2. Stabilization of Rice Bran

Rice bran is rich in nutrients and contains many bioactive compounds that benefit health. However, the rapid deterioration of rice bran limits its development and utilization. Hydrolytic rancidity is the main obstacle in rice bran extraction. The only way to prevent the hydrolytic rancidity of rice bran is to promote enzyme inactivation [34]. Microwave pretreatment stabilizes rice bran and increases its phytochemical content. In one study, when rice bran was treated with 440 W of microwave radiation for 2.5 min, increases were evident in antioxidant activity (0.5-fold), total phenolic contents (1.3-fold), total flavonoid contents (0.9-fold), total γ-oryzanols (1.6-fold), and total phytosterols (1.4-fold). Furthermore, phytochemicals were enhanced, especially trans-p-coumaric acid (10.3-fold) and kaempferol (8.6-fold) [35]. In the microwave-assisted stabilization of rice bran from the Ariete, Teti, and Luna varieties of rice (three commercial Carolino rice cultivars, Ariete, Luna, and Teti, grown in Portugal in the Mondego, Sado, and Tejo River valleys, respectively), oil was extracted, and the effects of rice bran stability on γ-aminobutyric acid (GABA) and γ-glutamate compounds were studied. The stability of LUNA rice bran did not significantly affect these two components. The content of γ-glutamate in the ARIETE and TETI varieties decreased by 34.4% and 24.2%, respectively, and GABA increased by 26.5% and 47.0%, respectively. However, the γ-glutamate content of rice bran oil is not affected by its stability of rice bran [36].

In another study, four methods of treating rice bran were compared, including steaming, hot air drying, microwave treatment, and a combination of microwave and heat treatment, and the rice bran was stored at room temperature for 50 d. The results showed that the combined microwave treatments had the best effects on the oxidation stability of rice bran; FFAs and PV were the lowest, as was the increment. Microwave-treated rice bran has the highest total phenol and γ-glutamate, antioxidant activity, and half-maximal inhibitory concentration (IC50) value [31]. The stabilization of rice bran using infrared radiation heating can improve the quality of rice bran oil. In one study, when rice bran was stabilized using infrared radiation at 125 and 135 V for 5 to 10 min, lipase activity was the lowest (inhibition rate 93~96%), γ-oryzanol and α-tocopherol contents were not significantly changed, and the increases in FFA content and PV were completely inhibited after storage at 38 °C for eight weeks [37].

5. Extraction and Refining of Rice Bran Oil

In industrial production, the traditional process for extracting rice bran oil is solvent extraction. The crude oil obtained is refined through processes such as deacidification, deodorization, decolorization, and degumming to obtain rice bran oil. The byproducts of the refining process can be used to extract various active substances, such as γ-oryzanol, tocopherol, and plant sterols (Figure 4).

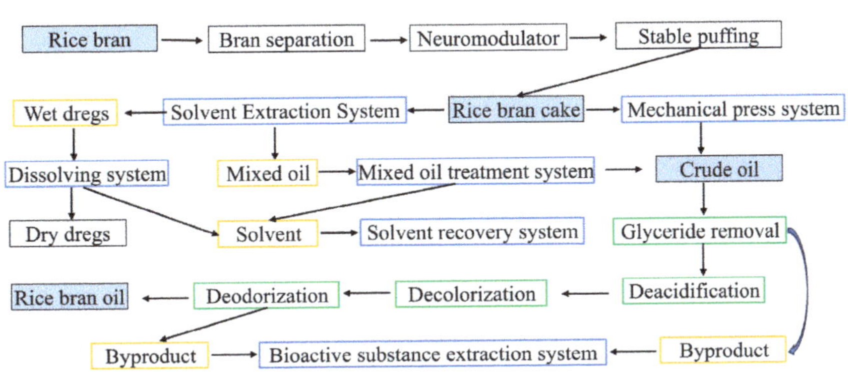

Figure 4. Two extraction processes for rice bran oil.

5.1. Extraction Technology of Rice Bran Oil

The extraction of rice bran oil is usually achieved through solvent extraction. N-hexane is the primary solvent because of its high solubility, low boiling point, and low cost. However, residual organic solvents can easily cause poisoning in humans. It was reported that with isopropyl alcohol and cyclohexane as extraction solvents instead of n-hexane, the peroxide, acid, iodine, and fatty acid compositions of rice bran oil were significantly lower than those of n-hexane. In contrast, the glutamine and total sterol contents increased to 2.7% and 5.1%, respectively [38]. The liquid–solid extraction of rice bran oil with ethyl acetate and ethanol was more compatible with the oil compounds in the rice bran and had a lower viscosity. Ethyl acetate had a higher extraction rate at the beginning and reached its highest yield within a short time [39] (Table 3).

Table 3. New rice bran oil extraction technology.

Extraction Technology	Advantages and Disadvantages
Enzyme-assisted extraction	The reaction conditions are relatively mild, do not damage the quality of rice bran oil, and can also improve the availability of physiologically active components. However, enzyme activity and stability need to be further improved.
Ultrasonic-assisted extraction	It has a low temperature, fast extraction rate, and no change in the chemical composition and structure of nutrients and can also promote the extraction of active substances.
Supercritical and subcritical CO_2 extraction	The extraction time is short, the solvent usage is low, and the oil yield and antioxidant activity are higher, but the equipment cost of this technology is relatively high.
Microwave-assisted extraction	Lipase activity inhibition is effective and does not damage nutritional components, making it a promising method for extracting rice bran oil, but be careful to control the microwave power and temperature.
Subcritical fluid extraction	It can simultaneously inactivate lipase and extract rice bran oil in a short period of time, but equipment costs are high.
CO_2 expansion liquid extraction	Mild conditions and the minimal use of organic solvents allow for the extraction of impurity-free rice bran oil without the need for refining, but equipment costs are high.

Mechanical pressing is a conventional extraction technique that does not involve heat treatment or any organic solvent and is less expensive than solvent extraction. In addition, the oil extracted using this method has better nutritional characteristics and is safer, thereby increasing the value of cold-pressed vegetable oil. New combined technologies have emerged to improve the efficiency and quality of rice bran oil. These include enzyme-assisted extraction, ultrasound-assisted water extraction, supercritical and subcritical CO_2 extraction, microwave-assisted extraction, and subcritical fluid technology. All these have broad application prospects [40].

5.1.1. Enzyme-Assisted Extraction

The addition of specific enzymes (cellulase, pectinase, and protease) to destroy the cell wall and release active substances significantly improves the extraction rate. Most of the properties are the same as those of the solvent extraction of rice bran oil, but the content of colored substances is lower and the acidity (FFA) is higher [41]. The highest yield of rice bran oil was 75% using alkaline protease, hemicellulase, pentosan complex enzyme, and papain [42]. Micron rice bran oil bodies were obtained by combining xylanase with the plant extraction enzyme at a ratio of 2:1 (*w*/*w*), an additional amount of 3%, a pH of 5.0, and an enzymolysis time of 1.5 h [43]. The grain size of rice bran oil extracted using enzyme-assisted $NaHCO_3$ was smaller, and the content of the β-folding structure was higher, which was conducive to the stretching and aggregation of protein spatial structure and better physical stability [44]. The specificity of the enzymatic reaction was high, the reaction conditions were mild and easy to control, and the extraction time was shortened. The surface protein of the rice bran oil body is not destroyed, the stability of the rice bran oil is good, its structure is complete, and the availability of physiologically active components can be improved.

5.1.2. Ultrasonic-Assisted Extraction

Ultrasonic waves transmit high-frequency signals and convert them into high-frequency mechanical oscillations. The intense flow of the solvent produces tiny bubbles, and the closure of the bubbles produces shock waves that improve the permeability of the material surface. The extraction rate of rice bran oil was close to that of Soxhlet extraction when the ultrasonic-assisted extraction technology was at pH 12, the temperature was 45 °C, the stirring speed was 800 r/min, the stirring time was 15 min, the ultrasonic treatment time was 70 min, and the ultrasonic treatment temperature was 25 °C [45]. However, the content of ω-6 polyunsaturated fatty acids (especially 9, 12-octadecadienoic acid (Z,Z)-fatty acids, namely, linoleic acid (64.19%)), in rice bran oil extracted through ultrasonic-assisted extraction was higher than that of hexane extraction [46]. Ultrasound-assisted solvent extraction shortened the homeostasis time from 15 min to 1 min and significantly increased the yield of rice bran oil and γ-oryzanol [47]. The technology has a low temperature, saves considerable energy, is a simple extraction process, improves the extraction rate, and reduces production costs. Low-temperature extraction does not change the chemical composition and structure of nutrients, promotes the extraction of active substances, and improves the nutritional value of rice bran oil.

5.1.3. Supercritical and Subcritical CO_2 Extraction

In the supercritical state, the supercritical fluid comes into contact with the substance to be separated, and the components of polarity, boiling point, and molecular weight are selectively and successively extracted. When the pressure exceeded the critical point, the CO_2 fluid significantly increased the conversion rate of phytosterols to 93.2% and the retention rate of vitamin E to 80.0% [48]. The yield of the subcritical CO_2 extraction of rice bran oil was lower than that of hexane extraction, but the tocopherol and glutarol contents in the rice bran oil extracted were approximately 10 times those of hexane extraction, and the FFA and peroxide values were significantly reduced [49]. The contents of total flavonoids and total polyphenols increased, reaching 61.28 μmol GAE/g and 1696.8 μmol EC/g, respectively [50]. This technology has a short extraction time, less solvent use, and higher safety, oil yield, and antioxidant activity. However, the technology has a high equipment cost and is not suitable for industrial applications. Subcritical CO_2 combined with Soxhlet extraction overcomes the disadvantage of supercritical CO_2 extraction by maintaining solubility under high pressure.

5.1.4. Microwave-Assisted Extraction

High-frequency microwaves produce dipole eddy currents, ion conduction, and high-frequency friction; generate a large amount of heat in a short time; and accelerate solvent penetration into the material. When the microwave power was 200–560 W, the microwave time was 30–120 s, the solvent bran ratio was 1.6–3 mL/g, the overall oil quality obtained using microwave-assisted extraction was better than that obtained using conventional solvent extraction, the α-tocopherol and antioxidant activities were significantly increased, the free fatty acids were significantly decreased, and the extraction speed was faster [51].

Using ethanol, d-limonene, and other green solvents as substitutes for n-hexane, the microwave-assisted extraction of rice bran oil afforded the highest yield of 24.64% and consumed less energy [52]. Microwaves can reduce or even deactivate lipase activity in rice bran and do not destroy fatty acids, cereals, and other nutrients in rice bran oil, making it a promising alternative to the traditional method.

5.1.5. Subcritical Fluid Extraction

Using the principle of phase dissolution, fat-soluble components are extracted via molecular diffusion between materials and subcritical fluids. Under the conditions of 40 °C, 20 MPa, and 120 min, the oil yield of rice bran was 17.19% [53]. The oxidative induction time and α-tocopherol content of safflower seed oil extracted using SFE were significantly increased [54]. It simultaneously inactivates lipase and extracts stable rice bran oil within a short residence time.

5.1.6. CO_2 Expansion Liquid Extraction

High-pressure CO_2 was added to a small amount of organic solvent to promote the rapid expansion of its volume and extraction of lipids. Under the conditions of a temperature of 25 °C, pressure of 5.1 MPa, CO_2 molar fraction of 0.87, and CO_2 expansion of hexane 0.2 mol, the oil yield (25%) of rice bran was significantly higher than that of n-hexane extraction (20%), the phosphorus content was 50 times lower than that of n-hexane extraction, and the FFA concentration was reduced by 17% [55]. At 25 °C, 5.0 MPa, and a CO_2 molar fraction of 0.76, a phosphorus concentration of 4.2 ppm, oil yield of 23.7%, and FFA of 9.60 wt% were obtained [56]. This extraction method has good gas solubility, mass transfer, low viscosity, low condition temperature, and a smaller amount of organic solvent. Thus, rice bran oil can be extracted without impurities or refining.

5.2. Refining Rice Bran Oil and Extraction of Bioactive Substances

The high acid value and dark color of rice bran crude oil make it difficult to refine rice oil. Nutrients are readily lost during the refining process. In addition, the production of trans-fatty acids, 3-chloropropanol ester, and other potentially harmful substances affects the quality and safety of rice oil products. The rice oil refining process generally involves deacidification, decolorization, deodorization, and triglyceride removal (Table 4). The byproducts of rice bran oil refinement can also be used to extract bioactive substances, such as γ-oryzanol and tocopherol.

Table 4. Refinement and extraction of active substances.

Processing	Functions
Deacidification	Remove free fatty acids
Decoloration	Adsorption decolorization, bleaching
Deodorization	Nitrogen removes odors
Glyceride removal	Remove triglycerides
Extraction of γ-oryzanol and tocopherol	Extracting active ingredients from refined byproducts

5.2.1. Deacidification

Rice bran oil usually contains many FFAs. Their removal has always been a challenge in the refining process. Alkali refining and water steam distillation deacidification are widely used in industry, but glycerolysis side reactions can easily occur. In one study, highly acidic rice bran oil was deacidified with a glyceryl eutectic solvent. The deacidification rate reached 94.5%, and the glycerolysis side reaction was prevented [57]. Magnetic carrier starch nanoparticles (MDSNP) were obtained by combining dialdehyde starch nanoparticles (DSNP) with modified Fe_3O_4. Magnetically immobilized Candida antarctic lipase B (MDSNPCALB) was obtained by cross-linking the enzyme with the carrier. MDSNPCALB was added to the degummed rice bran oil. Ethanolamine was used as an acyl receptor in the acylation and deacidification reactions. After ten repeated uses, MDSNPCALB maintained high activity and could be effectively used to deacidify rice bran oil [58]. When CALB was fixed on hydrophobic ordered mesoporous silicon, its catalytic activity increased by 6.6 times, temperature tolerance increased by 20 °C, 91.4% FFAs were removed in a series continuous flow enzyme reactor, and plant sterol esters and diglycerides were increased by 9 and 12 times. The reported retention rate of γ-oryzanol was at least 40% higher than that of traditional alkali refining [59]. Polyedopamine/hydroxyethyl methacrylate/acrylamide hydrogel microspheres were deacidified through esterification with a fixed lipase as the catalyst. A deacidification rate of up to 98.19% was achieved in 2 h, which was economical and sustainable [60].

5.2.2. Decolorization

In one study, the decolorization rate of rice bran oil was 97.08%; the retention rates were 89.62% for glutamic acid, 90.16% for sterol, and 79.91% for vitamin E; and the adsorption rate of benzo (a) pyrene was 95.98% with a 5% compound decolorization agent (active clay/activated carbon = 5:1) [61]. Rice bran oil is rich in nutrients that benefit the human body. However, the loss of nutrients is serious because of the difficulty of bleaching. The color and quality of rice bran oil were reportedly significantly improved using ultrasonic-assisted adsorbents (floroli silica soil, magnesium trisilicate, and active blended soil), whereas the composition of micronutrients and fatty acids was maintained within an acceptable range [62].

5.2.3. Deodorization

Rice bran oil contains a large amount of phytosterols, which are naturally active substances with various active functions that are beneficial to the human body. Deodorization during refining affects the content of phytosterols. A comparison of the phytosterol content in rice bran oil after deodorization with nitrogen and water vapor showed that nitrogen as a stripping gas promoted the formation of phytosterol esters, reduced the production of phytosterol oxidation products, and was more suitable for deodorizing rice bran oil [63]. The addition of sesamol to rice bran oil during deodorization inhibited glycerol formation. When 0.05% sesamol was added, the inhibition rate of glycerol was close to that of 0.02% tert-butylhydroquinone [64].

5.2.4. Glyceride Removal

Rice furoic acid oil is a byproduct of rice bran oil extraction. It comprises FFAs, triglycerides, and a large amount of the natural antioxidant γ-oryzanol. Before the recovery of γ-glutamate, enzymatic hydrolysis is an alternative to the traditional base-catalyzed green triglyceride removal method with a removal rate of 99%. Subcritical water hydrolysis at 220 °C for 10 min can achieve a high triglyceride removal rate of >95%, and the content of γ-oryzanol is still acceptable (50%) [65].

5.2.5. Extraction of γ-Oryzanol and Tocopherol

γ-Oryzanol and tocopherol can be extracted from the fatty acid distillation residue of the rice bran oil refining process. The concentration of tocopherol was reportedly increased by approximately six times, with a recovery rate of 51.50% using methanol and ethyl ketone (1:1). The recovery rates of tocopherol and γ-oryzanol were 92.15% and 84.12%, respectively, and the initial concentrations of these phytochemicals were increased by approximately three times [66]. γ-Tocotrienol and delta-tocotrienol are the most effective natural radiation protection agents, which can be extracted from the distillate of rice bran oil deodorant with 100% purity through the gradient elution of hexane-acetic acid (99.1:0.9) and ethyl acetoacetic acid (99.1:0.9) [67]. The byproduct of the deodorized distillate from rice bran oil contains squalene (approximately 8%) and phytosterol (approximately 4%) as unsaponifiable substances. Phytosterol and squalene can be obtained with 97% purity through supercritical fluid extraction combined with a solvent [68].

6. Application of Rice Bran Oil in Food

Rice bran oil has high stability, nutritional value, and gel properties, Therefore, it is widely used in food(Table 5). The structure, concentration, and synergism of γ-glutenin, polyphenols, phospholipids, phytosterols, and squalene give rice bran oil its unique properties [69].

Table 5. Application of rice bran oil in food.

Application	Advantages and Disadvantages
Add to other oils	Enhance stability
Fat replacement	Preparation of gel instead of saturated fat
Nutritional supplementation	Supplement plant sterols, etc.

6.1. Improved Quality of Other Edible Oils

The addition of 25% to 75% rice bran oil to soybean oil can inhibit oxidation and degradation. FFA and PV remain within the recommended range for human consumption during 12 months of storage, and thiobarbituric acid shows a similar trend to PV [70]. The thermal stability of rice bran oil and soybean oil in deep-oil frying of French fries has been studied. In the first four heating cycles, the peroxide value of the two oils remained below 10 meq/kg, and the acid value (<0.6 mg KOH/100 g) in the fifth cycle of rice bran oil and the third cycle of soybean oil met the requirements. FFA and total polar material showed significant stability after repeated heating, indicating the recoverability of the two oils. The total saturated fatty acid content increased in rice bran oil, the unsaturated fatty acid content decreased, and the stability was enhanced, indicating that rice bran oil was more stable during the frying process [71]. When rice bran oil was added to flaxseed oil at concentrations of 500, 1000, and 1500 ppm, the antioxidant capacity of flaxseed oil also increased with the increase in rice bran oil [72].

6.2. Preparation of Gel to Replace Fat

With the development of gel technology, liquid vegetable oils have been converted into solid fats, which have the functional and textural properties of solid fats and can be used as potential substitutes for saturated and trans-fats. The mixture of β-sitosterol and γ-oryzanol in rice bran oil can be used as a gel to construct an oil gel, which self-assembles to form a fibrous network structure in the oil phase mainly through hydrogen bonding and van der Waals forces (Figure 5). In the case of sterols only, the crystal structure of the oil gel sample consists mainly of large flakes, and excessive sterols interfere with the self-assembly of the three-dimensional gel network structure (Figure 5a). With the addition of γ-oryzanol, the liquid oil is fixed by the spherular-shaped aggregate structure formed by the two, and the network structure is composed of sterol flake crystals (Figure 5b). As the proportion of γ-glutamate is increased, at a certain proportion, the two can form one-dimensional tubules through self-assembly. These crystals grow radially to form a three-dimensional network space dominated by a spherulite shape with a dense structure, and liquid oil is deposited in an orderly manner in the lattice to form a solid oil gel (Figure 5c).

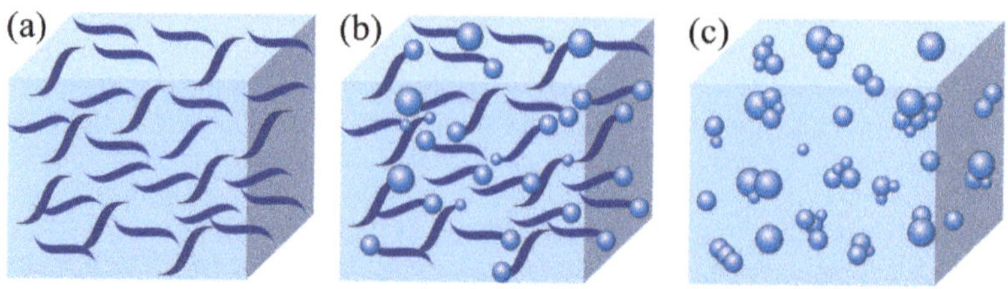

Figure 5. Preparation principle of sterol- and glutamate-mixed oil gels: (**a**) sterols only; (**b**) with the addition of γ-oryzanol; and (**c**) from a solid oil gel.

Oil gels have been studied and applied to many types of foods. In a study on the processing of cookies with soy wax/rice bran oil gel instead of butter, the dough became firmer as the amount of added oil gel increased. The prepared cookies had brown edges and light-colored centers. In addition, with an increase in oil gel content, the surface cracks of cookies increased correspondingly. Replacing up to 50% of butter with oil gel, which was feasible, had no significant effect on the physicochemical properties of cookie dough and cookies [73]. Replacing animal fat with oil gels composed of 25%, 50%, and 75% rice bran wax and rice bran oil can improve the textural, sensory, and nutritional properties of Thai sweet sausages. The higher the replacement ratio, the lower the cholesterol level ($p < 0.05$). There was no significant difference in total unsaturated fatty acid levels between the 50% and 75% oil gels. The optimal replacement ratio of oil gel was determined to be 50% [74]. The new oil gel was developed using sesame oil, rice bran oil, beeswax, and stearic acid and had similar properties to margarine. The binding rate of oil was 99.99%, showing higher oxidation stability than margarine, and a 3:1 ratio of beeswax and stearic acid synergistically affected the properties of the oil gel [75]. By adding 50% rice bran oil instead of shortening, the physical properties and sensory acceptability of the bread maximally increased the quality of the bread [76]. The PV of oil gel prepared from mixed oil, such as rice bran oil and rice bran wax, increased only slightly after 60 days of storage at 6 to 7 °C. Therefore, mixed oil-based gels can be used as new structural oil substitutes for chocolate sauce [77].

6.3. Supplementary Nutrition

As one of the main byproducts of the rice bran oil refining industry, rice bran gum is mainly composed of phospholipids, which can be made into powdered rice bran lecithin after bleaching with sodium chlorite to relieve the marketplace pressure of phospholipid demand [78]. Rice bran oil and mango kernel fat were used to make a cocoa butter substitute, with a 20% improved overall sensory acceptance and taste similar to that of control chocolate [79]. A double emulsion prepared from rice starch, rice protein isolate, and rice bran oil was reportedly used as a carrier, filler, and binder to enhance the physical, functional, and sensory properties of food [80]. The viscosity, permeability, and water-holding capacity of yogurt supplemented with rice bran oil and soybean protein nanoparticles decreased with decreasing size of the nanoparticles. Compared with control yogurt, the yogurt supplemented with soybean protein nanoparticles showed stronger anti-free radical scavenging ability and reduced iron antioxidant properties [81]. Rice bran oil, curcumin, and medium-chain triglycerides were ultrasonically treated into rice bran oil bodies, which can effectively transfer hydrophobic nutrients. The 1.5 wt% preparation had the highest encapsulation rate (87.67%), particle size (190 nm), storage stability, and bioavailability (61.04%) [82].

7. Physiological Functions of Rice Bran Oil

Rice bran oil is rich in tocopherol, γ-glutamic acid, phytosterol, and unsaponifiable substances, which are essential in antioxidant, anti-inflammatory, cancer prevention, and nerve protection functions. These constituents improve the nutritional quality and promote the value-added utilization of rice bran oil (Figure 6).

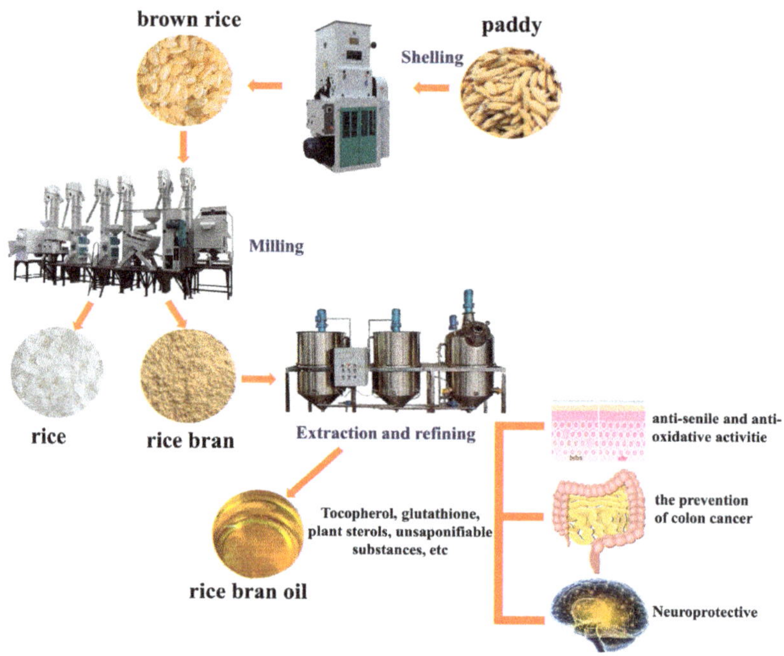

Figure 6. Extraction process and physiological function of rice bran oil.

7.1. Antioxidant and Anti-Inflammatory Activities

The antioxidant and anti-aging ability of physically refined rice bran oil was reported to be significantly higher than that of dissolved γ-oryzanol, α-tocopherol, and sitosterol, and was superior to the mixture, indicating that the antioxidant and anti-aging effects of physically refined rice bran oil were superior to those of the antioxidants derived from rice bran oil [83]. The extraction of rice bran oil by hot pretreatment and cold pressing can improve oil yield, oxidation stability, bioactive compounds, and antioxidant activity, with a stronger inhibitory effect on cell oxidative stress induced by hydrogen peroxide [84].

Using purified rice bran oil as a lipid matrix model, different concentrations of α-tocopherol, γ-oryzanol, and phytosterol significantly affected oxidation stability and free radical scavenging ability. The inhibitory effect of phytosterol on α-tocopherol and hydrogen bond formation between γ-oryzanol and phytosterol was suggested to be mainly due to the influence on the degree of synergism or antagonism of interaction [85].

Rice bran oil is rich in tocotrienol. The scavenging of tocotrienol on 2,2-diphenyl-1-picrylhydrazyl free radicals was 53% at 100 mg/mL. Tocotrienol can effectively inhibit the production of reactive oxygen species, inhibit the migration of neutrophils to inflammatory sites, and upregulate the expression of the antioxidant genes superoxide dismutase and glutathione peroxidase. Inhibiting the expression of pro-inflammatory factors, tumor necrosis factor-alpha (TNF-α), interleukin (IL)-8, and cupric sulfate-induced inflammation have significant antioxidant and anti-inflammatory effects [86]. Rice bran oil is rich in polyphenols. Supplementation with 0.02% rice bran oil in lipopolysaccharide (LPS)-contaminated feed can improve the antioxidant capacity of piglets exposed to LPS, increase the concentrations of catalase and superoxide dismutase, increase the total antioxidant capacity, and decrease the plasma concentrations of diamine oxidase and malondialdehyde [87]. In another study, when elderly individuals with prehypertension were supplemented with 1000 mg of rice bran oil daily for eight weeks, the plasma levels of malondialdehyde, glutathione disulfide, and tumor necrosis factor-alpha (TNF-α) were significantly reduced, and the ratio of reduced glutathione to glutathione disulfide was significantly increased (all $p < 0.05$). The results indicate that the long-term consumption of rice bran oil can relieve oxidative stress and inflammation in older adults with prehypertension [88].

In vitro, rice bran oil treatment was demonstrated to significantly reduce the levels of the pro-inflammatory cytokines (IL-6 and TNF-α) in the culture supernatant of mouse cells, upregulate the secretion of the anti-inflammatory cytokine IL-10, and inhibit the transcription and activation markers of genes encoding the cyclooxygenase-2 and inducible nitric oxide synthase inflammatory mediators in CD80 and INOS cells. The expression of CD86 and MHC-II and the increase in mitochondrial respiration demonstrated that rice bran oil can regulate the inflammatory response of mouse macrophages by upregulating mitochondrial respiration [89]. Rice bran oil in male Wister rats and female Sprague–Dawley rats also showed effective anti-inflammatory, analgesic, and antiarthritic activities, demonstrating the therapeutic value of various bioactive ingredients, including γ-oryzanol [90].

7.2. *Reduction of Blood Lipids and Prevention of Cardiovascular Diseases*

Hyperlipidemia can lead to high blood pressure, high blood sugar levels, cardiovascular disease, liver damage, and atherosclerosis. The main components of rice bran oil (unsaturated fatty acids, triterpene alcohols, phytosterols, tocotrienols, and tocopherols) can lower cholesterol levels. The consumption of rice bran oil can significantly reduce serum total cholesterol, low-density lipoprotein cholesterol, and triglyceride levels [91]. After consumption, rice bran oil mainly enhances the expression of the low-density lipoprotein receptor cytochrome P450 family 7 subfamily member 1 and sterol regulatory element-binding protein 2, promotes the excretion of fecal cholesterol and bile acid, and decreases β-hydroxy β-methylglutaryl-CoA reductase and fatty acid synthase, thereby reducing cholesterol absorption [92]. Tocotrienol from rice bran oil extract can significantly upregulate pparγ and cyp7a1 mRNA levels, significantly down-regulate cpt1a mRNA levels ($p < 0.01$), increase the protein expression levels of Pparγ and Rxrα by two times, and reduce the synthesis and storage of adipose tissue of zebrafish [93]. The active polyphenols in rice bran oil can prevent cardiovascular diseases by improving the oxidative stress mediated by free radicals [94]. Rice bran oil extract inhibits lipid uptake in vivo. It reduces lipid accumulation in HepG2 cells, leading to the activation of AMPK and inhibition of the signal transducer and activator of transcription 3, thereby inhibiting hyperlipidemia and hepatic steatosis caused by a high-fat diet [95].

7.3. *Cancer Prevention*

After injecting diethylnitrosamine and 1,2-dimethylhydrazine into rats, rice bran oil equivalent to 100 mg/kg body weight of γ-oryzanol 5 days per week for 10 weeks inhibited precancerous lesions, including hepatic glutathione S-transferase placental positive lesions and abnormal rectum lesions. Inducing apoptosis of hepatocytes and colorectal cells and reducing the expression of pro-inflammatory cytokines also reportedly promotes diethylnitrosamine- and 1,2-dimethylhydrazine-induced changes in rat gut microbiota, such as an increase in the Firmicutes/Bacteroidetes ratio [96]. After the occurrence of colon cancer induced by 1,2-dimethylhydrazine and sodium glucan sulfate, superoxide dismutase activity and expression of the pro-inflammatory protein cyclooxygenase-2 in the colon of rats were increased when 0.8% γ-oryzanol and unsaponifiable matter extracted from rice bran oil were added to the diet. Chemically induced colon cancer in rats may be inhibited by antioxidant and anti-inflammatory mechanisms, and unsaponifiable compounds may be more effective than gamma-glutamate [97].

7.4. *Neuroprotective Effect*

1,3-Dipalmitoyl-2-oleyl glycerol (POP) is a triacylglycerol ester present in rice bran oil. Oral administration of 1 5 mg/kg POP prevents middle cerebral artery occlusion/reperfusion-induced glutathione depletion and lipid oxidative degradation in rat brains. It also significantly inhibited the phosphorylation of p38 in the ischemic brain up-regulation of MAPKs, inflammatory factors (inducible nitric oxide synthase (i-NOS) and cyclooxygenase-2 (COX-

2)), and pro-apoptotic proteins (B-cell lymphoma-2 (Bcl-2) associated X protein (Bax) and lysed caspase 3) and the down-regulation of anti-apoptotic protein (Bcl-1). The down-regulation of phosphatidylinositol 3′-kinase (PI3K), phosphorylated protein kinase B (Akt), and phosphorylated ring (adenosine monophosphate) AMP response element-binding protein (CREB) in an ischemic brain was inhibited, suggesting that POP may play a neuroprotective role by inhibiting p38 MAPK and activating the PI3K/Akt/CREB pathway [69]. Phytosterols are important bioactive compounds in rice bran and rice bran oil and are found in rotenone-treated Caenorhabditis elegans. Sitosterol, ferulic acid, 24-methylene-ferulic acid, and rapeseed ester in rice bran oil activate the DAF-16/FOXO pathway and increase the neuroprotective effect of oxidative stress resistance induced by the overexpression of the cell death protein 3 apoptosis protein [98].

8. Challenges, Limitations, and Future Research Directions

Rice bran, which is a byproduct of rice processing, has a large annual output. However, because of its low cost and high susceptibility to spoilage (3–5 days), it is mainly used to feed animals, with only 2–5% used as a raw material for rice bran oil processing. With increasing economic development and health awareness, individuals are increasingly paying attention to the health benefits and nutritional properties of vegetable oil, which is recommended for its rich nutritional content. Different types of vegetable oils can be selected according to individual physical conditions and nutritional needs to supplement more unsaturated fatty acids and plant sterols. In terms of the nutritional quality evaluation of rice bran oil, the analysis of bioactive components has been limited mainly to the analysis of γ-oryzanol, phytosterols, and lipids. Analyses and the identification of hitherto unknown small molecule active substances are lacking. Therefore, the health benefits of rice bran oil require further investigation. Rice bran oil is rich in unsaturated fatty acids and various bioactive components, which have been confirmed by many researchers. However, because it has a higher price than soybean oil, a unique odor, and insufficient consumer awareness, its consumption is limited. Therefore, research on low-cost and high-quality rice bran oil processing technologies and supporting equipment, as well as consumer awareness promotion and education, are effective ways to increase rice bran oil consumption.

N-hexane solvent extraction has become a mainstream production technology in the rice bran processing industry due to its low cost. However, from the perspective of nutritional characteristics and safety, under the premise of ensuring the yield and quality of rice bran oil, increased research of emerging extraction and auxiliary technologies is needed, as is an exploration of the potential combined use of the technologies. With technological developments, new rice bran oil extraction technologies have emerged. The efficiency and superiority of the extraction were mainly investigated in the experimental research stage. Owing to the immaturity of the production equipment and higher equipment costs, it is difficult to apply new technologies in production. Therefore, future research should focus on efficient and low-cost emerging technologies that support equipment production. Ultrasound-assisted extraction is very promising and highly recommended. This technology is characterized by low temperatures and fast extraction, does not change the chemical composition and structure of nutrients, and more importantly, promotes the extraction of active substances.

With the gradual penetration of Western eating habits, foods containing light cream are becoming increasingly preferred by Chinese people. In 2022, the output of China's margarine industry was approximately 192.75 million tons, with a demand of approximately 2.5297 million tons. Commercial butter is primarily plant-based cream, accounting for approximately 80% of the total. Traditional vegetable cream (hydrogenated oil) has a high content of trans-fatty acids, which increases the risk of cardiovascular disease, diabetes, and other diseases. The use of trans-fatty acids has been restricted in many countries worldwide. The mixture of β-sitosterol and γ-oryzanol in rice bran oil can be used as a gel to construct an oil gel, which can convert liquid vegetable oil into solid fat and can be used as an ideal alternative to traditional vegetable cream and improve the nutritional properties of the oil gel. Other applications of rice bran oil in the food sector have yet to be further explored.

Rice bran oil is rich in tocopherol, γ-glutamate, phytosterol, and other unsaponifiable substances, which have remarkable antioxidant, anti-inflammatory, cancer prevention, and nerve protective effects. Future studies should prioritize reducing blood sugar and lipid levels while also enhancing the intestinal environment. Correlation analyses of the mechanism of action are needed. This study provides a scientific basis for the functional evaluation and application of rice bran oil.

Author Contributions: Conceptualization, W.H. and B.L.; investigation, D.S. and A.C.; resources, G.C. and S.Z.; data curation, F.L. and J.D.; writing—original draft preparation, W.H.; writing—review and editing, C.R.; visualization, J.L. and B.H.; funding acquisition, C.R. All authors have read and agreed to the published version of the manuscript.

Funding: This study was funded by the National Key Research and Development Program Project (2021YFD2100902), the Heilongjiang Province Research Business Project (CZKYF2023-1-C015), the National Rice Industry Technology System (CARS-01-50), and the Heilongjiang Province Key R&D Project (Innovation Base) (JD2023GJ04).

Institutional Review Board Statement: Not applicable.

Informed Consent Statement: Not applicable.

Data Availability Statement: No new data were created or analyzed in this study. Data sharing is not applicable to this article.

Conflicts of Interest: The authors declare no conflicts of interest.

References

1. Tang, L.; Risalat, H.; Cao, R.; Hu, Q.; Pan, X.; Hu, Y.; Zhang, G. Food Security in China: A Brief View of Rice Production in Recent 20 Years. *Foods* **2022**, *11*, 3324. [CrossRef] [PubMed]
2. Saito, K.; Dossou-Yovo, E.R.; Ibrahim, A. Ratoon rice research: Review and prospect for the tropics. *Field Crops Res.* **2024**, *314*, 109414. [CrossRef] [PubMed]
3. Huang, Y.; Du, L.; Wang, M.; Ren, M.; Yu, S.; Yang, Q. Multifaceted roles of zinc finger proteins in regulating various agronomic traits in rice. *Front. Plant Sci.* **2022**, *13*, 974396. Available online: https://www.frontiersin.org/journals/plant-science/articles/10.3389/fpls.2022.974396 (accessed on 25 July 2022). [CrossRef] [PubMed]
4. Yahya, A.B.; Usaku, C.; Daisuk, P.; Shotipruk, A. Enzymatic hydrolysis as a green alternative for glyceride removal from rice bran acid oil before γ-oryzanol recovery: Statistical process optimization. *Biocatal. Agric. Biotechnol.* **2023**, *50*, 102727. [CrossRef]
5. Spaggiari, M.; Righetti, L.; Folloni, S.; Ranieri, R.; Dall'Asta, C.; Galaverna, G. Impact of air classification, with and without micronisation, on the lipid component of rice bran (*Oryza sativa* L.): A focus on mono-, di- and triacylglycerols. *Int. J. Food Sci. Technol.* **2020**, *55*, 2832–2840. [CrossRef]
6. Go, A.W.; Pham, T.Y.N.; Truong, C.T.; Quijote, K.L.; Angkawijaya, A.E.; Agapay, R.C.; Gunarto, C.; Ju, Y.H.; Santoso, S.P. Improved solvent economy and rate of rice bran lipid extraction using hydrolyzed rice bran with hexane as solvent. *Biomass Bioenergy* **2020**, *142*, 105773. [CrossRef]
7. Tan, B.L.; Norhaizan, M.E.; Chan, L.C. Rice Bran: From Waste to Nutritious Food Ingredients. *Nutrients* **2023**, *15*, 2503. [CrossRef]
8. Punia, S.; Kumar, M.; Sandhu, K.S.; Whiteside, W.S. Rice-bran oil: An emerging source of functional oil. *J. Food Process. Preserv.* **2021**, *45*, e15318. [CrossRef]
9. Ghosh, M. Review on Recent Trends in Rice Bran Oil Processing. *J. Am. Oil Chem. Soc.* **2007**, *84*, 315–324. [CrossRef]
10. Bhatnagar, A.S.; Prasanth Kumar, P.K.; Hemavathy, J.; Gopala Krishna, A.G. Fatty Acid Composition, Oxidative Stability, and Radical Scavenging Activity of Vegetable Oil Blends with Coconut Oil. *J. Am. Oil Chem. Soc.* **2009**, *86*, 991–999. [CrossRef]
11. Orsavova, J.; Misurcova, L.; Ambrozova, J.V.; Vicha, R.; Mlcek, J. Fatty Acids Composition of Vegetable Oils and Its Contribution to Dietary Energy Intake and Dependence of Cardiovascular Mortality on Dietary Intake of Fatty Acids. *Int. J. Mol. Sci.* **2015**, *16*, 12871–12890. [CrossRef] [PubMed]
12. Fox, M.; Newcomb, K.; Oliveira, C.; Shakiba, E.; Nawarathne, I.N. Facile analysis of rice bran oil to compare free unsaturated fatty acid compositions of parental and hybrid rice lines. *J. Am. Oil Chem. Soc.* **2022**, *99*, 1103–1111. [CrossRef] [PubMed]
13. Alauddin, M.; Sultana, A.; Faruque, M.O.; Islam, F.; Kabir, M.A.; Shozib, H.B.; Siddiquee, M.A.; Howlader, M.Z.H. Functional Evaluation of Fermented Rice Bran and Extracted Rice Bran Oil Addressing for Human Health Benefit. *J. Oleo Sci.* **2024**, *73*, 467–477. [CrossRef] [PubMed]
14. Pranowo, D.; Savira, T.D.; Sukardi. Rice bran stabilization using autoclave and optimization of crude rice bran oil recovery using ultrasound-assisted extraction. *BioResources* **2023**, *18*, 8341–8361. [CrossRef]

15. Kumar, R.M.; Yamanura, B. Rice Bran Oil-A Silver Lining to Indian Edible Oil Economy: A Review. *Agric. Rev.* **2022**, *43*, 469–474. [CrossRef]
16. Sawada, K.; Nakagami, T.; Rahmania, H.; Matsuki, M.; Ito, J.; Mohri, T.; Ogura, Y.; Kuwahara, S.; Hashimoto, H.; Nakagawa, K. Isolation and structural elucidation of unique γ-oryzanol species in rice bran oil. *Food Chem.* **2021**, *337*, 127956. [CrossRef]
17. Ali, M.A.; Islam, M.A.; Othman, N.H.; Noor, A.M.; Hossen, J.; Ibrahim, M. Effect of Heating on Compositional Characteristics and Oxidative Stability of Crude and Refined Rice Bran Oil. *J. Oleo Sci.* **2019**, *68*, 1085–1097. [CrossRef]
18. Gingras, L. Refining of rice bran oil. *Inf. Int. News Fats Oils Relat. Mater.* **2019**, *68*, 1085–1097.
19. Hoffmann Bruscatto, M.; Murowaniecki Otero, D.; Ribeiro Pestana-Bauer, V.; Lorini, A.; Rosane Barboza Mendonça, C.; Carlos Zambiazi, R. Study of the thermal stability of γ-oryzanol present in rice bran oil over time. *J. Sci. Food Agric.* **2021**, *101*, 5715–5720. [CrossRef]
20. Zhu, B.; Zhang, Q.; Lu, L.; Bao, J.; Rong, X.; Wang, J.R.; Mei, X. Cocrystals to tune oily vitamin E into crystal vitamin E. *Int. J. Pharm.* **2021**, *592*, 120057. [CrossRef]
21. Sookwong, P.; Yuenyong, J.; Bennett, C. Bioactive Constituents in Cold-Pressed Plant Oils: Their Structure, Bioactivity and Chromatographic Analysis. *J. Oleo Sci.* **2024**, *73*, 393–409. [CrossRef] [PubMed]
22. Yu, K.; Miao, H.; Liu, H.; Zhou, J.; Sui, M.; Zhan, Y.; Xia, N.; Zhao, X.; Han, Y. Genome-wide association studies reveal novel QTLs, QTL-by-environment interactions and their candidate genes for tocopherol content in soybean seed. *Front. Plant Sci.* **2022**, *13*, 1026581. Available online: https://www.frontiersin.org/journals/plant-science/articles/10.3389/fpls.2022.1026581 (accessed on 27 October 2022). [CrossRef] [PubMed]
23. Xu, D.; Gao, Q.; Ma, N.; Hao, J.; Yuan, Y.; Zhang, M.; Cao, Y.; Ho, C.T. Structures and physicochemical characterization of enzyme extracted oil bodies from rice bran. *LWT* **2021**, *135*, 109982. [CrossRef]
24. Rogowska, A.; Szakiel, A. The role of sterols in plant response to abiotic stress. *Phytochem. Rev.* **2020**, *19*, 1525–1538. [CrossRef]
25. Michellod, D.; Bien, T.; Birgel, D.; Violette, M.; Kleiner, M.; Fearn, S.; Zeidler, C.; Gruber-Vodicka, H.R.; Dubilier, N.; Liebeke, M. De novo phytosterol synthesis in animals. *Science* **2023**, *380*, 520–526. [CrossRef]
26. Zhao, B.; Zhang, Y.; Li, H.; Deng, J.; Gong, H.; Chen, Z. Nutritional Component and Chemical Characterization of Chinese Highland Barley Bran Oil. *J. Oleo Sci.* **2020**, *69*, 1339–1347. [CrossRef]
27. Tufail, T.; Ain, H.B.; Chen, J.; Virk, M.S.; Ahmed, Z.; Ashraf, J.; Shahid, N.U.; Xu, B. Contemporary Views of the Extraction, Health Benefits, and Industrial Integration of Rice Bran Oil: A Prominent Ingredient for Holistic Human Health. *Foods* **2024**, *13*, 1305. [CrossRef]
28. Ardiansyah; Nada, A.; Rahmawati, N.T.; Oktriani, A.; David, W.; Astuti, R.M.; Handoko, D.D.; Kusbiantoro, B.; Budijanto, S.; Shirakawa, H. Volatile Compounds, Sensory Profile and Phenolic Compounds in Fermented Rice Bran. *Plants* **2021**, *10*, 1073. [CrossRef]
29. Wu, X.; Li, F.; Wu, W. Effect of rice bran rancidity on the structure and antioxidant properties of rice bran soluble dietary fiber. *J. Cereal Sci.* **2022**, *105*, 103469. [CrossRef]
30. Yuan, C.; Pu, W.F.; Ifticene, M.A.; Zhao, S.; Varfolomeev, M.A. Crude Oil Oxidation in an Air Injection Based Enhanced Oil Recovery Process: Chemical Reaction Mechanism and Catalysis. *Energy Fuels* **2022**, *36*, 5209–5227. [CrossRef]
31. Aluthge, D.S.U.; Ranaweera, K.K.D.S.; Gunathilake, I.A.D.S.R. The effect of stabilization heat treatment on rice bran quality parameters, including total phenolic content, gamma oryzanol content, antioxidant potential, oxidative stability and extraction yield during storage. *Food Chem. Adv.* **2023**, *3*, 100531. [CrossRef]
32. Rohfritsch, Z.; Canelli, G.; Pollien, P.; Bel-Rhlid, R. Wheat and Rice Bran as Natural Additives for the Protection of Fish Oil from Oxidation. *ACS Food Sci. Technol.* **2021**, *1*, 1160–1168. [CrossRef]
33. Liu, X.; Li, Z.; OuYang, B.; Wang, W.; Lan, D.; Wang, Y. Lipidomics analysis of rice bran during storage unveils mechanisms behind dynamic changes in functional lipid molecular species. *Food Chem.* **2024**, *447*, 138946. [CrossRef] [PubMed]
34. Yılmaz Tuncel, N. Stabilization of Rice Bran: A Review. *Foods* **2023**, *12*, 1924. [CrossRef] [PubMed]
35. Pokkanta, P.; Yuenyong, J.; Mahatheeranont, S.; Jiamyangyuen, S.; Sookwong, P. Microwave treatment of rice bran and its effect on phytochemical content and antioxidant activity. *Sci. Rep.* **2022**, *12*, 7708. [CrossRef]
36. Reis, N.; Castanho, A.; Lageiro, M.; Pereira, C.; Brites, C.M.; Vaz-Velho, M. Rice Bran Stabilisation and Oil Extraction Using the Microwave-Assisted Method and Its Effects on GABA and Gamma-Oryzanol Compounds. *Foods* **2022**, *11*, 912. [CrossRef]
37. Duangsi, R.; Krongyut, W. Stabilization of Rice Bran by Infrared Radiation Heating for Increased Resilience and Quality of Rice Bran Oil Production. *Prev. Nutr. Food Sci.* **2023**, *28*, 189–199. [CrossRef]
38. Wang, Z.; Li, S.; Zhang, M.; Yang, H.; Li, G.; Ren, X.; Liang, S. Optimization of Oil Extraction from Rice Bran with Mixed Solvent Using Response Surface Methodology. *Foods* **2022**, *11*, 3849. [CrossRef]
39. Ribas, F.B.T.; Gasparetto, H.; Salau, N.P.G. Sustainable extraction of rice bran Oil: Assessing renewable solvents, kinetics, and thermodynamics. *Chem. Eng. Res. Des.* **2023**, *197*, 342–354. [CrossRef]
40. Modupalli, N.; Krisshnan, A.; CK, S.; DV, C.; Natarajan, V.; Koidis, A.; Rawson, A. Effect of novel combination processing technologies on extraction and quality of rice bran oil. *Crit. Rev. Food Sci. Nutr.* **2024**, *64*, 1911–1933. [CrossRef]
41. Hanmoungjai, P.; Pyle, D.L.; Niranjan, K. Enzyme-assisted water-extraction of oil and protein from rice bran. *J. Chem. Technol. Biotechnol.* **2002**, *77*, 771–776. [CrossRef]
42. Hanmoungjai, P.; Pyle, D.L.; Niranjan, K. Enzymatic process for extracting oil and protein from rice bran. *J. Am. Oil Chem. Soc.* **2001**, *78*, 817–821. [CrossRef]

43. Mounika, E.; Smith, D.D.D.; Edukondalu, D.L.; Raja, D.S. Aqueous enzymatic extraction of rice bran oil. *J. Pharmacogn. Phytochem.* **2020**, *9*, 3149–3151. [CrossRef]
44. Hao, J.; Wang, Q.; Li, X.; Xu, D. Extraction of structurally intact and well-stabilized rice bran oil bodies as natural pre-emulsified O/W emulsions and investigation of their rheological properties and components interaction. *Food Res. Int.* **2023**, *164*, 112457. [CrossRef] [PubMed]
45. Khoei, M.; Chekin, F. The ultrasound-assisted aqueous extraction of rice bran oil. *Food Chem.* **2016**, *194*, 503–507. [CrossRef]
46. Gopala Krishna, A.G.; Hemakumar, K.H.; Khatoon, S. Study on the composition of rice bran oil and its higher free fatty acids value. *J. Am. Oil Chem. Soc.* **2006**, *83*, 117–120. [CrossRef]
47. Fraterrigo Garofalo, S.; Demichelis, F.; Mancini, G.; Tommasi, T.; Fino, D. Conventional and ultrasound-assisted extraction of rice bran oil with isopropanol as solvent. *Sustain. Chem. Pharm.* **2022**, *29*, 100741. [CrossRef]
48. Yu, D.; Wang, T.; Chen, J.; Tang, H.; Li, D.; Zhang, X.; Geng, H.; Wang, L.; Elfalleh, W.; Jiang, L. Enzymatic esterification of rice bran oil and phytosterol in supercritical CO_2. *J. Food Process. Preserv.* **2019**, *43*, e14066. [CrossRef]
49. Fraterrigo Garofalo, S.; Cavallini, N.; Destefano, R.; Micera, M.; Cavagnero, C.; Botto, A.; Savorani, F.; Tommasi, T.; Fino, D. Optimization of Supercritical Carbon Dioxide Extraction of Rice Bran Oil and γ-Oryzanol Using Multi-Factorial Design of Experiment. *Waste Biomass Valorization* **2023**, *14*, 3327–3337. [CrossRef]
50. Coelho, J.P.; Robalo, M.P.; Fernandes, I.S.; Stateva, R.P. Optimization of Oil Recovery from Japonica Luna Rice Bran by Supercritical Carbon Dioxide Applying Design of Experiments: Characterization of the Oil and Mass Transfer Modeling. *ChemEngineering* **2022**, *6*, 63. [CrossRef]
51. Pandey, R.; Shrivastava, S.L. Comparative evaluation of rice bran oil obtained with two-step microwave assisted extraction and conventional solvent extraction. *J. Food Eng.* **2018**, *218*, 106–114. [CrossRef]
52. Gotama, B.; Rahman, A.K.; Ahmad, A.; Hariyadi, A. Extraction of rice bran oil using microwave-assisted extraction and green solvents. *IOP Conf. Ser. Earth Environ. Sci.* **2022**, *1105*, 012052. [CrossRef]
53. Moreira, B.P.; Draszewski, C.P.; Rosa, N.C.; Tres, M.V.; Zabot, G.L.; Pereira, F.C.; Abaide, E.R.; Castilhos, F. Integrated rice bran processing by supercritical CO2 extraction and subcritical water hydrolysis to obtain oil, fermentable sugars, and platform chemicals. *J. Supercrit. Fluids* **2023**, *192*, 105786. [CrossRef]
54. Hou, N.C.; Gao, H.H.; Qiu, Z.J.; Deng, Y.H.; Zhang, Y.T.; Yang, Z.C.; Gu, L.B.; Liu, H.M.; Zhu, X.L.; Qin, Z.; et al. Quality and active constituents of safflower seed oil: A comparison of cold pressing, hot pressing, Soxhlet extraction and subcritical fluid extraction. *LWT* **2024**, *200*, 116184. [CrossRef]
55. Okajima, I.; Ly, L.T.T.; Yi, K.C.; Sako, T. Phosphorus-free oil extraction from rice bran using CO2-expanded hexane. *Chem. Eng. Process. Process Intensif.* **2021**, *166*, 108502. [CrossRef]
56. Mathias, M.G.; Okajima, I.; Aoki, Y.; Kong, C.Y.; Itika, A.; Elisante, E.; Sako, T. Extraction of high-quality rice bran oil with CO_2-expanded acetone for biofuel production. *Fuel* **2024**, *356*, 129491. [CrossRef]
57. Zhou, P.; Liu, W.; Zhang, Y.; Liu, G.; Zhao, Z.; Zhang, M.; Wei, Z.; Zhang, Y.; Tang, X.; Li, P.; et al. Tuning multi-purpose glycerol-based deep eutectic solvents for biocatalytic deacidification of high-acid rice bran oil: Kinetics, mechanism, and inhibition of side glycerolysis. *LWT* **2023**, *189*, 115535. [CrossRef]
58. Dong, T.; Zhou, X.; Dai, Y.; Yang, X.; Zhang, W.; Yu, D.; Liu, T. Application of magnetic immobilized enzyme of nano dialdehyde starch in deacidification of rice bran oil. *Enzym. Microb. Technol.* **2022**, *161*, 110116. [CrossRef]
59. Xu, L.; Zhang, Y.; Zivkovic, V.; Zheng, M. Deacidification of high-acid rice bran oil by the tandem continuous-flow enzymatic reactors. *Food Chem.* **2022**, *393*, 133440. [CrossRef]
60. Yin, Y.; Fei, X.; Tian, J.; Xu, L.; Li, Y.; Wang, Y. Synthesis of lipase-hydrogel microspheres and their application in deacidification of high-acid rice bran oil. *New J. Chem.* **2022**, *46*, 21287–21300. [CrossRef]
61. Sun, B.; Gao, P.; Yu, H.; Dong, Z.; Yin, J.; Zhong, W.; Hu, C.; He, D.; Wang, X. Optimization of Composite Decolorizer Efficacy Based on Decolorization Efficiency, Toxicity, and Nutritional Value of Rice Bran Oil. *J. Oleo Sci.* **2023**, *72*, 755–765. [CrossRef] [PubMed]
62. Zhang, F.; Wu, Z.; Huang, H.; Wang, P.; Yu, Y.; Wu, C.; Li, T.; Li, X.; Zhou, D.; Fan, G. Ultrasound bleaching and quality evaluation of rice bran oil. *J. Food Process Eng.* **2024**, *47*, e14557. [CrossRef]
63. Yu, D.; Dong, T.; Zhang, L.; Zhou, X.; Wang, L.; Yang, F.; Liu, T. Effects of different deodorization methods on the oxidation of sterol components in rice bran oil. *Food Chem.* **2023**, *404*, 134568. [CrossRef] [PubMed]
64. Han, L.; Li, J.; Wang, S.; Cheng, W.; Ma, L.; Liu, G.; Han, D.; Niu, L. The inhibitory effects of sesamol and sesamolin on the glycidyl esters formation during deodorization of vegetables oils. *J. Sci. Food Agric.* **2021**, *101*, 3605–3612. [CrossRef] [PubMed]
65. Meedam, A.; Usaku, C.; Daisuk, P.; Shotipruk, A. Comparative study on physicochemical hydrolysis methods for glycerides removal from rice bran acid oil for subsequent γ-oryzanol recovery. *Biomass Convers. Biorefinery* **2022**, *12*, 245–252. [CrossRef]
66. Pestana-Bauer, V.R.; Mendonça, C.R.; Bruscatto, M.H.; Krumreich, F.D.; Otero, D.M.; Costa, I.H.; Zambiazi, R.C. Fatty acid distillation residues from rice oil bran refining as a source of phytochemicals. *Int. J. Food Sci. Technol.* **2023**, *58*, 4627–4637. [CrossRef]
67. Kordsmeier, M.; Howard, L.R.; Brownmiller, C.; Proctor, A.; Hauer-Jensen, M. Isolation of Gamma and Delta Tocotrienols from Rice Bran Oil Deodorizer Distillate Using Flash Chromatography. *J. Am. Oil Chem. Soc.* **2015**, *92*, 1243–1252. [CrossRef]
68. Sugihara, N.; Kanda, A.; Nakano, T.; Nakamura, T.; Igusa, H.; Hara, S. Novel Fractionation Method for Squalene and Phytosterols Contained in the Deodorization Distillate of Rice Bran Oil. *J. Oleo Sci.* **2010**, *59*, 65–70. [CrossRef]

69. Ali, M.A.; Chew, S.C.; Majid, F.A.A. Contribution of endogenous minor components in the oxidative stability of rice bran oil. *J. Food Meas. Charact.* **2022**, *17*, 187–210. [CrossRef]
70. Mohammed, A.S.; Hashem, H.A.A.; Abdel Maksoud, B.S. Improving the quality properties of soybean oil by using rice bran oil. *Sci. Rep.* **2024**, *14*, 2723. [CrossRef]
71. Siddiqua, T.; Uddin, I.; Hasan, M.R.; Begum, R. Effect of heating and Re-heating on physico-chemical properties of rice bran oil and soybean oil. *J. Agric. Food Res.* **2024**, *15*, 100979. [CrossRef]
72. Waseif, M.A.E.; Badr, S.A.; Fahmy, H.M.; Sabry, A.M.; Abd-Eazim, E.I.; Shaaban, H.A. Improving Stability of Flaxseed Oil by Rice Bran Oil as Source of γ-Oryzanol. *Pak. J. Biol. Sci.* **2022**, *25*, 698–704. [CrossRef] [PubMed]
73. Pradhan, A.; Anis, A.; Alam, M.A.; Al-Zahrani, S.M.; Jarzebski, M.; Pal, K. Effect of Soy Wax/Rice Bran Oil Oleogel Replacement on the Properties of Whole Wheat Cookie Dough and Cookies. *Foods* **2023**, *12*, 3650. [CrossRef] [PubMed]
74. Issara, U. Improvement of Thai Sweet Sausage (Goon Chiang) Properties by Oleogel Made of Rice Bran Wax and Rice Bran Oil: A Textural, Sensorial, and Nutritional Aspect. *IOP Conf. Ser. Earth Environ. Sci.* **2022**, *995*, 012045. [CrossRef]
75. Sivakanthan, S.; Fawzia, S.; Mundree, S.; Madhujith, T.; Karim, A. Optimization and characterization of new oleogels developed based on sesame oil and rice bran oil. *Food Hydrocoll.* **2023**, *142*, 108839. [CrossRef]
76. Eng, H.Y.; Mohd Rozalli, N.H. Rice bran and its constituents: Introduction and potential food uses. *Int. J. Food Sci. Technol.* **2022**, *57*, 4041–4051. [CrossRef]
77. Prakansamut, N.; Adulpadungsak, K.; Sonwai, S.; Aryusuk, K.; Lilitchan, S. Application of functional oil blend-based oleogels as novel structured oil alternatives in chocolate spread. *LWT* **2024**, *203*, 116322. [CrossRef]
78. Dhara, O.; Reddy, J.R.C.; Eanti, A.; Karuna, M.S.L.; Chakrabarti, P.P. Valorization of Rice Bran Gums: An Untapped Source of Natural Phospholipids. *Waste Biomass Valorization* **2024**, *15*, 2937–2945. [CrossRef]
79. Mokbul, M.; Cheow, Y.L.; Siow, L.F. Physical properties, sensory profile and storage stability of compound chocolates made with cocoa butter replacer consisting of mango kernel fat and rice bran oil. *Food Chem. Adv.* **2023**, *3*, 100515. [CrossRef]
80. Bako, H.K.; Ibeogu, H.I.; Bassey, A.P.; Yar, M.S.; Zhou, T.; Li, C. Optimisation and characterization of double emulsion derived from rice starch, rice protein isolates and rice bran oil. *Int. J. Biol. Macromol.* **2024**, *258*, 128966. [CrossRef]
81. Sengupta, S.; Bhattacharyya, D.K.; Goswami, R.; Bhowal, J. Emulsions stabilized by soy protein nanoparticles as potential functional non-dairy yogurts. *J. Sci. Food Agric.* **2019**, *99*, 5808–5818. [CrossRef]
82. Wang, H.; Chen, L.; Cai, Q.; Wu, S.; Shen, W.; Hu, Z.; Huang, W.; Jin, W. Formation, digestion properties, and physicochemical stability of the rice bran oil body carrier system. *Food Chem.* **2023**, *409*, 135283. [CrossRef] [PubMed]
83. Liu, X.T.; Yang, C.; Liu, J.; Liu, Y.X.; Shang, S.; Hua, F.; Lv, X.X. The antioxidant and antisenescence activities of physically refined rice bran oil surpass those of the combination of γ-oryzanol, α-tocopherol and sitosterol. *CyTA J. Food* **2023**, *21*, 410–417. [CrossRef]
84. Junyusen, T.; Chatchavanthatri, N.; Liplap, P.; Junyusen, P.; Phan, V.M.; Nawong, S. Effects of Extraction Processes on the Oxidative Stability, Bioactive Phytochemicals, and Antioxidant Activity of Crude Rice Bran Oil. *Foods* **2022**, *11*, 1143. [CrossRef]
85. Liu, R.; Xu, Y.; Chang, M.; Tang, L.; Wang, X. Antioxidant interaction of α-tocopherol, γ-oryzanol and phytosterol in rice bran oil. *Food Chem.* **2020**, *343*, 128431. [CrossRef]
86. Liu, N.; Zhang, P.; Xue, M.; Zhang, M.; Xiao, Z.; Xu, C.; Fan, Y.; Liu, W.; Wu, Y.; Wu, M.; et al. Anti-inflammatory and antioxidant properties of rice bran oil extract in copper sulfate-induced inflammation in zebrafish (Danio rerio). *Fish Shellfish. Immunol.* **2023**, *136*, 108740. [CrossRef]
87. Huang, J.; Qin, W.; Xu, B.; Sun, H.; Jing, F.; Xu, Y.; Zhao, J.; Chen, Y.; Ma, L.; Yan, X. Rice bran oil supplementation protects swine weanlings against diarrhea and lipopolysaccharide challenge. *J. Zhejiang Univ. Sci. B* **2023**, *24*, 430–441. [CrossRef]
88. Prasertsri, P.; Boonla, O.; Vierra, J.; Yisarakun, W.; Koowattanatianchai, S.; Phoemsapthawee, J. Effects of Riceberry Rice Bran Oil Supplementation on Oxidative Stress and Cardiovascular Risk Biomarkers in Older Adults with Prehypertension. *Prev. Nutr. Food Sci.* **2022**, *27*, 365–375. [CrossRef]
89. Gallyas, F.; Lee, S.; Yu, S.; Park, H.J.; Jung, J.; Go, G.W.; Kim, W. Rice bran oil ameliorates inflammatory responses by enhancing mitochondrial respiration in murine macrophages. *PLoS ONE* **2019**, *14*, e0222857. [CrossRef]
90. Ashraf, S.; Saeed, S.M.G.; Saify, Z.S.; Haider, S.; Simji, S. Potential nutraceutical benefits of basmati rice bran oil as analgesic, anti-inflammatory and anti-arthritis. *Pak. J. Pharm. Sci.* **2019**, *32*, 2545–2551.
91. Pourrajab, B.; Sohouli, M.H.; Amirinejad, A.; Fatahi, S.; Găman, M.A.; Shidfar, F. The impact of rice bran oil consumption on the serum lipid profile in adults: A systematic review and meta-analysis of randomized controlled trials. *Crit. Rev. Food Sci. Nutr.* **2022**, *62*, 6005–6015. [CrossRef] [PubMed]
92. Sen, S.; Chakraborty, R.; Kalita, P. Rice—Not just a staple food: A comprehensive review on its phytochemicals and therapeutic potential. *Trends Food Sci. Technol.* **2020**, *97*, 265–285. [CrossRef]
93. Liu, N.; Zhang, P.; Xue, M.; Zhang, M.; Huang, Z.; Xu, C.; Meng, Y.; Fan, Y.; Liu, W.; Zhang, F.; et al. Hypolipidemic Effect of Rice Bran Oil Extract Tocotrienol in High-Fat Diet-Induced Hyperlipidemia Zebrafish (Danio Rerio) Induced by High-Fat Diet. *Int. J. Mol. Sci.* **2024**, *25*, 2954. [CrossRef] [PubMed]
94. Saji, N.; Francis, N.; Schwarz, L.J.; Blanchard, C.L.; Santhakumar, A.B. Rice Bran Derived Bioactive Compounds Modulate Risk Factors of Cardiovascular Disease and Type 2 Diabetes Mellitus: An Updated Review. *Nutrients* **2019**, *11*, 2736. [CrossRef] [PubMed]

95. Son, J.E.; Jo, J.Y.; Kim, S.; Park, M.J.; Lee, Y.; Park, S.S.; Park, S.Y.; Jung, S.M.; Jung, S.K.; Kim, J.Y.; et al. Rice Bran Extract Suppresses High-Fat Diet-Induced Hyperlipidemia and Hepatosteatosis through Targeting AMPK and STAT3 Signaling. *Nutrients* **2023**, *15*, 3630. [CrossRef]
96. Phannasorn, W.; Pharapirom, A.; Thiennimitr, P.; Guo, H.; Ketnawa, S.; Wongpoomchai, R. Enriched Riceberry Bran Oil Exerts Chemopreventive Properties through Anti-Inflammation and Alteration of Gut Microbiota in Carcinogen-Induced Liver and Colon Carcinogenesis in Rats. *Cancers* **2022**, *14*, 4358. [CrossRef]
97. Ke-Chen, H.; Hsu, W.F.; Li, S.C.; Shih, C.K. P-0179 Effects of γ-Oryzanol and Unsaponifiable Matter from Rice Bran Oil on Colon Carcinogenesis in Rats. *Ann. Oncol.* **2012**, *23*, iv82. [CrossRef]
98. Zhang, J.; Zhang, L.; Wu, Z.; Zhang, P.; Liu, R.; Chang, M.; Wang, X. The dopaminergic neuroprotective effects of different phytosterols identified in rice bran and rice bran oil. *Food Funct.* **2021**, *12*, 10538–10549. [CrossRef]

MDPI AG
Grosspeteranlage 5
4052 Basel
Switzerland
Tel.: +41 61 683 77 34

Foods Editorial Office
E-mail: foods@mdpi.com
www.mdpi.com/journal/foods

www.ingramcontent.com/pod-product-compliance
Lightning Source LLC
Chambersburg PA
CBHW061453210926
51850CB00008B/1702
9783725856954